Control and State Estimation for Dynamical Network Systems with Complex Samplings

This book focuses on the control and state estimation problems for dynamical network systems with complex samplings subject to various network-induced phenomena. It includes a series of control and state estimation problems tackled under the passive sampling fashion. Further, it explains the effects from the active sampling fashion, i.e., event-based sampling is examined on the control/estimation performance, and novel design technologies are proposed for controllers/estimators. Simulation results are provided for better understanding of the proposed control/filtering methods. By drawing on a variety of theories and methodologies such as Lyapunov function, linear matrix inequalities, and Kalman theory, sufficient conditions are derived for guaranteeing the existence of the desired controllers and estimators, which are parameterized according to certain matrix inequalities or recursive matrix equations.

- Covers recent advances of control and state estimation for dynamical network systems with complex samplings from the engineering perspective
- Systematically introduces the complex sampling concept, methods, and application for the control and state estimation
- Presents unified framework for control and state estimation problems of dynamical network systems with complex samplings
- Exploits a set of the latest techniques such as linear matrix inequality approach, Vandermonde matrix approach, and trace derivation approach
- Explains event-triggered multi-rate fusion estimator, resilient distributed sampled-data estimator with predetermined specifications

This book is aimed at researchers, professionals, and graduate students in control engineering and signal processing.

Control and State Estimation for Dynamical Network Systems with Complex Samplings

Bo Shen
Zidong Wang
Qi Li

CRC Press
Taylor & Francis Group
Boca Raton London New York

CRC Press is an imprint of the
Taylor & Francis Group, an **informa** business

First edition published 2023
by CRC Press
6000 Broken Sound Parkway NW, Suite 300, Boca Raton, FL 33487-2742

and by CRC Press
4 Park Square, Milton Park, Abingdon, Oxon, OX14 4RN

CRC Press is an imprint of Taylor & Francis Group, LLC

Library of Congress Cataloging-in-Publication Data

Names: Shen, Bo, 1981- author. | Wang, Zidong, 1966- author. | Li, Qi (Professor of information science and engineering), author.
Title: Control and state estimation for dynamical network systems with complex samplings / Bo Shen, Zidong Wang, Qi Li.
Description: First edition. | Boca Raton : CRC Press, 2023. | Includes bibliographical references and index.
Identifiers: LCCN 2022009046 (print) | LCCN 2022009047 (ebook) | ISBN 9781032309965 (hardback) | ISBN 9781032310206 (paperback) | ISBN 9781003307648 (ebook)
Subjects: LCSH: Adaptive control systems. | Sensor networks. | Neural networks (Computer science) | Parameter estimation. | Observers (Control theory) | Sampling (Statistics)
Classification: LCC TJ217 .S49 2023 (print) | LCC TJ217 (ebook) | DDC 629.8/36--dc23/eng/20220606
LC record available at https://lccn.loc.gov/2022009046
LC ebook record available at https://lccn.loc.gov/2022009047

ISBN: 978-1-032-30996-5 (hbk)
ISBN: 978-1-032-31020-6 (pbk)
ISBN: 978-1-003-30764-8 (ebk)

DOI: 10.1201/9781003307648

Typeset in CMR10
by KnowledgeWorks Global Ltd.

Publisher's note: This book has been prepared from camera-ready copy provided by the authors.

This book is dedicated to the Dream Dynasty consisting of a group bright people who have been enamored with the challenging research on the control and estimation for dynamical network systems with complex samplings ......

Contents

List of Figures

List of Tables

Preface

In response to the ongoing advances of digital medias, the sampling issue becomes vitally important in the control systems. Over the past decades, periodic sampling has received considerable research attention due to its advantages of easy implementation and analysis. The main feature of periodic sampling is that the state or output signals of a continuous-time system are sampled and then transmitted according to a fixed sampling period. However, in practical applications, it is often the case that the sampling period is time-varying and/or uncertain. This is referred to as complex samplings. Usually, the complex samplings can be classified into two categories, i.e., passive sampling and active sampling. The passive sampling relates to the case that the samplings occur in an uncertain way owing to some undesirable physical constraints such as aperiodic faults in the samplers, fluctuated network loads, intermittent signal quantization/saturation and unwanted changes of some components of the system itself. As for the active sampling, the sampling instants are determined only according to the specific engineering requirements such as energy saving. It is worth mentioning that the utilization of complex sampling mechanisms renders the traditional periodic-sampling-based control and estimation algorithm is no longer effective. Therefore, it makes both theoretical and practical sense to develop new techniques to address the control and estimation issues for various systems with complex samplings such that the required performance can still be guaranteed.

In this book, we focus on the control and state estimation problems for dynamical network systems with complex samplings subject to various network-induced phenomena. The network-induced phenomena under consideration include state/sensor saturations, quantization effects and sensor degradations. The dynamical network systems cover the general networked control systems, sensor networks, neural networks and complex networks. The main content of this book is identified mainly into two parts. In the first part (Chapters 2-3), a series of control and state estimation problems are tackled under the passive sampling mechanism, where sufficient conditions are derived in terms of matrix inequalities to guarantee the existence of the desired controllers and estimators. In the second part (Chapters 4-9), the effects from the active sampling mechanism, i.e., ET sampling, are examined on the control and estimation performance, where novel design technologies are proposed for controllers and estimators with the hope of saving energy while the acceptable control and estimation performances are still guaranteed.

Also, simulation results are provided to have a better understanding of the proposed control/estimation methods.

The compendious frame and description of the book are given as follows. Chapter 1 presents background on complex samplings as well as the recent advances of these topics. Chapter 2 is concerned with the stabilization and control problem for sampled-data systems under noisy sampling intervals. Chapter 3 studies the distributed H_∞ state estimation problem over sensor networks with nonuniform samplings under infinite-distributed delays, where the resilient issue is taken into account in order to accommodate the potential estimator implementation errors. For ET sampling, Chapter 4 addresses the ET control problem for two classes of switched systems with exogenous disturbances. In Chapters 5 and 6, the ET state estimation issues are investigated, respectively, for state-saturated systems and neural networks subject to various network-induced phenomena. In addition, the ET robust fusion estimation problem is investigated in Chapter 7 for multi-rate systems with stochastic nonlinearities, coloured measurement noises and sensor degradations. Chapter 8 is concerned with the dynamic ET synchronization controller design problem for delayed complex networks. In this chapter, a new discrete-time version of the dynamic ET mechanism is put forward to further save energy. Chapter 9 examines the filtering and state estimation problems for several classes of dynamical systems under dynamic ET mechanisms. Chapter 10 presents the conclusion and some possible future research directions.

This book is a research monograph whose intended audiences are graduate and postgraduate students as well as researchers.

Bo Shen
Shanghai, China

Zidong Wang
London, U.K.

Qi Li
Hangzhou, China

Author Biographies

Bo Shen received the B.Sc. degree in mathematics from Northwestern Polytechnical University, Xian, China, in 2003, and the Ph.D. degree in control theory and control engineering from Donghua University, Shanghai, China, in 2011.

From 2009 to 2010, he was a Research Assistant with the Department of Electrical and Electronic Engineering, The University of Hong Kong, Hong Kong. From 2010 to 2011, he was a Visiting Ph.D. Student with the Department of Information Systems and Computing, Brunel University London, London, U.K. From 2011 to 2013, he was a Research Fellow (Scientific Co-Worker) with the Institute for Automatic Control and Complex Systems, University of Duisburg-Essen, Duisburg, Germany. He is currently a Professor at the College of Information Science and Technology, Donghua University. He has published around 80 articles in refereed international journals. His research interests include nonlinear control and filtering, stochastic control and filtering, as well as complex networks and neural networks.

Professor Shen is a program committee member for many international conferences. He serves (or has served) as an Associate Editor or Editorial Board Member for nine international journals, including Systems Science and Control Engineering, Journal of The Franklin Institute, Asian Journal of Control, Circuits, Systems, and Signal Processing, Neurocomputing, Assembly Automation, Neural Processing Letters, IEEE Access, and Mathematical Problems in Engineering.

Zidong Wang was born in Jiangsu, China, in 1966. He received the B.Sc. degree in mathematics from Suzhou University, Suzhou, China, in 1986, and the M.Sc. degree in applied mathematics and the Ph.D. degree in electrical engineering from the Nanjing University of Science and Technology, Nanjing, China, in 1990 and 1994, respectively.

He is currently a Professor of Dynamical Systems and Computing in the Department of Computer Science, Brunel University London, Uxbridge, U.K. From 1990 to 2002, he held teaching and research appointments in universities in China, Germany and the U.K. He has published more than 600 papers in international journals. He is a holder of the Alexander von Humboldt Research Fellowship of Germany, the JSPS Research Fellowship of Japan, and William Mong Visiting Research Fellowship of Hong Kong. His research interests include dynamical systems, signal processing, bioinformatics, and control theory and applications.

Professor Wang serves (or has served) as the Editor-in-Chief for the International Journal of Systems Science, Neurocomputing, and Systems Science & Control Engineering, and an Associate Editor for 12 international journals including IEEE Transactions on Automatic Control, IEEE Transactions on Control System Technology, IEEE Transactions on Neural Networks, IEEE Transactions on Signal Processing, and IEEE Transactions on Systems, Man, and Cybernetics-Part C. He is a Member of the Academia Europaea, a Member of the European Academy of Sciences and Arts, an Academician of the International Academy for Systems and Cybernetic Sciences, a Fellow of IEEE, a Fellow of the Royal Statistical Society and a member of program committee for many international conferences.

Qi Li received her B.Eng. degree in electrical engineering and automation from Jiangsu University of Technology, Changzhou, China, in 2013 and the Ph.D. degree in control science and engineering from Donghua University, Shanghai, China, in 2018. She is currently an Associate Professor at the School of Information Science and Engineering, Hangzhou Normal University, Hangzhou, China. From June 2016 to July 2016, she was a Research Assistant in the Department of Mathematics, Texas A&M University at Qatar, Qatar. From November 2016 to November 2017, she was a Visiting Ph.D. Student in the Department of Computer Science, Brunel University London, U.K. Her current research interests include network communication, complex networks and sensor networks. She is a very active reviewer for many international journals.

Acknowledgements

We would like to express our sincere appreciation to many people who have been directly involved in various aspects of the research leading to this book. Particular thanks go to Professor Huisheng Shu from Donghua University, Shanghai, Professor Lifeng Ma from Nanjing University of Science and Technology, Nanjing, Dr. Hailong Tan from Anhui Polytechnic University, Wuhu, Dr. Jinghui Suo from Donghua University, Shanghai, and Dr. Cong Huang from Nantong University, Nantong. Special thanks are also given to Yufei Liu, Jie Sun, Xuelin Wang, and Haijing Fu for their considerable help in the editorial and proofreading work.

The writing of this book was supported in part by the National Natural Science Foundation of China under Grants 61873059, 61922024, 61873148, 61933007, and 62003121, the Program of Shanghai Academic/Technology Research Leader of China under Grant 20XD1420100, the Zhejiang Provincial Natural Science Foundation of China under Grant LQ20F030014, the Royal Society of the UK, and the Alexander von Humboldt Foundation of Germany.

Symbols

Symbol Description

Symbol	Description
$\mathbb{R}^n$	The n-dimensional Euclidean space.
$\mathbb{R}^{n\times m}$	The set of all $n \times m$ real matrices.
$\mathbb{Z}_-$	The set of all nonpositive integers.
$\mathbb{Z}_+$	The set of all nonnegative integers.
$\mathbb{N}_+$	The set of positive integers.
$\mathbb{R}_+$	The set of all nonnegative real numbers.
1_n	The n-dimensional vector whose elements are all 1.
$\|A\|$	The norm of matrix A defined by $\|A\| = \sqrt{\text{trace}(A^TA)}$.
$\|A\|_\infty$	The infinite norm of matrix A.
A^T	The transpose of the matrix A.
I	The identity matrix of compatible dimension.
ID	The identity function.
$\otimes$	The Kronecker product.
$\circ$	The Hadamard product.
$\diamond$	The composition of two functions.
$L_2[0,\infty)$	The space of square integrable vector functions.
$l_2[0,\infty)$	The space of square summable vector functions.
$\mathcal{C}([-d,0];\mathbb{R}^n)$	The family of all continuous $\mathbb{R}^n$-value functions ξ on $[-d,0]$ with norm $\|\xi\|_d \triangleq \sup\{\|\xi(s)\| : -d \leq s \leq 0\}$.
$\mathcal{K}$-function	The function $\gamma : \mathbb{R}_+ \to \mathbb{R}_+$ is continuous, strictly increasing and satisfies $\gamma(0) = 0$.
$\mathcal{K}_\infty$-function	The function $\gamma : \mathbb{R}_+ \to \mathbb{R}_+$ is a $\mathcal{K}$-function and also $\gamma(s) \to \infty$ as $s \to \infty$.
$\mathcal{KL}$-function	The function $\gamma(s,k) : \mathbb{R}_+ \times \mathbb{R}_+ \to \mathbb{R}_+$ is a $\mathcal{K}$-function for each fixed k and also decreasing with $\gamma(s,t) \to 0$ as $t \to \infty$.
$\mathcal{L}_\infty^n$	The set of all locally essentially bounded sequences $\nu : \mathbb{Z}_+ \to \mathbb{R}_+$ with norm $\|\nu\|_\infty = ess\sup_{k\geq 0} \|v(k)\| < \infty$.
$D^+\varphi(t)$	The upper right-hand Dini derivative of function $\varphi(t)$ with $D^+\varphi(t) = \limsup_{h\to 0^+} \frac{\varphi(t+h)-\varphi(t)}{h}$.
Prob$(\cdot)$	The occurrence probability of the event "$\cdot$".
$\mathbb{E}\{x\}$	The expectation of the stochastic variable x.
$\mathbb{E}\{x\|y\}$	The expectation of x conditional on y.

Var$\{x\}$ The covariance of stochastic variable x.

$(\Omega, \mathcal{F}, \mathrm{Prob})$ A complete probability space.

$\lambda_{\min}(A)$ The smallest eigenvalue of a square matrix A.

$\lambda_{\max}(A)$ The largest eigenvalue of a square matrix A.

$*$ The ellipsis for terms induced by symmetry, in symmetric block matrices.

tr(A) The trace of a matrix A.

$X > Y$ The $X - Y$ is positive definite, where X and Y are real symmetric matrices.

$X \geq Y$ The $X - Y$ is positive semi-definite, where X and Y are real symmetric matrices.

diag$\{\cdots\}$ The block-diagonal matrix.

$\mathrm{diag}_N\{A\} = \mathrm{diag}\{\underbrace{A, \cdots, A}_{N}\}$.

$\mathrm{diag}_N^i\{A_i\} = \mathrm{diag}\{A_1, \cdots, A_N\}$.

$\mathrm{vec}_N\{A\} = \underbrace{\begin{bmatrix} A & \cdots & A \end{bmatrix}}_{N}$.

$\mathrm{vec}_N^i\{A_i\} = \begin{bmatrix} A_1 & \cdots & A_N \end{bmatrix}$.

$\mathrm{vec}_N^i\{A_{ij}\} = \begin{bmatrix} A_{i1} & \cdots & A_{iN} \end{bmatrix}$.

$\mathrm{col}_N\{A\} = \underbrace{\begin{bmatrix} A^T & \cdots & A^T \end{bmatrix}^T}_{N}$.

$\mathrm{col}_N^i\{A_i\} = \begin{bmatrix} A_1^T & \cdots & A_N^T \end{bmatrix}^T$.

List of Acronyms

ET	event-triggered
LMIs	linear matrix inequalities
SPs	stochastic parameters
DDSs	delayed differential systems
ZOH	zero-order hold
ROMDs	randomly occurring mixed delays
IMs	incomplete measurements
MJPs	Markovian jumping parameters
TVDs	time-varying delays
GRNs	genetic regulatory networks
mRNAs	messenger ribonucleic acids
PECs	prediction error covariances
FECs	filtering error covariances
CMs	censored measurements
PUs	parameter uncertainties
CI	covariance intersection

1

Introduction

1.1 Background

In recent years, in response to the rapid development of digital technology, sampled-data control systems have attracted recurring research attention both from academic research and industrial application. Typically, a sampled-data control system consists of a continuous-time plant and a discrete-time controller, which are connected together in feedback by the sampler and hold devices. A general architecture of a sampled-data control system is illustrated in Fig. 1.1. In this system, the sensor first converts a physical stimulus of the plant into a readable measurement $y(t)$ and then the sampler takes samples of $y(t)$ to yield the corresponding discrete-time measurement $y(t_k)$, where t_k ($k = 0, 1, \ldots$) are sampling instants. Subsequently, the controller uses the transmitted measurement $y(t_k)$ to generate a discrete-time control input $u(t_k)$, which is further converted back into a continuous-time control input signal $u(t)$ by using the hold. Finally, the input signal $u(t)$ would be sent to the actuator to control the plant. In reality, the functions of the sampler and the hold are, respectively, implemented with the help of an analogue-to-digital (A/D) converter and a digital-to-analogue (D/A) converter. In fact, due to its practical backgrounds and wide applications, the analysis/synthesis of sampled-data control systems has long been a hot topic of research with a number of excellent results reported in the existing literature, see [16, 73, 193, 209] and the reference therein.

In classic sampled-data control theory, the sampling intervals defined by $T_k \triangleq t_{k+1} - t_k$ are usually expected to be an invariant constant (i.e. $T_k = T > 0$) for its simplicity in analysis, design, and implementation. However, under certain undesirable physical environments such as the unsteady power voltage, the sampler may tremble slightly, and the practically implemented sampling intervals often fluctuate around a nominal sampling period. This is referred to as nonuniform sampling. It is worth mentioning that the nonuniform sampling is essentially performed in a *deterministic* manner. Actually, in many cases such as seismic data extraction, it is quite common that the sampling occurs in a *probabilistic* way, and accordingly, the sampling intervals are stochastic due to unavoidable noises. In such cases, it seems natural to describe the sampling intervals T_k by a random variable obeying certain probability

DOI: 10.1201/9781003307648-1

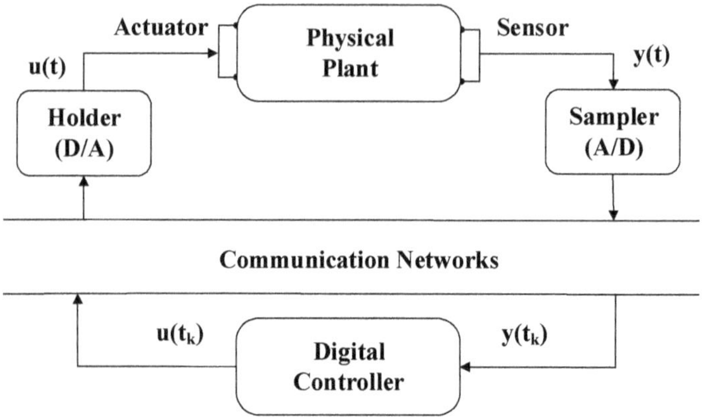

FIGURE 1.1
A general architecture of a sampled-data control system.

distribution. Note that both nonuniform sampling and stochastic sampling can be regarded as passive sampling, which occurs in an uncertain way owing to some undesirable physical constraints such as aperiodic faults in the samplers, fluctuated network loads, and unwanted changes of some components of the system itself.

On the other hand, when resource limit becomes a concern, the so-called ET (or event-based) sampling method has served as an ideal way for energy-saving purposes. The main idea of the ET strategy lies in that the sampling/transmission is executed only if a predefined triggering condition is satisfied, thereby effectively reducing unnecessary transmissions. In view of such an advantage, ET sampling has become more and more popular with wide application potentials [22, 28, 101, 191, 197, 215]. Roughly speaking, the existing triggering conditions can be typically divided into two categories, one is the absolute change triggering condition and the other is the relative change triggering condition. It is noticeable that, in recent years, the ET sampling scheme is under continuous improvement with aim to further reduce the energy consuming, and thus some improved versions of ET strategy have been proposed. For example, a dynamic ET sampling method initiated in [50], which introduces an additional dynamical variable in the typical triggering condition, has recently aroused particular research interests due to its capability of saving resource even further without significantly degrading the system performance. In addition, it also has been shown in [50] that the dynamic ET sampling is more general, which contains the traditional ET one as a special case. So far, a large number of research results have been reported on the investigation of dynamic ET sampling; see e.g. [46, 60, 174] for some latest literature.

Several typical mathematical models for the complex samplings mentioned above, including nonuniform sampling, stochastic sampling, ET sampling, dynamic ET sampling, are shown in Table 1.1.

Types	Mathematical models
Nonuniform sampling	The sampling interval is variable but bounded, i.e., $T_k \in [T_m, T_M]$, where T_m and T_M are known constants.
Stochastic sampling	The sampling interval $T_k = T + v_k$, where T is a constant and v_k is a random variable.
ET sampling	• Absolute change triggering condition: $t_{k+1} = \inf\{t \mid t > t_k \wedge (y(t) - y(t_k))^T(y(t) - y(t_k)) \geq \sigma\}$; • Relative change triggering condition: $t_{k+1} = \inf\{t \mid t > t_k \wedge (y(t) - y(t_k))^T(y(t) - y(t_k)) \geq \sigma y^T(t)y(t)\}$; where σ is a given positive scalar.
Dynamic ET sampling	• Absolute change triggering condition: $t_{k+1} = \inf\{t \mid t > t_k \wedge (y(t) - y(t_k))^T(y(t) - y(t_k)) \geq \frac{1}{\theta}\eta(t) + \sigma\}$ with $\eta(t)$ satisfying $\dot{\eta}(t) = -\lambda\eta(t) + \sigma - \big(y(t) - y(t_k)\big)^T\big(y(t) - y(t_k)\big)$; • Relative change triggering condition: $t_{k+1} = \inf\{t \mid t > t_k \wedge \big(y(t) - y(t_k)\big)^T\big(y(t) - y(t_k)\big) \geq \frac{1}{\theta}\eta(t) + \sigma y^T(t)y(t)\}$ with $\eta(t)$ satisfying $\dot{\eta}(t) = -\lambda\eta(t) + \sigma y^T(t)y(t) - \big(y(t) - y(t_k)\big)^T\big(y(t) - y(t_k)\big)$; where λ, θ and σ are given positive scalars, $\eta(0) \geq 0$ is the initial condition.

TABLE 1.1
Several typical mathematical models of complex samplings.

Nowadays, with the ongoing advances of network technologies, more and more data information of the system components (i.e., sensor, sampler, controller, and actuator) are exchanged through communication networks. Note that the usage of networks has many distinctive advantages such as low installation and maintenance costs, easy manipulation, increased system flexibility, and reduced hardwire. Such advantages have given a great impetus to extensive applications of communication networks in various practical areas such as modern automobiles, wastewater treatment processes, power systems, and transportation systems. Nevertheless, owing primarily to the inherent network constraints such as limited bandwidth, some networked-induced phenomena arise inevitably during the signal transmission. These networked-induced phenomena include, but are not limited to, packet dropouts, saturations, quantizations, and sensor degradations. It is generally acknowledged that these phenomena, if they are not appropriately tackled, could deteriorate the system performance or even result in the instability of the overall dynamical systems. Bearing these in mind, it makes practical sense to look into the effects from the networked-induced phenomena on the sampled-data system (see Fig. 1.1), where the closed-loop sampled-data control is implemented with the help of communication networks to transmit the measured output and the control input signals. Therefore, the aim of this paper is to deal with the control and state estimation problems for complex sampled systems subject to various networked-induced phenomena.

1.2 Recent Advances

In this section, we shall give an overview of the results on complex samplings including nonuniform sampling, stochastic sampling, ET sampling and dynamic ET sampling.

1.2.1 Nonuniform Sampling

In networked and embedded control systems, it is often the case that the sampling intervals are uncertain and/or vary with time due to unpredictable network-induced phenomena. Recently, the analysis/design issues for such nonuniform sampled-data control systems have received considerable research attention. Generally speaking, there have been three different approaches categorized according to the types of the transformed equivalent models of the sampled-data control systems, i.e., 1) continuous-time systems with a delayed control input, 2) discrete-time systems, and 3) hybrid modelling of sampled-data systems.

Up to now, the first approach has been extensively investigated for nonunifrom sampled-data systems. The main idea of this approach is to transform the sampled-data system to a continuous-time counterpart with a bounded TVD. Specifically, if the current sampling instant is t_k and the next sampling instant is t_{k+1}, then the control input between the sequel sampling instants is $u(t) = u(t_k)$, $t \in [t_k, t_{k+1})$. Setting $u(t) = u(t - (t - t_k))$, the underlying sampled-data system can be equivalently transformed to a time-varying input delay system with the delay defined by $\tau(t) = t - t_k$. This approach was proposed in [37], where an LMI-based criterion of the stability for the sampled-data systems has been derived. By using this approach, a number of significant results can be referred to [41, 199, 200], where the stability analysis and control problems have been studied for sampled-data control systems with nonuniform samplings. After that, in [40,116], the scaled small gain theorem has been applied to establish a new simple stability condition which improves the results of [37]. Moreover, the input delay approach has been refined in [36, 100, 131] by introducing a proper time-dependent Lyapunov-Krasovskii functional, where the major contribution lies in that much less conservative stability criteria have been obtained than those [40, 116]. Inspiringly, recent years also have seen a quite frequent use of such an input delay approach for the filtering (or state estimation) issue for nonuniformly sampled systems, see e.g. [4, 8, 44, 45, 57, 71, 83, 109] and the reference therein. More recently, the sampled-data state estimation problem with nonuniform samplings has been examined in [71,109] for a class of Takagi-Sugeno fuzzy systems.

In the past decade, the second approach with discrete-time transformation has also received particular research interests because of the exact integration over a sampling interval leading to less conservative stability conditions. In this approach, a discrete-time controller design would be performed based on a discretized model of the continuous-time system. For example, in [39,150], the stabilization problem of nonuniform sampling systems has been investigated by modelling the considered plant as a linear discrete-time system, where the sampling interval variations are treated as norm bounded matrix uncertainty which is handled by using robust control techniques. Following these works, in [124], an aperiodic sampling has been modeled as parametric uncertainty, rather than matrix uncertainty for less conservative stability analysis, and an LMI-based sufficient condition has been presented accordingly for all possible parameter values. Further improvement can also be found in [69], where the major contribution is to refine the stability conditions proposed in [39,150] by taking into account the positive real property of the time-varying uncertainty in the system. Moreover, the studies in [159] have extended the results in [39,150] to the realm of nonlinear systems with the consideration of large TVDs and packet dropouts. It is worth pointing out that, in [205,210], a novel approach to dealing with the sampled-data filtering problems has been proposed by converting the nonuniform sampling system into a discrete-time switched system with a finite number of subsystems, and then the corresponding distributed H_∞ filtering problem has been studied by restoring to the average dwell time method for switched systems.

A lot of work has been done by following the third approach as well. This approach is based on the representation of the aperiodic sampled-data system in the form of hybrid discrete/continuous model, or more precisely, the impulsive model. In the seminal work [119], by modelling the nonuniformly sampling system with an impulsive system, some sufficient conditions have been derived for asymptotic and exponential stability of such impulsive systems by using a Lyapunov functional with discontinuity at impulsive instants. Later, these results have been extended in [120, 158] to the case that the network TVDs are considered; in [10] to the case that the try-once-discard/round-robin protocols are considered; and in [145] to the case that the input saturations are considered. Subsequent work in [19] has provided a less conservative stability condition than the one in [119] by using a piecewise differentiable Lyapunov functional, which is continuous at impulse times but not necessarily positive definite inside the impulse intervals. Besides, in [74, 82], by using the conception of average impulse interval, new stability criteria with no restrictions on the bounds of sampling intervals have been obtained for the nonuniform sampling systems. Similarly, the impulsive system approach has also been applied to study the sampled-data filtering issues. For example, the design problem of the impulsive observer has been investigated in [34] for a class of continuous-time dynamical systems with nonuniformly sampled measurements. In [98], the performance index in terms of H_∞ has been proposed, and the corresponding sampled-data filtering issue has been solved. Some other typical results along this approach can be found in [3,12,18,32,113].

1.2.2 Stochastic Sampling

In practice, due to some undesirable physical constraints, the aperiodic samplings often occur in a probabilistic way. In this regard, it seems natural to model the sampling intervals by a stochastic process. For example, [42] has considered two different sampling intervals whose occurrence probabilities are given to be constant and satisfy Bernoulli distribution. In this work, an H_∞ control method has been proposed to ensure both the mean-square exponential stability and H_∞ performance of the sampled-data systems. Following the idea in [42], much effort has been devoted to the system analysis/design problems with such stochastic samplings and many results have been obtained, see e.g. [5, 27, 29, 52, 80, 90, 103, 111, 139, 140, 144, 148, 180, 181, 211]. In [181], the H_∞ filtering issue has been investigated for continuous-time systems under sampled measurements with stochastic sampling, where the input delay approach and Lyapunov theory have been utilized to obtain the stability conditions in the form of LMI. In [139], the distributed H_∞ filtering problem for sensor networks has been investigated by using a stochastic sampled-data approach. The sampled-data synchronization control for complex networks with stochastic sampling has been discussed in [140]. The authors in [52] have considered the consensus under stochastic sampling for multiagent systems and have shown that the results are independent of the number of agents thus facilitating its application to large-scale networked agents.

In recent years, many researchers have attempted to further extend the two different sampling intervals to a more general case of multiple stochastic sampling intervals, see e.g. [56, 75, 76, 127, 129, 202]. The so-called multiple stochastic sampling intervals mainly refer to that the sampling intervals take values in a finite set with m different values (rather than two different values) and switch among these values in a random way with given probability. Based on such a stochastic sampling model, the stochastic sampled-data synchronization issue has been studied in [128] for complex dynamical networks with control packet loss and additive TVDs. For consensus issue of multi-agent systems with stochastic sampling, in [162], a sufficient condition for the mean square node-to-node consensus has been presented in terms of LMIs. A study of exponential synchronization of chaotic Lur'e systems with stochastic sampling has been presented in [208], where a unified probability framework has been proposed to design the synchronization controllers. In [220], a consensus protocol has been proposed in which each Euler-Lagrange system employs the stochastic sampled information to communicate with its neighboring nodes, and it has been proved that the consensus can be reached without utilizing relative coordinate derivatives.

Another way to describe the sampling intervals is treating them as random variables obeying certain probability distributions. For example, in [2], the sampling interval has been modeled by a stochastic variables following Erlang distribution, and a stochastic dynamic programming framework has been used for the solution of the optimal control problems. Following a similar line, in [68], the H_2 performance constraint has been introduced to the optimal control context, and an H_2 state feedback controller has been designed for the sampled-data system with randomly sampled measurement. Recently, in [134], a new quantized/saturated control problem has been addressed for a class of sampled-data systems with noisy sampling intervals obeying Erlang distribution, where the confluent Vandermonde matrix approach has been exploited to deal with multiple control inputs. Later, the random variable obeying categorical distribution has been introduced in [61] to describe the stochastic sampling phenomenon, and a probability-distribution-dependent control scheme has been proposed to guarantee the mean-square exponential synchronization of the error dynamical network. On the other hand, in [136], the sampling intervals have been represented by an arbitrary stochastic variable with known probability density function, and a fundamental stabilization problem has been investigated for such stochastic sampled-data systems.

Different from the previous work, a Markov chain has been utilised in [118] to characterize the process of random sampling, which effectively reflects the statistical correlation between adjacent sampling intervals. By transforming the sampled-data systems into Markov jump systems, a mode-dependent filter with desired H_∞ performance has been designed. Inspired by that, in [193], the output feedback stabilization issue has been considered for Markov-based sampled-data networked control systems, where sufficient and necessary LMI-based conditions of the stability have been established for the closed-loop system. These results have been extended in [206] to more general nonlinear systems, which can be approximated by Takagi-Sugeno fuzzy systems. Moreover, the authors in [156] have proposed a hidden Markov model with conditional probability matrix to make full use of the partial information of sampling interval. It should be pointed out that, in all the papers mentioned above, the transition probabilities of the Markov chain have been assumed to be completely known. When the transition probabilities are partially unknown, some initial effort has been paid recently in [104] and [216] to investigate the sampled-data state estimation and fault detection problems, respectively.

1.2.3 Event-Triggered Sampling

As has been mentioned earlier, ET sampling is an effective strategy capable of reducing the number of sampling, and thus saving the computation and communication resources. The direct consideration of event-based sampling was firstly reported in the work of [9], where such a scheme is integrated with a PID controller, and the experiments have shown that the ET sampling could achieve large reductions in CPU utilization compared with conventional time-triggered one. Since then, the ET sampling has received a great deal of research interest in the field of networked control systems, see e.g. [1, 28, 43, 54, 55, 108, 152, 153, 168, 188, 198]. Representatively, for the state-feedback control system, the event-based scheduling policy has been employed in [152] to determine the execution time of the control task, where a Lyapunov method has been utilized to analyze the asymptotic behaviour of the nonlinear system and a lower bound on the inter-execution times has been calculated based on the Lipschitz continuous condition. Note that it is sometimes the case that the information of the system states is not always available for feedback. As such, a dynamical output-based control approach with decentralized ET mechanism has been proposed in [28], where both the stability and $\mathcal{L}_\infty$-performance have been analyzed. In addition, a novel periodic ET control strategy has been developed in [54], in which the triggering condition is verified only periodically, instead of continuously in most existing ET schemes. Recently, in the presence of unknown disturbances, the robust stability issue has been addressed in [1] for nonlinear systems using ET output feedback laws.

In the aforementioned papers, the common assumption is that the controller is known as a prior. In fact, in many practical applications, it is often necessary to find the desired controller such that certain performance requirements could be achieved. To fill the research gap, [201] has been concerned with the ET state feedback controller design problem for networked control systems and derived the criteria for co-designing both the feedback gain and the trigger parameters. After that, an increasing number of excellent results concerning on the ET controller design problem of networked control systems have been reported, see e.g. [59, 81, 93, 125, 133, 149, 172, 203, 213, 217, 219]. For instance, in [125], a parallel distribution compensation fuzzy controller has been designed for a class of networked Takagi-Sugeno fuzzy systems under a discrete ET communication scheme. Furthermore, in [149], the dissipativity has been taken into account in designing an ET fuzzy controller, which ensures the resulting closed-loop system is asymptotically stable with required dissipative performance. In addition to the above-mentioned ET state feedback controller design work, there also have been a variety of results on the design of the ET output feedback controllers in the existing literature. Recently, in [93], the H_∞ static output feedback tracking controller design approach has been developed for a class of nonlinear networked systems subject to quantization effects and asynchronous ET constraints. For continuous-time switched nonlinear systems with asymmetric input saturation, the ET output feedback controller has been constructed in [81] via the common Lyapunov functional approach and the dynamic gain control design approach.

It is notable that the ET sampling method is particularly beneficial in saving communication cost for state estimation issue whose main concern often faced with is the limited energy supply of sensors. In this case, recently, the ET filtering problem has stirred particular research interests and there have been a number of relevant results reported in the existing literature, see e.g. [51,62,63,85,143,163–165,169,178,182,186,187,204]. For example, in [163], the ET H_∞ filtering problem has been addressed for networked Markovian jump systems, where the time-delay modelling method has been employed to describe the ET scheme. By selecting an appropriate ET threshold, a desired balance between the sensor-to-estimator communication rate and the estimation quality has been achieved in [182]. In [143], by using the probability measure change approach, the remote state estimation problem has been investigated for a class of discrete-state hidden Markov systems subject to event-based measurement updates. It is worth pointing out that a novel randomized ET transmission schedule has been proposed in [51], where the minimum mean squared error estimates have been obtained for both the open-loop and closed-loop scenarios. Further studies in [178] have extended the results in [51] to a more general multi-sensor networked system, where the correlated sensor noises are taken into account.

For the large-scale network systems, the energy-saving issue is especially important since the communications among a large number of nodes may consume a significant amount of energy. As such, it is practically crucial to take advantage of the ET sampling to look into the network control/estimation problems, and much research has recently been conducted for various network systems, see e.g. [20, 23, 24, 47, 79, 101, 151, 166, 185, 221, 222] for some representative results. For complex networks, an ET pinning feedback control strategy has been adopted in [79] to achieve the expected synchronization behaviour of stochastic complex networks with nonlinear inner-coupling and exogenous disturbances. In [166], by solving certain convex optimization problem, a series of event-based state estimators has been designed for a class of delayed complex networks subject to nonlinearities and stochastic noises. In the context of multiagent systems, in [20], the ET consensus control issue has been solved for linear multiagent systems subject to nonuniform time-varying communication delays. A similar problem has been solved in [222] for a class of switched stochastic nonlinear multiagent systems. When it comes to sensor networks, an ET distributed filtering problem has been considered in [101] for linear time-varying systems, and a matrix simplification technique has been firstly proposed to deal with the sparsity issue of the sensor network topology. In [47], a distributed H_∞ consensus filtering algorithm has been proposed for sensor networks, where each sensor's communication action with its neighboring sensors is determined by the ET strategy whose threshold parameter is time-varying.

1.2.4 Dynamic Event-Triggered Sampling

In order to further reduce the communication cost, a dynamic ET mechanism has been proposed in [50] by introducing an additional dynamical variable with its dynamics related to the state of the underlying system. It has been shown in [50] that, comparing to the general ET approach, the dynamic one typically generates much fewer triggering times while still guaranteeing the same system performance. Inspired by the idea in [50], much research has been done on system analysis issue with the dynamically ET method. To mention a few, based on the framework of [50], a dynamically ET control method has been developed in [174] under the stochastic scenario to guarantee the asymptotical stability of the nonlinear systems, and a lower bound on the intersample times has been constructed to exclude the Zeno phenomenon. Subsequently, these results have been extended in [110] to the case that the sporadic measurements and varying communication delays are taken into account simultaneously. In [207], a time-delay approach has been utilized to analyze the exponential mean-square stability issue for a dynamic ET stochastic system under a given static output-feedback controller. Instead of continuous monitoring, a two-stage dynamic ET scheme, which only needs to monitor the state information after an admissible sampling period, has been put forward in [11,92] to achieve the exponential stabilization of the underlying continuous-time stochastic systems. In addition, due to the wide applications of discrete-time systems, the dynamic ET controller implementation paradigms have been proposed in [60] under the discrete-time setting, and a unifying framework for the stability analysis has been provided for discrete-time systems.

In spite of the above stability analysis issue, another important issue concerned is to design the desired controller/estimator under a specific dynamic ET scheme such that the required performance is achieved. When it comes to the dynamic ET controller design issue, results obtained along this line of research include [35, 112, 114, 194, 218] and the reference therein. For example, the problem of state feedback H_∞ control with dynamic ET communication strategy has been investigated for various types of systems such as singularly perturbed systems [112], switched systems [35, 114], and fuzzy system [194, 218], in which LMIs-based conditions have been derived for the design of dynamic ET controllers guaranteeing a certain H_∞ performance. Following that, the problem of output feedback H_∞ control has been solved in [223], where the outputs of sensor and controller are transmitted asynchronously based on two independent dynamic ET strategies. Moreover, in [154], the problem of dynamic ET observer-based H_∞ control has been solved for a class of body slip angle of electric vehicles subject to parameter variations of vehicle controllers. Besides, [170, 171] have focused on the design of dynamic ET dissipative asynchronous controller for a class of Markovian jump systems with unknown transition probability. Additionally, in [13, 195], the dynamic ET method has also been integrated into the sliding mode control scheme to stabilize Markovian jump fuzzy systems and interval Type-2 fuzzy systems, respectively.

On the other hand, with respect to dynamical network systems (e.g., multiagent systems, complex networks, and neural networks), some initial research has also been recently conducted on the networked control problems by taking advantage of the dynamic ET mechanism. In the context of multiagent systems, a few preliminary research results have been reported on the centralized or decentralized dynamic ET consensus control problem for first- or second-order systems with fixed topologies or switching topologies subject to cyber attacks, denial-of-service attacks or time-varying communication delays, see [7, 30, 72, 91, 95, 179] and the reference therein.

When it comes to complex networks, [86] has presented a dynamic ET synchronization control approach which achieves the synchronization of all the states of the network nodes. In this work, a new discrete-time version of the dynamic ET mechanism has been proposed in terms of the absolute errors between control input updates for the efficiency of energy utilization. Later, in [99], the cluster synchronization problem has been discussed for a class of complex networks with switching signal, which is characterized by average dwell-time constraint. For neural networks, some initial efforts have been made in [196] on the synchronization problem for inertial memristive neural networks with TVDs via a dynamic ET control scheme.

As for the dynamic ET filter design problem, up to now, there are many effective strategies developed on the basis of variance-constraint filtering, which designs an optimal filter to guarantee the minimum upper bound of estimation error variance; H_∞ filtering which aims to ensure the H_∞ performance requirement of the estimation error dynamics; and set-membership filtering which determines a geometric set enclosing all possible states. Let us summarize some representative work here. Within the variance-constraint framework, a recursive distributed dynamic ET filter has been designed in [87], where each sensor node in the network shares its local measurement with its neighbors according to a time-varying topology, which is connected via Gilbert-Elliott channels. After that, the recursive dynamic ET filtering problems have been thoroughly studied for complex networks [88,167] and multirate systems [142], where their corresponding gain parameters have been derived by solving a set of Riccati-like difference equations. Based on the H_∞ filtering, a general framework has been established in [89] for designing the dynamic ET H_∞ state estimator to reach the required H_∞ performance constraint in the presence of estimator gain perturbations and randomly occurring sensor saturations. A similar problem has been solved in [84] for the case of randomly occurring nonlinearities. In [96,105], the LMI approach has been developed to discuss the resilient H_∞ state estimation problem for delayed systems with dynamic ET mechanisms. In the set-membership framework, the distributed dynamic ET estimation issue has been investigated in [33,46], and the solvability has been presented in terms of the feasibility of a convex optimization problem with linear inequality constraints.

Till now, we have systematically reviewed the research on various kinds of samplings and a great number of results on the analysis and synthesis issues for each sampling have been carefully discussed. We have been working in the control and state estimation areas for more than ten years and would like to share our research results with the audience. Our results on control and state estimation for dynamical network systems with complex samplings will be introduced in detail in this book.

1.3 Outline

The outline of this book is given as follows.

- In Chapter 1, the research background, motivations and research advances are first introduced, which mainly involve nonuniform sampling, stochastic sampling, ET sampling, and dynamic ET sampling. Then, the outline of the book is listed.

- Chapter 2 addresses the stabilization problem for a class of sampled-data systems under noisy sampling intervals. A random variable obeying certain probability distributions is used to describe the noisy sampling intervals. By converting the addressed sampled-data control system into an equivalent discrete-time stochastic system, the stabilization controller is designed such that the resulting discrete-time stochastic system is stochastically stable. Then, by using similar analysis techniques, the stabilization problem is also studied for a general class of stochastic sampled-data control systems with multiple inputs under quantization and/or saturation effects, and a set of parallel results is derived.

- In Chapter 3, the distributed sampled-data H_∞ state estimation problem is investigated for a class of continuous-time systems with infinite-distributed delays and estimator parameter variations. A set of sensors is deployed to acquire the plant output by collaborating with their neighbors according to a given network topology. By utilizing the input delay approach, the effect of the sampling intervals is transformed into an equivalent bounded TVD. A sufficient condition is first established to ensure both the asymptotical stability and the H_∞ performance requirement of the estimation error dynamics. Subsequently, the parameters of the desired distributed estimators are obtained by resorting to the solutions to some matrix inequalities.

- In Chapter 4, the ET control problem is addressed for switched systems. In order to reduce the communication burden, the ET strategy is adopted in the controller design of switched systems. We firstly consider a kind of switched DDSs with exogenous disturbances. The notions of boundedness and input-to-state practical stability are employed to characterize the control objectives in the presence of exogenous disturbances, switching signals, and ET schemes. Some upper bounds on the system states are obtained and the Zeno phenomenon is shown to be excluded under the proposed ET scheme. Then, the gain of the feedback controller, the parameters of the ET function, and the switching signals are jointly designed for the underlying DDSs. Moreover, we also consider the ET control problem for the synchronization of a class of switched stochastic complex networks with identical nodes. The notion of bounded synchronization in probability here is introduced to characterize the performance of the controlled dynamical networks in the presence of exogenous disturbances, switching rules, and ET schemes. Upper bounds of the states of the switched stochastic complex networks are provided, and then the controller gain, the ET parameters, and the average dwell time are co-designed for switched subsystems.

- Chapter 5 is concerned with the ET H_∞ state estimation problem for state-saturated systems. In this system under consideration, a saturation function is introduced to constrain the state variables to stay within a bounded set. Firstly, the ET distributed H_∞ state estimation problem is investigated for a class of state-saturated systems with ROMDs over sensor networks. The mixed delays, which comprise both discrete and distributed delays, are allowed to occur in a random manner governed by two mutually independent Bernoulli distributed random variables. Moreover, the ET H_∞ state estimation problem is also studied for a class of state-saturated complex networks subject to quantization effects as well as randomly occurring distributed delays. The main purpose is to design ET state estimators such that the error dynamics of state estimation is exponentially mean-square stable with a prescribed H_∞ performance index. Sufficient conditions are derived via intensive stochastic analysis to guarantee the existence of the desired estimators, and the parameters of the desired estimators are then obtained in light of the feasibility of certain sets of matrix inequalities.

- In Chapter 6, the ET state estimation problem is investigated for discrete-time neural networks. For the purpose of energy saving, the ET mechanism is adopted, and the measurement outputs are only transmitted to the estimator when a certain triggering condition is met. Firstly, we develop a new ET estimation technique for the delayed neural networks with SPs and IMs. In order to cater for more realistic transmission process of the neural signals, we make the first attempt to introduce a set of stochastic variables to characterize the random fluctuations of system parameters. The incomplete information under consideration includes randomly occurring sensor saturations and quantizations. A Lyapunov functional is constructed to obtain sufficient conditions under which the estimation error dynamics is exponentially ultimately bounded in the mean square. It is worth noting that the ultimate boundedness of the error dynamics is explicitly estimated. The other research focus is to design the ET H_∞ state estimators for a class of discrete-time stochastic GRNs (that can be reviewed as a special case of neural networks) such that, in the presence of MJPs and TVDs, the estimation error dynamics is stochastically stable with a prescribed H_∞ performance level.

- In Chapter 7, the ET fusion estimation problem is studied for multi-rate systems. The multi-rate systems under consideration include several sensor nodes with different sampling rates. The ET robust fusion estimation problem is firstly considered for a class of uncertain multi-rate sampled-data systems with stochastic nonlinearities and coloured measurement noises. A new augmentation approach is proposed by which the multi-rate system is transformed into the single-rate system. The main purpose is to design a set of ET local filters for each sensor node such that the upper bound of each local FEC is guaranteed and minimized at each sampling instant. Sufficient conditions are established for the existence of upper bounds of the local FECs, and then the local filter gains are parameterized according to certain matrix recursions. Subsequently, for the local state estimates, a new fusion estimation scheme is proposed with the help of CI method, and the consistency of the proposed CI-based fusion estimation scheme is shown. Moreover, the ET fusion estimation problem is also studied for a class of multi-rate systems subject to sensor degradations. A set of random variables obeying known probability distributions are used to characterize the phenomenon of the sensor degradations. By using similar analysis techniques, the corresponding ET filters are designed.

- In Chapter 8, the synchronization control problem is considered for a class of discrete time-delay complex dynamical networks under a dynamic ET mechanism. For the efficiency of energy utilization, we make the first attempt to introduce a dynamic event-triggering strategy into the design of synchronization controllers for complex dynamical networks. By constructing an appropriate Lyapunov functional, the dynamics of each network node combined with the introduced event-triggering mechanism are first analyzed, and a sufficient condition is then provided under which the synchronization error dynamics is exponentially ultimately bounded. Subsequently, a set of the desired synchronization controllers is designed by solving a matrix inequality.

- In Chapter 9, the filtering and state estimation problems are studied by using the dynamic ET mechanisms. For the sake of energy saving, a dynamic ET mechanism is employed in the design of filters and state estimators. Firstly, the dynamic ET filtering problem is investigated for a class of discrete time-varying systems with CMs and PUs. The CMs under consideration are described by the Tobit measurement model. By means of the mathematical induction, an upper bound is derived for the FEC in terms of recursive matrix equations and such an upper bound is then minimized by designing the filter gain properly. Furthermore, the boundedness is discussed for the minimized upper bound of the FEC. Secondly, we consider the system outputs are collected through a sensor network subject to a time-varying topology that is connected via Gilbert-Elliott channels and governed by a set of Markov chains. By using similar analysis techniques, the corresponding dynamic ET distributed filtering problem is studied, and a set of parallel results is derived. Thirdly, the finite-time resilient H_∞ state estimation problem is discussed for delayed neural networks under dynamic ET mechanisms. In order to handle the possible fluctuation of the estimator gain parameters when the state estimator is implemented, a resilient state estimator is adopted. Lyapunov functional approach is carried out to obtain sufficient conditions for the existence of desired estimators ensuring both the finite-time boundedness and the H_∞ performance of the estimation error system.

- In Chapter 10, we sum up the results of the book and discuss some related topics for future research work.

2

Stabilization and Control under Noisy Sampling Intervals

In this chapter, a fundamental stabilization problem is investigated for a class of sampled-data systems under noisy sampling intervals. The stochastic sampled-data control system under consideration is converted into a discrete-time system whose system matrix is represented as an equivalent yet tractable form via the matrix exponential computation. By introducing a Vandermonde matrix, the mathematical expectation of the quadratic form of the system matrix is computed. By recurring to the Kronecker product operation, the sampled-data stabilization controller is designed such that the closed-loop system is stochastically stable in the presence of noisy sampling intervals. Subsequently, by using similar analysis techniques, the stabilization problem is also studied for a general class of stochastic sampled-data control systems with multiple inputs under quantization and/or saturation effects, and a set of parallel results is derived. Finally, some numerical simulation examples are provided to demonstrate the effectiveness of the proposed design approach.

2.1 Stabilization with Single Input

In this section, a general framework is developed to deal with the stabilization problem for a class of sampled-data systems with signal input under noisy sampling intervals.

2.1.1 Problem Formulation

Consider the following continuous-time system

$$\dot{x}(t) = Ax(t) + Bu(t) \tag{2.1}$$

where $x(t) \in \mathbb{R}^n$ is the state vector, $u(t) \in \mathbb{R}^m$ is the control input, A is the system matrix and B is the input matrix. The initial value is given by x_0.

DOI: 10.1201/9781003307648-2

The control input $u(t)$ is generated by a ZOH function with a sequence of hold times $0 = t_0 < t_1 < \cdots < t_k < \cdots$

$$u(t) = Kx(t_k), \quad t_k \leq t < t_{k+1} \tag{2.2}$$

where K is the gain matrix to be determined and t_k denotes the sampling instant satisfying $\lim_{k\to\infty} t_k = \infty$.

By substituting (2.2) into (2.1), the closed-loop system is obtained as follows

$$\dot{x}(t) = Ax(t) + BKx(t_k), \quad t_k \leq t < t_{k+1}. \tag{2.3}$$

Integrating the above equation from t_k to t_{k+1}, one has

$$x(t_{k+1}) = \left(\mathrm{e}^{A(t_{k+1}-t_k)} + \int_{t_k}^{t_{k+1}} \mathrm{e}^{A(t_{k+1}-\theta)} d\theta BK\right) x(t_k). \tag{2.4}$$

Letting the sampling interval be $T_k = t_{k+1} - t_k$ and denoting $x(t_k)$ by x_k, we come up with a discrete-time system of the following form

$$x_{k+1} = \left(\mathrm{e}^{AT_k} + \int_0^{T_k} \mathrm{e}^{As} dsBK\right) x_k. \tag{2.5}$$

The sampling interval T_k under consideration is subject to noisy perturbations and consists of two parts, i.e., $T_k = T + v_k$, where T is a constant which stands for the nominal sampling interval and v_k is a random variable which accounts for the sampling errors/drifts/deviations resulting from unpredictable environmental phenomena. The probability density function of the random variable v_k is denoted by $f(v)$, where the argument v satisfies $T + v > 0$.

Remark 2.1 *It is worth mentioning that, to date, very little attention has been paid to noisy sampling interval issue despite its practical significance in engineering. In [42, 68], the noisy sampling interval has been taken into account, where the sampling intervals have been simply assumed to obey Bernoulli or Erlang distribution. In this section, the probability distributions of the sampling intervals are general that include the above probability distributions as special cases and the aim of this section is to develop a design approach for the sampled-data control systems with such a general noisy sampling model.*

Note that the system (2.5) is a discrete-time stochastic system due to the random nature of the sampling error, and therefore the following notion of stochastic stability is needed.

Definition 2.1 *[26, 190] The discrete-time stochastic system (2.5) is said to be stochastically stable if*

$$\mathbb{E}\left\{\sum_{k=0}^{\infty} \|x_k\|^2\right\} < \infty. \tag{2.6}$$

In this section, we aim to investigate the sampled-data stabilization problem for the system (2.1) in presence of the random sampling error, that is, we are interested in finding the controller gain matrix K such that the discrete-time stochastic system (2.5) is stochastically stable with respect to noisy sampling intervals obeying certain probability distributions.

2.1.2 Main Results

To start with, a sufficient condition is provided in the following well-known lemma for the stochastic stability of the system (2.5).

Lemma 2.1 *Given the controller gain matrices K, the stochastic system (2.5) is stochastically stable if there exists a positive definite matrix Q such that the following inequality holds:*

$$\mathbb{E}\left\{\left(e^{AT_k}+\int_0^{T_k}e^{As}dsBK\right)^T Q\left(e^{AT_k}+\int_0^{T_k}e^{As}dsBK\right)\right\}-Q<0. \tag{2.7}$$

Proof *The proof of this lemma is straightforward and is therefore omitted.*

It can be seen from (2.7) that the random variable T_k appears in the upper boundary of integral, and this makes it difficult to directly calculate the mathematical expectation of the integral. In what follows, our main efforts will be made towards the computation of this mathematical expectation by recurring to the matrix theory.

Firstly, it follows from Theorem 1 in [107] that

$$\mathrm{e}^{CT_k}=\begin{bmatrix}\mathrm{e}^{AT_k} & \int_0^{T_k}\mathrm{e}^{As}dsB\\ 0 & I\end{bmatrix} \tag{2.8}$$

where

$$C=\begin{bmatrix}A & B\\ 0 & 0\end{bmatrix}. \tag{2.9}$$

Then, $\mathrm{e}^{AT_k}+\int_0^{T_k}\mathrm{e}^{As}dsBK$ can be rewritten as

$$\begin{aligned}&\mathrm{e}^{AT_k}+\int_0^{T_k}\mathrm{e}^{As}dsBK\\ &=\begin{bmatrix}I & 0\end{bmatrix}\begin{bmatrix}\mathrm{e}^{AT_k} & \int_0^{T_k}\mathrm{e}^{As}dsB\\ 0 & I\end{bmatrix}\begin{bmatrix}I\\ K\end{bmatrix}\\ &=\begin{bmatrix}I & 0\end{bmatrix}\mathrm{e}^{CT}\mathrm{e}^{Cv_k}\begin{bmatrix}I\\ K\end{bmatrix}.\end{aligned} \tag{2.10}$$

Secondly, we transform the matrix exponential e^{Cv_k} into a tractable form. In order to avoid unnecessary mathematical complexity, we make the following assumption.

Assumption 2.1 *Suppose that the matrix A has different eigenvalues $\lambda_i \neq 0$ $(i = 1, 2, \cdots, n)$ and the control input $u(t)$ is a scalar, i.e., $m = 1$.*

Under Assumption 2.1, it is easily known that the eigenvalues of matrix C (denoted by $\lambda_{C,i}$) are $\lambda_{C,i} = \lambda_i$ $(i = 1, 2, \cdots, n)$ and $\lambda_{C,n+1} = 0$. Then, the Vandermonde matrix generated by the eigenvalues of matrix C is given by

$$V = \begin{bmatrix} 1 & 1 & \cdots & 1 \\ \lambda_{C,1} & \lambda_{C,2} & \cdots & \lambda_{C,n+1} \\ \vdots & \vdots & \ddots & \vdots \\ \lambda_{C,1}^n & \lambda_{C,2}^n & \cdots & \lambda_{C,n+1}^n \end{bmatrix}. \tag{2.11}$$

By following the same analysis in [117], the matrix exponential e^{Cv_k} can be expressed by

$$\mathrm{e}^{Cv_k} = \sum_{i=1}^{n+1} \mathrm{e}^{\lambda_{C,i} v_k} C_i \tag{2.12}$$

where

$$C_i = \sum_{j=1}^{n+1} \nu_{ij} C^{j-1} \tag{2.13}$$

and ν_{ij} is the (i, j) entry of V^{-1}.

Remark 2.2 *Under Assumption 2.1, it is ensured that the matrix C has different eigenvalues. In this case, the matrix exponential e^{Cv_k} can be computed according to (2.12) and (2.13).*

Introduce the matrix Π_f defined by $\Pi_f = [\pi_{ij}^f]_{(n+1)\times(n+1)}$ with

$$\pi_{ij}^f = \begin{cases} \int_{-\infty}^{+\infty} \mathrm{e}^{2\lambda_{C,i} v} f(v) dv, & i = j, \\ \int_{-\infty}^{+\infty} \mathrm{e}^{(\lambda_{C,i} + \lambda_{C,j}) v} f(v) dv, & i \neq j. \end{cases} \tag{2.14}$$

Lemma 2.2 *For the matrix Π_f defined in (2.14), there exists a matrix U_f such that $\Pi_f = U_f^T U_f$.*

Proof *From (2.14), it is easily seen that the matrix Π_f is symmetric. For any $\eta \neq 0 \in \mathbb{R}^{n+1}$, we have*

$$\eta^T \Pi_f \eta = \int_{-\infty}^{+\infty} \eta^T \Pi_v \eta f(v) dv \tag{2.15}$$

where $\Pi_v = [\pi_{ij}^v]_{(n+1)\times(n+1)}$ with

$$\pi_{ij}^v = \begin{cases} e^{2\lambda_{C,i} v}, & i = j, \\ e^{(\lambda_{C,i} + \lambda_{C,j}) v}, & i \neq j. \end{cases} \tag{2.16}$$

Note that, for any $v \in \mathbb{R}$, the matrix Π_v defined in (2.16) is always positive semi-definite and, therefore, for all $\eta \in \mathbb{R}^{n+1}$, we have $\eta^T \Pi_v \eta \geq 0$. By considering that the probability density function $f(v)$ is positive for all $v \in \mathbb{R}$, it follows from (2.15) that $\eta^T \Pi_f \eta \geq 0$ for all $\eta \in \mathbb{R}^{n+1}$. This implies that the matrix Π_f is also positive semi-definite and hence we can find a matrix U_f such that $\Pi_f = U_f^T U_f$. The proof of Lemma 2.2 is complete.

Now, our main results are given as follows.

Theorem 2.1 *For the continuous-time system (2.1) with the sampled-data controller given by (2.2), the sampled-data stabilization problem with noisy sampling interval is solvable if there exist matrices $P > 0$ and Y such that the following inequality holds:*

$$\begin{bmatrix} -P & \Phi_f \\ * & -(I \otimes P) \end{bmatrix} < 0 \tag{2.17}$$

where

$$\Phi_f = \begin{bmatrix} P & Y^T \end{bmatrix} e^{C^T T} \bar{C}^T (U_f^T \otimes \begin{bmatrix} I \\ 0 \end{bmatrix}), \quad \bar{C} = \begin{bmatrix} C_1^T & C_2^T & \cdots & C_{n+1}^T \end{bmatrix}^T.$$

Furthermore, if inequality (2.17) is feasible, the desired controller gain is given by

$$K = YP^{-1}. \tag{2.18}$$

Proof *According to (2.12), one has*

$$\begin{aligned}
&\mathbb{E}\left\{ e^{C^T v_k} \begin{bmatrix} I \\ 0 \end{bmatrix} Q \begin{bmatrix} I & 0 \end{bmatrix} e^{C v_k} \right\} \\
&= \mathbb{E}\left\{ \left(\sum_{i=1}^{n+1} e^{\lambda_{C,i} v_k} C_i \right)^T \begin{bmatrix} I \\ 0 \end{bmatrix} Q \begin{bmatrix} I & 0 \end{bmatrix} \sum_{i=1}^{n+1} e^{\lambda_{C,i} v_k} C_i \right\} \\
&= \bar{C}^T \left(\left(\int_{-\infty}^{+\infty} \Pi_v f(v) dv \right) \otimes \left(\begin{bmatrix} I \\ 0 \end{bmatrix} Q \begin{bmatrix} I & 0 \end{bmatrix} \right) \right) \bar{C} \\
&= \bar{C}^T \left(\Pi_f \otimes \left(\begin{bmatrix} I \\ 0 \end{bmatrix} Q \begin{bmatrix} I & 0 \end{bmatrix} \right) \right) \bar{C} \\
&= \bar{C}^T (U_f^T \otimes \begin{bmatrix} I \\ 0 \end{bmatrix})(I \otimes Q)(U_f \otimes \begin{bmatrix} I & 0 \end{bmatrix}) \bar{C}
\end{aligned} \tag{2.19}$$

where Π_f and Π_v are defined in (2.14) and (2.16), respectively. It then follows from (2.10) that

$$\mathbb{E}\left\{ \left(e^{AT_k} + \int_0^{T_k} e^{As} ds BK \right)^T Q \left(e^{AT_k} + \int_0^{T_k} e^{As} ds BK \right) \right\}$$

$$
\begin{aligned}
&= \begin{bmatrix} I & K^T \end{bmatrix} e^{C^T T} \mathbb{E}\left\{ e^{C^T v_k} \begin{bmatrix} I \\ 0 \end{bmatrix} Q \begin{bmatrix} I & 0 \end{bmatrix} e^{C v_k} \right\} e^{CT} \begin{bmatrix} I \\ K \end{bmatrix} \\
&= \bar{\Phi}_f (I \otimes Q) \bar{\Phi}_f^T
\end{aligned} \tag{2.20}
$$

where $\bar{\Phi}_f = \begin{bmatrix} I & K^T \end{bmatrix} e^{C^T T} \bar{C}^T (U_f^T \otimes \begin{bmatrix} I \\ 0 \end{bmatrix})$.

Subsequently, by using the Schur Complement Lemma, the condition (2.7) in Lemma 2.1 is true if and only if

$$
\begin{bmatrix} -Q & \bar{\Phi}_f (I \otimes Q) \\ * & -(I \otimes Q) \end{bmatrix} < 0 \tag{2.21}
$$

which is equivalent to inequality (2.17) by performing a congruence transformation of $\mathrm{diag}\{Q^{-1}, I \otimes Q^{-1}\}$ *and noting the relations* $P = Q^{-1}$ *and* $Y = KP$. *That is, inequality (2.17) implies that condition (2.7) in Lemma 2.1 holds, and the rest of the proof follows directly from Lemma 2.1.*

In Theorem 2.1, the existence condition of the sampled-data controller is obtained and, if the probability density function of the sampling error is determined, the desired sampled-data controller can be designed by finding a solution to inequality (2.17). The feature of the design approach developed is that we have exploited the matrix theories including matrix exponential computation, Vandermonde matrix, and Kronecker product operation which deal well with the difficulties from both the random nature and the nonlinear characteristics.

Remark 2.3 *Note that the design approach developed can cope with the case when the sampling intervals follow certain probability distributions. For other sampling disturbance, so long as the probability distributions of their sampling intervals are known, the design approach developed is always effective.*

In what follows, we discuss a special case, where the sampling error obeys the continuous uniform distribution.

Suppose that the sampling error v_k follows the continuous uniform distribution between $-\delta$ and δ, i.e, the probability density function of the random variable v_k is given by

$$
f(v) = \begin{cases} \frac{1}{2\delta}, & -\delta \leq v \leq \delta, \\ 0, & \text{elsewhere.} \end{cases} \tag{2.22}
$$

where $0 < \delta < T$.

By the same definition of Π_f as before, we obtain $\Pi_\delta = [\pi_{ij}^\delta]_{(n+1)\times(n+1)}$ with

$$
\pi_{ij}^\delta = \begin{cases} \lim_{\lambda \to \lambda_{C,i}} \frac{1}{4\delta\lambda} (\mathrm{e}^{2\lambda\delta} - \mathrm{e}^{-2\lambda\delta}), & i = j, \\ \lim_{\lambda \to (\lambda_{C,i} + \lambda_{C,j})} \frac{1}{2\delta\lambda} (\mathrm{e}^{\lambda\delta} - \mathrm{e}^{-\lambda\delta}), & i \neq j. \end{cases} \tag{2.23}
$$

In virtue of Lemma 2.2, we can find a matrix U_δ such that $\Pi_\delta = U_\delta^T U_\delta$. Then, by Theorem 2.1, the following corollary is easily obtained.

Corollary 2.1 *Suppose that the sampling error v_k obeys the continuous uniform distribution between $-\delta$ and δ $(0 < \delta < T)$. If there exist matrices $P > 0$ and Y satisfying the following inequality:*

$$\begin{bmatrix} -P & \Phi_\delta \\ * & -(I \otimes P) \end{bmatrix} < 0 \tag{2.24}$$

where

$$\Phi_\delta = \begin{bmatrix} P & Y^T \end{bmatrix} e^{C^T T} \bar{C}^T (U_\delta^T \otimes \begin{bmatrix} I \\ 0 \end{bmatrix}), \tag{2.25}$$

then, the sampled-data controller (2.2) with gain matrix given by (2.18) achieves the stochastic stability of the stochastic systems (2.5).

Proof *The proof follows directly from Theorem 2.1.*

Until now, the desired sampled-data stabilization controller has been designed such that the sampled-data control system is stochastically stable when the sampling intervals follow certain probability distributions. As a special case, the sampled-data stabilization controller has been designed when the sampling intervals follow continuous uniform distribution.

2.2 Quantized/Saturated Control with Multiple Inputs

In this section, we consider the stabilization problem for a general class of stochastic sampled-data control systems with multiple inputs under quantization and/or saturation effects.

2.2.1 Problem Formulation

Consider the following system:

$$\dot{x}(t) = Ax(t) + Bg(u(t)) \tag{2.26}$$

where $x(t) \in \mathbb{R}^n$ is the state vector, $u(t) \in \mathbb{R}^p$ is the control input, $g(\cdot)$ is the nonlinear function, A and B are the known matrices with appropriate dimensions.

The control input $u(t)$ is assumed to be piecewise constant, i.e.,

$$u(t) = Kx(t_k), \quad t_k \leq t < t_{k+1} \tag{2.27}$$

where K is the gain matrix to be determined and the set of sampling instants $\{t_k\}_{k \in N}$ satisfies $0 = t_0 < t_1 < \cdots < t_k < \cdots$ and $\lim_{k\to\infty} t_k = \infty$.

Denote by $T_k = t_{k+1} - t_k$ the sampling interval and suppose that $T_k = T + v_k$, where T is the nominal sampling period and v_k represents the sampling error. Let v_k obey the Erlang distribution with the following probability density function

$$f(s,\mathcal{K},\mu) = \frac{\mu^{\mathcal{K}} s^{\mathcal{K}-1} e^{-\mu s}}{(\mathcal{K}-1)!} \quad \text{for} \quad s > 0 \tag{2.28}$$

where $\mathcal{K} \in N$ is the shape parameter and $\mu > 0$ is the rate parameter.

Remark 2.4 *In [2], the Erlang distribution model has been proposed to describe the stochastic sampling phenomenon. The Erlang distribution is more general that includes the exponential distribution as a special case. Therefore, in this section, we choose the Erlang distributed random variable to represent the stochastic sampling error.*

From (2.26) and (2.27), we have the following closed-loop system

$$\dot{x}(t) = Ax(t) + Bg(Kx(t_k)), \quad t_k \le t < t_{k+1}. \tag{2.29}$$

Integrating (2.29) from t_k to t_{k+1} and denoting $x(t_k)$ by x_k, an equivalent discrete-time stochastic system is given by

$$x_{k+1} = e^{AT_k} x_k + \int_0^{T_k} e^{As} ds Bg(Kx_k). \tag{2.30}$$

We are interested in the following two cases.

Case 1: The nonlinear function $g(\cdot)$ is a logarithmic type quantization mapping, i.e., $g(u) = q(u)$, which is defined as

$$q(u) = \begin{bmatrix} q_1(u_1) & q_2(u_2) & \dots & q_p(u_p) \end{bmatrix}^T,$$

with $u = \begin{bmatrix} u_1 & u_2 & \dots & u_p \end{bmatrix}^T$. For each j ($j = 1, 2, \dots, p$), the set of quantization levels is described by

$$\mathcal{U}_j = \{\pm\chi_i^{(j)},\ \chi_i^{(j)} = \rho_j^i \chi_0^{(j)},\ i = 0, \pm 1, \dots\} \cup \{0\}, \quad 0 < \rho_j < 1,\ \chi_0^{(j)} > 0,$$

and the logarithmic quantizer $q_j(\cdot)$ is defined as

$$q_j(u_j) = \begin{cases} \chi_i^{(j)}, & \frac{1}{1+\delta_j}\chi_i^{(j)} < u_j \le \frac{1}{1-\delta_j}\chi_i^{(j)}, \\ 0, & u_j = 0, \\ -q_j(u_j), & u_j < 0, \end{cases}$$

with $\delta_j = \frac{1-\rho_j}{1+\rho_j}$.

Case 2: The nonlinear function $g(\cdot)$ is a saturation mapping, i.e., $g(u) = s(u)$, which is defined as

$$s(u) = \begin{bmatrix} s_1(u_1) & s_2(u_2) & \dots & s_p(u_p) \end{bmatrix}^T, \tag{2.31}$$

where $s_j(u_j) = \text{sign}(u_j)\min\{u_{j,\max}, |u_j|\}$ for each $j = 1, 2, \ldots, p$.

The objective of this section is to design a controller gain K ($K = K_q$ in case of quantized control and $K = K_s$ in case of saturated control) for sampled-data system (2.26) such that, when the sampling error follows the Erlang distribution, the equivalent discrete-time stochastic system (2.30) is stochastically stable.

2.2.2 Main Results

A: Confluent Vandermonde Matrix Approach

Let $C \in \mathbb{R}^{n_c \times n_c}$ be a square matrix and denote the eigenvalues of matrix C by $\lambda_1, \lambda_2, \cdots, \lambda_f$ with their multiplicities $m_1, m_2, \cdots, m_f$, respectively, where $m_1 + m_2 + \cdots + m_f = n_c$. The confluent Vandermonde matrix V_c generated by the eigenvalues λ_d $(d = 1, 2, \cdots, f)$ is given by

$$V_c = \begin{bmatrix} \Lambda_1 & \Lambda_2 & \cdots & \Lambda_f \end{bmatrix} \tag{2.32}$$

where, for each d $(d = 1, 2, \cdots, f)$,

$$\Lambda_d = \begin{bmatrix} 1 & 0 & 0 & 0 & \cdots & 0 \\ \lambda_d & 1 & 0 & 0 & \cdots & 0 \\ \lambda_d^2 & 2\lambda_d & 2 & 0 & \cdots & 0 \\ \vdots & \vdots & \vdots & \vdots & \cdots & \vdots \\ \lambda_d^{m_d-1} & (m_d-1)\lambda_d^{m_d-2} & \frac{(m_d-1)!}{(m_d-3)!}\lambda_d^{m_d-3} & \frac{(m_d-1)!}{(m_d-4)!}\lambda_d^{m_d-4} & \cdots & (m_d-1)! \\ \vdots & \vdots & \vdots & \vdots & \cdots & \vdots \\ \lambda_d^{n_c-1} & (n_c-1)\lambda_d^{n_c-2} & \frac{(n_c-1)!}{(n_c-3)!}\lambda_d^{n_c-3} & \frac{(n_c-1)!}{(n_c-4)!}\lambda_d^{n_c-4} & \cdots & \frac{(n_c-1)!}{(n_c-m_d)!}\lambda_d^{n_c-m_d} \end{bmatrix}.$$

Lemma 2.3 *For matrix $C \in \mathbb{R}^{n_c \times n_c}$ and a scalar v, we have*

$$e^{Cv} = \left(\left(\pi(v) V_c^{-1} \right) \otimes I \right) \bar{C} \tag{2.33}$$

where

$$\bar{C} = \begin{bmatrix} I & C^T & \cdots & C^{(n_c-1)T} \end{bmatrix}^T, \quad \pi(v) = \begin{bmatrix} \pi_1(v) & \pi_2(v) & \cdots & \pi_f(v) \end{bmatrix},$$
$$\pi_d(v) = \begin{bmatrix} e^{\lambda_d v} & v e^{\lambda_d v}, & \cdots & , v^{m_d-1} e^{\lambda_d v} \end{bmatrix}.$$

Proof *It follows from [117] that*

$$e^{Cv} = \sum_{j=1}^{n_c} \alpha_j(v) C^{j-1} \tag{2.34}$$

where $\alpha_j(v)$ $(j = 1, 2, \cdots, n_c)$ are functions of v.

From (2.34), it can be obtained that, for each eigenvalue λ_d ($d = 1, 2, \cdots, f$)

$$e^{\lambda_d v} = \sum_{j=1}^{n_c} \alpha_j(v)\lambda_d^{j-1}. \tag{2.35}$$

Derivatives of both sides of the above equation with respect to λ_d from first to $(m_d - 1)$th order yield

$$\begin{aligned} ve^{\lambda_d v} &= \sum_{j=2}^{n_c}(j-1)\alpha_j(v)\lambda_d^{j-2}, \\ &\cdots\cdots \\ v^{m_d-1}e^{\lambda_d v} &= \sum_{j=m_d}^{n_c} \frac{(j-1)!}{(j-m_d)!}\alpha_j(v)\lambda_d^{j-m_d}. \end{aligned} \tag{2.36}$$

By setting $\alpha(v) = \begin{bmatrix}\alpha_1(v) & \alpha_2(v) & \cdots & \alpha_{n_c}(v)\end{bmatrix}$, the set of equations given by (2.35) and (2.36) is then rewritten as the compact form $\pi_d(v) = \alpha(v)\Lambda_d$, from which, we further have

$$\pi(v) = \alpha(v)V_c. \tag{2.37}$$

By substituting (2.37) into (2.34), the equation (2.33) is obtained immediately and hence the proof of Lemma 2.3 is complete.

Remark 2.5 *With the help of confluent Vandermonde matrix, the matrix exponential is transformed into a general matrix form in Lemma 2.3, which is convenient for the design of controller in the case of multiple inputs. Comparing to the representation of the integral of matrix exponential proposed in [48], the method developed in this section is capable of handling the case of stochastic sampling interval.*

Define matrix $\Pi = \mathbb{E}\{\pi^T(v_k)\pi(v_k)\}$. By using the definition of the mathematical expectation, the matrix Π is obtained immediately in the following lemma.

Lemma 2.4 *For a random variable e_k obeying the probability density function (2.28), under the condition $\mu > 2\rho(A)$ ($\rho(A)$ is the largest eigenvalue of matrix A), matrix $\Pi = [\pi_{ij}]_{n_c \times n_c}$ is given by*

$$\pi_{ij} = \frac{(\mathcal{K}+\theta-1)!\mu^{\mathcal{K}}}{(\mathcal{K}-1)!(\mu-(\lambda_a+\lambda_b))^{\mathcal{K}+\theta}} \tag{2.38}$$

where

$$a = \min\{d | i / \sum_{l=1}^{d} m_l \le 1,\ 1 \le d \le f\},$$

$$b = \min\{d|j/\sum_{\tau=1}^{d} m_\tau \leq 1,\ 1 \leq d \leq f\},$$

$$\theta = i + j - \sum_{l=1}^{a} m_l - \sum_{\tau=1}^{b} m_\tau + m_a + m_b - 2.$$

Remark 2.6 *The existence of the scalar π_{ij} given by (2.38) is guaranteed under the assumption $\mu > 2\rho(A)$.*

It is easily verified that matrix Π is positive semi-definite and hence there exists a matrix U such that $\Pi = U^T U$.

In what follows, we select $C = \begin{bmatrix} A & B \\ 0 & 0 \end{bmatrix}$. According to Theorem 1 in [107], it is easily known that

$$e^{CT_k} = \begin{bmatrix} e^{AT_k} & \int_0^{T_k} e^{As} ds B \\ 0 & I \end{bmatrix}. \tag{2.39}$$

B: Design of Quantized Sampled-Data Controller

According to analysis in [38], the quantization effect of the logarithmic type quantizer can be expressed by $q_j(u_j) = (1 + \Delta_j)u_j$ with $|\Delta_j| \leq \delta_j$. Defining $\Delta = \mathrm{diag}\{\Delta_1, \Delta_2, \ldots, \Delta_p\}$, $\Xi = \mathrm{diag}\{\delta_1, \cdots, \delta_p\}$ and $F = \Delta\Xi^{-1}$, one has $q(u) = (I + F\Xi)u$ and $F^T F = FF^T \leq I$.

By using the confluent Vandermonde matrix approach, the quantized sampled-data controller is designed in the following theorem.

Theorem 2.2 *Suppose that $\mu > 2\rho(A)$. For the continuous-time system (2.26) with the sampled-data controller given by (2.27), the discrete-time stochastic system (2.30) is stochastically stable if there exist matrices $P_q > 0$, Y_q and a constant $\varepsilon_q > 0$ such that the following inequality holds:*

$$\begin{bmatrix} -P_q & \begin{bmatrix} P_q & Y_q^T \end{bmatrix} G^T & 0 & Y_q^T \Xi^T \\ * & -I \otimes P_q & \varepsilon_q G \begin{bmatrix} 0 \\ I \end{bmatrix} & 0 \\ * & * & -\varepsilon_q I & 0 \\ * & * & * & -\varepsilon_q I \end{bmatrix} < 0 \tag{2.40}$$

where $G = \left((UV_c^{-1}) \otimes \begin{bmatrix} I & 0 \end{bmatrix} \right) \bar{C} e^{CT}$. Furthermore, if inequality (2.40) is feasible, the desired quantized controller gain is given by

$$K_q = Y_q P_q^{-1}. \tag{2.41}$$

Proof *Let the Lyapunov function candidate be*

$$V(x_k) = x_k^T Q_q x_k \tag{2.42}$$

where Q_q is a positive definite matrix and the difference of the Lyapunov function is defined by

$$\Delta V(x_k) = \mathbb{E}\{V(x_{k+1})|x_k\} - V(x_k). \tag{2.43}$$

It then follows from (2.30), (2.39), and (2.43) that

$$\begin{aligned}
&\mathbb{E}\{\Delta V(x_k)\} \\
&= \mathbb{E}\Bigg\{ \left(e^{AT_k} x_k + \int_0^{T_k} e^{As} ds B q(K_q x_k) \right)^T Q_q \\
&\quad \times \left(e^{AT_k} x_k + \int_0^{T_k} e^{As} ds B q(K_q x_k) \right) - x_k^T Q_q x_k \Bigg\} \\
&= \mathbb{E}\Bigg\{ x_k^T \Bigg(\begin{bmatrix} I & K_q^T (I + F\Xi)^T \end{bmatrix} e^{C^T T} e^{C^T v_k} \begin{bmatrix} I \\ 0 \end{bmatrix} Q_q \\
&\quad \times \begin{bmatrix} I & 0 \end{bmatrix} e^{C v_k} e^{CT} \begin{bmatrix} I \\ (I + F\Xi) K_q \end{bmatrix} - Q_q \Bigg) x_k \Bigg\}.
\end{aligned} \tag{2.44}$$

From Lemmas 2.3 and 2.4, it is obtained that

$$\begin{aligned}
&\mathbb{E}\left\{ e^{C^T v_k} \begin{bmatrix} I \\ 0 \end{bmatrix} Q_q \begin{bmatrix} I & 0 \end{bmatrix} e^{C v_k} \right\} \\
=& \bar{C}^T (V_c^{-T} \otimes I) \mathbb{E}\left\{ (\pi^T(v_k)\pi(v_k)) \otimes \left(\begin{bmatrix} I \\ 0 \end{bmatrix} Q_q \begin{bmatrix} I & 0 \end{bmatrix} \right) \right\} (V_c^{-1} \otimes I) \bar{C} \\
=& \bar{C}^T (V_c^{-T} \otimes I) \left(\Pi \otimes \left(\begin{bmatrix} I \\ 0 \end{bmatrix} Q_q \begin{bmatrix} I & 0 \end{bmatrix} \right) \right) (V_c^{-1} \otimes I) \bar{C} \\
=& \bar{C}^T \left((U V_c^{-1})^T \otimes \begin{bmatrix} I \\ 0 \end{bmatrix} \right) (I \otimes Q_q) \left((U V_c^{-1}) \otimes \begin{bmatrix} I & 0 \end{bmatrix} \right) \bar{C}.
\end{aligned} \tag{2.45}$$

Substituting (2.45) into (2.44), we have

$$\mathbb{E}\{\Delta V(x_k)\} = \mathbb{E}\{x_k^T \Omega_q x_k\} \tag{2.46}$$

where

$$\Omega_q = \begin{bmatrix} I & K_q^T (I + F\Xi)^T \end{bmatrix} G^T (I \otimes Q_q) G \begin{bmatrix} I \\ (I + F\Xi) K_q \end{bmatrix} - Q_q. \tag{2.47}$$

By using Schur complement formula, $\Omega_q < 0$ if and only if

$$M_q =: \begin{bmatrix} -Q_q & \begin{bmatrix} I & K_q^T (I + F\Xi)^T \end{bmatrix} G^T \\ * & -I \otimes Q_q^{-1} \end{bmatrix} < 0. \tag{2.48}$$

Rewrite matrix M_q as follows

$$M_q = \begin{bmatrix} -Q_q & [I \ \ K_q^T]\, G^T \\ * & -I \otimes Q_q^{-1} \end{bmatrix} + \begin{bmatrix} 0 \\ G \begin{bmatrix} 0 \\ I \end{bmatrix} \end{bmatrix} F \begin{bmatrix} \Xi K_q & 0 \end{bmatrix}$$
$$+ \begin{bmatrix} K_q^T \Xi^T \\ 0 \end{bmatrix} F^T \begin{bmatrix} 0 & [0 \ \ I]\, G^T \end{bmatrix}$$
$$\leq \begin{bmatrix} -Q_q & [I \ \ K_q^T]\, G^T \\ * & -I \otimes Q^{-1} \end{bmatrix} + \varepsilon_q^{-1} \begin{bmatrix} K_q^T \Xi^T \\ 0 \end{bmatrix} \begin{bmatrix} \Xi K_q & 0 \end{bmatrix}$$
$$+\varepsilon_q \begin{bmatrix} 0 \\ G \begin{bmatrix} 0 \\ I \end{bmatrix} \end{bmatrix} \begin{bmatrix} 0 & [0 \ \ I]\, G^T \end{bmatrix}. \tag{2.49}$$

By using Schur complement formula again, it can be seen that $M_q < 0$ if the following inequality holds

$$\begin{bmatrix} -Q_q & [I \ \ K_q^T]\, G^T & 0 & K_q^T \Xi^T \\ * & -I \otimes Q_q^{-1} & \varepsilon_q G \begin{bmatrix} 0 \\ I \end{bmatrix} & 0 \\ * & * & -\varepsilon_q I & 0 \\ * & * & * & -\varepsilon_q I \end{bmatrix} < 0. \tag{2.50}$$

Performing a congruence transformation of $\mathrm{diag}\{Q_q^{-1}, I, I, I\}$ *on inequality (2.50) and setting $P_q = Q_q^{-1}$ and $Y = K_q P_q$ yield (2.40) immediately. This means that inequality (2.40) implies $\Omega_q < 0$ and we can obtain a positive scalar $\rho = -\lambda_{\max}(\Omega_q)$ such that $\mathbb{E}\{\Delta V(x_k)\} \leq -\rho \mathbb{E}\{\|x_k\|^2\}$. Then, it can be shown that the discrete-time stochastic system (2.30) is stochastically stable by following the same lines in [190] and the proof of Theorem 2.2 is complete.*

C: Design of Saturated Sampled-Data Controller

It has been originally shown in [157] that the saturation function $s(u)$ can be written as

$$s(u) = Hu + \psi(u) \tag{2.51}$$

where H is a diagonal matrix satisfying $0 < H < I$ and the nonlinear function $\psi(u)$ satisfies

$$\psi^T(u)\big(\psi(u) - (I - H)u\big) \leq 0. \tag{2.52}$$

Similar to what has been done in subsection 2.2.2-B, the design method of the saturated sampled-data controller is given in Theorem 2.3.

Theorem 2.3 *Suppose that $\mu > 2\rho(A)$. For the continuous-time system (2.26) with the sampled-data controller given by (2.27), the discrete-time* *stochastic system (2.30) is stochastically stable if there exist matrices $P_s > 0$, Y_s and a positive constant ε_s such that the following inequality holds:*

$$\begin{bmatrix} -P_s & Y_s^T(I-H)/2 & G_1^T \\ * & -\varepsilon_s I & \varepsilon_s G_2^T \\ * & * & -(I \otimes P_s) \end{bmatrix} < 0 \tag{2.53}$$

where $G_1^T = \begin{bmatrix} P_s & Y_s^T H \end{bmatrix} G^T$ and $G_2^T = \begin{bmatrix} 0 & I \end{bmatrix} G^T$. Furthermore, if inequality (2.53) is feasible, the desired saturated controller gain is given by

$$K_s = Y_s P_s^{-1}. \tag{2.54}$$

Proof *Choose $V(x_k) = x_k Q_s x_k$, where Q_s is a positive definite matrix. By using Lemmas 2.3 and 2.4, it follows from (2.30), (2.39), (2.43), and (2.51) that*

$$\begin{aligned}
&\mathbb{E}\{\Delta V(x_k)\} \\
&= \mathbb{E}\Big\{\Big(e^{AT_k}x_k + \int_0^{T_k} e^{As}dsBs(K_s x_k)\Big)^T Q_s\Big(e^{AT_k}x_k + \int_0^{T_k} e^{As}ds \\
&\quad \times Bs(K_s x_k)\Big) - x_k^T Q_s x_k\Big\} \\
&= \mathbb{E}\Big\{ \begin{bmatrix} x_k^T & \psi^T(K_s x_k)\end{bmatrix} \begin{bmatrix} I & K_s^T H \\ 0 & I \end{bmatrix} \Big(e^{C^T T} e^{C^T v_k} \begin{bmatrix} I \\ 0 \end{bmatrix} Q_s \begin{bmatrix} I & 0 \end{bmatrix} e^{C v_k} e^{CT} \Big) \\
&\quad \times \begin{bmatrix} I & 0 \\ HK_s & I \end{bmatrix} \begin{bmatrix} x_k \\ \psi(K_s x_k) \end{bmatrix} - x_k^T Q_s x_k \Big\} \\
&= \mathbb{E}\Big\{ \begin{bmatrix} x_k^T & \psi^T(K_s x_k)\end{bmatrix} \begin{bmatrix} I & K_s^T H \\ 0 & I \end{bmatrix} G^T (I \otimes Q_s) G \begin{bmatrix} I & 0 \\ HK_s & I \end{bmatrix} \\
&\quad \times \begin{bmatrix} x_k \\ \psi(K_s x_k) \end{bmatrix} - x_k^T Q_s x_k \Big\}.
\end{aligned} \tag{2.55}$$

Considering (2.52), for a positive scalar $\tilde{\varepsilon}_s$, we have

$$\begin{aligned}
&\mathbb{E}\{\Delta V(x_k)\} \\
&\leq \mathbb{E}\Big\{ \begin{bmatrix} x_k^T & \psi^T(K_s x_k)\end{bmatrix} \begin{bmatrix} I & K_s^T H \\ 0 & I \end{bmatrix} G^T (I \otimes Q_s) G \begin{bmatrix} I & 0 \\ HK_s & I \end{bmatrix} \begin{bmatrix} x_k \\ \psi(K_s x_k) \end{bmatrix} \\
&\quad - \tilde{\varepsilon}_s \psi^T(K_s x_k)\big(\psi(K_s x_k) - (I-H)K_s x_k\big) - x_k^T Q_s x_k \Big\} \\
&= \mathbb{E}\{\xi_k^T \Omega_s \xi_k\}
\end{aligned} \tag{2.56}$$

where

$$\xi_k = \begin{bmatrix} x_k^T & \psi^T(K_s x_k) \end{bmatrix}^T,$$

$$\Omega_s = \begin{bmatrix} -Q_s & \tilde{\varepsilon}_s K_s^T(I-H)/2 \\ * & -\tilde{\varepsilon}_s I \end{bmatrix} + \begin{bmatrix} I & K_s^T H \\ 0 & I \end{bmatrix} G^T(I \otimes Q_s) G \begin{bmatrix} I & 0 \\ HK_s & I \end{bmatrix}.$$

By Schur complement formula, one has $\Omega_s < 0$ *if and only if*

$$\begin{bmatrix} -Q_s & \tilde{\varepsilon}_s K_s^T(I-H)/2 & \tilde{G}_1^T \\ * & -\tilde{\varepsilon}_s I & G_2^T \\ * & * & -(I \otimes Q_s^{-1}) \end{bmatrix} < 0 \tag{2.57}$$

where $\tilde{G}_1^T = \begin{bmatrix} I & K_s^T H \end{bmatrix} G^T$. *By setting* $P_s = Q_s^{-1}$, $Y_s = K_s P_s$ *and* $\varepsilon_s = \tilde{\varepsilon}_s^{-1}$ *and performing a congruence transformation of* $\mathrm{diag}\{Q_s^{-1}, \tilde{\varepsilon}_s^{-1} I, I\}$ *on the above inequality, inequality (2.53) is obtained easily. The rest of the proof of Theorem 2.3 follows from the same lines in Theorem 2.2 and is therefore omitted.*

Remark 2.7 *Note that the approach proposed in this section is always effective as long as the probability density function of sampling error is given. For example, if the sampling error* v_k *follows the continuous uniform distribution between* $-\delta$ *and* δ $(0 < \delta < T)$*, the corresponding results can be derived by modifying* π_{ij} *given by (2.38).*

To this end, the desired quantized/saturated controller has been designed such that, when the sampling interval follows the Erlang distribution, the discrete-time stochastic system (2.30) is stochastically stable. The noisy sampling intervals, quantization and saturations concerned are all motivated by engineering practice.

2.3 Illustrative Examples

In this section, some numerical examples are presented to demonstrate the sampled-data control approach presented in this chapter.

2.3.1 Example 1

In this example, we aim to show the effectiveness and applicability of the proposed method in Section 2.1.

Consider a continuous-time system described by (2.1) with the following parameters

$$A = \begin{bmatrix} 1.8 & -1.3 \\ 0 & 1.2 \end{bmatrix}, \quad B = \begin{bmatrix} 1 \\ -1.2 \end{bmatrix}.$$

It is easy to see that the system matrix A has eigenvalues $\lambda_1 = 1.8$ and $\lambda_2 = 1.2$, which implies that the system without control input is unstable.

From the definition of matrix C in (2.9), it is known that the eigenvalues of matrix C are $\lambda_{C,1} = 1.8$, $\lambda_{C,2} = 1.2$, and $\lambda_{C,3} = 0$. Then, according to (2.11) and (2.13), we have

$$C_1 = \begin{bmatrix} 1 & -2.1667 & 2 \\ 0 & 0 & 0 \\ 0 & 0 & 0 \end{bmatrix}, C_2 = \begin{bmatrix} 0 & 2.1667 & -2.1667 \\ 0 & 1 & -1 \\ 0 & 0 & 0 \end{bmatrix}, C_3 = \begin{bmatrix} 0 & 0 & 0.1667 \\ 0 & 0 & 1 \\ 0 & 0 & 1 \end{bmatrix}.$$

Suppose that the nominal sampling interval is $T = 0.8$ and the sampling error v_k obeys continuous uniform distribution between -0.3 and 0.3. By (2.23), matrix Π_δ is obtained as

$$\Pi_\delta = \begin{bmatrix} 1.2061 & 1.1406 & 1.0493 \\ 1.1406 & 1.0887 & 1.0217 \\ 1.0493 & 1.0217 & 1.0000 \end{bmatrix},$$

from which we can find matrix

$$U_\delta = \begin{bmatrix} 0.0041 & -0.0063 & 0.0021 \\ 0.1334 & 0.0273 & -0.1768 \\ 1.0901 & 1.0430 & 0.9843 \end{bmatrix}$$

such that $\Pi_\delta = U_\delta^T U_\delta$.

With the above parameters, the inequality (2.24) can be solved by using the Matlab software and we can obtain a set of feasible solutions as follows

$$P = \begin{bmatrix} 1.2477 & -1.5093 \\ -1.5093 & 15.2625 \end{bmatrix}, \quad Y = \begin{bmatrix} -3.3913 & 19.7073 \end{bmatrix}.$$

Then, according to (2.18), the desired controller gain matrix is designed as

$$K = YP^{-1} = \begin{bmatrix} -1.3132 & 1.1614 \end{bmatrix}.$$

In the simulation, the initial value is set as $x_0 = \begin{bmatrix} 1 & 0.8 \end{bmatrix}^T$. Simulation results are presented in Figs. 2.1–2.2. Fig. 2.1 plots the state trajectories without control input. The state trajectories with sampled-data control input are depicted in Fig. 2.2, where the asterisks represent the sampling instants. From simulation results, it can be seen that the designed sampled-data controller stabilizes the unstable system very well, which has confirmed that the design approach proposed in Section 2.1 is effective.

2.3.2 Example 2

This example considers the design of both the quantized and the saturated controllers with multiple control inputs.

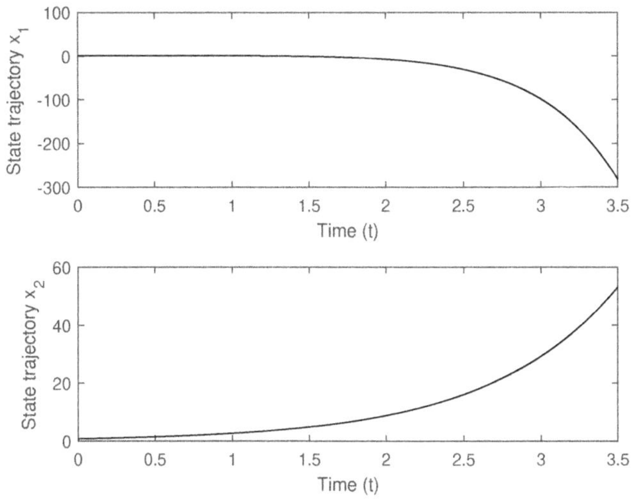

FIGURE 2.1
State trajectories without control inputs.

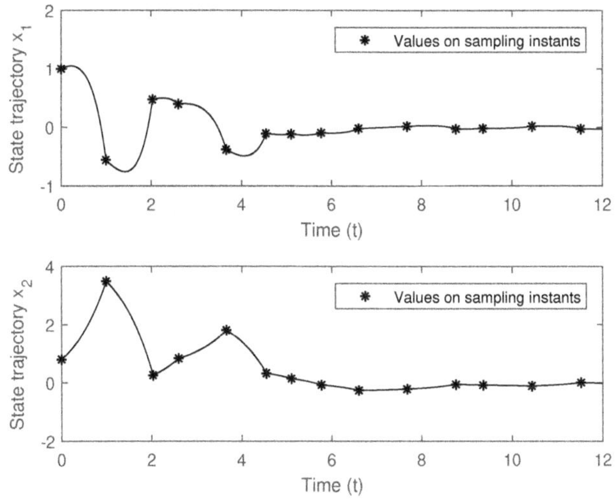

FIGURE 2.2
State trajectories with control inputs.

Consider a system of form (2.26) with the following parameters

$$A = \begin{bmatrix} -0.5 & 0.8 & 0.4 \\ 0 & -0.5 & 0.12 \\ 0 & 0 & 0.3 \end{bmatrix}, \quad B = \begin{bmatrix} 0.2 & 0.15 \\ 0.11 & 0.6 \\ -0.3 & 0.6 \end{bmatrix}$$

from which it is seen that the eigenvalues of matrix C are $\lambda_1 = -0.5$, $\lambda_2 = 0.3$, and $\lambda_3 = 0$, with multiplicities $m_1 = 2$, $m_2 = 1$, and $m_3 = 2$, respectively. Then, the confluent Vandermonde matrix V_c can be obtained according to (2.32).

For each sampling interval, the nominal sampling period is taken as $T = 0.6$ and the sampling error v_k is assumed to obey the Erlang distribution with parameter $\mathcal{K} = 1$ and $\mu = 4$. By formula (2.38), it is obtained that

$$\Pi = \begin{bmatrix} 0.8000 & 0.1600 & 0.9524 & 0.8889 & 0.1975 \\ 0.1600 & 0.0640 & 0.2268 & 0.1975 & 0.0878 \\ 0.9524 & 0.2268 & 1.1765 & 1.0811 & 0.2922 \\ 0.8889 & 0.1975 & 1.0811 & 1.0000 & 0.2500 \\ 0.1975 & 0.0878 & 0.2922 & 0.2500 & 0.1250 \end{bmatrix},$$

and matrix U can be computed as

$$U = \begin{bmatrix} 0.0001 & 0.0000 & 0.0000 & 0.0000 & 0.0000 \\ -0.0001 & 0.0001 & 0.0008 & -0.0008 & -0.0004 \\ -0.0002 & -0.0280 & 0.0068 & -0.0054 & 0.0150 \\ 0.1289 & -0.1491 & -0.0508 & 0.0334 & -0.2410 \\ 0.8851 & 0.2025 & 1.0834 & 0.9994 & 0.2583 \end{bmatrix}.$$

Case 1: The design of quantized sampled-data controller

The parameters of the logarithmic quantizer $q(\cdot)$ are taken as $\chi_0^1 = \chi_0^2 = 0.009$, and $\rho_1 = \rho_2 = 0.9$. With the above parameters, the inequality (2.40) can be solved by using the Matlab software and the feasible solution is derived as follows

$$P_q = \begin{bmatrix} 155.0915 & 13.8373 & -9.4461 \\ 13.8373 & 154.0177 & 9.6401 \\ -9.4461 & 9.6401 & 157.5231 \end{bmatrix},$$

$$Y_q = \begin{bmatrix} -171.8875 & -185.7690 & 326.4018 \\ -48.2920 & -131.8642 & -136.0331 \end{bmatrix}.$$

Then, according to (2.41), the desired quantized controller gain is given by

$$K_q = Y_q P_q^{-1} = \begin{bmatrix} -0.8682 & -1.2594 & 2.0971 \\ -0.2928 & -0.7777 & -0.8335 \end{bmatrix}.$$

Case 2: The design of saturated sampled-data controller

The parameters of saturation function are chosen as $u_{1,max} = u_{2,max} =$

0.5, $H = \begin{bmatrix} 0.01 & 0 \\ 0 & 0.19 \end{bmatrix}$. Then, the inequality (2.53) is solved and the feasible solution is obtained as follows

$$P_s = \begin{bmatrix} 25.1506 & -22.0987 & -19.6683 \\ -22.0987 & 62.2328 & 34.4212 \\ -19.6683 & 34.4212 & 23.9222 \end{bmatrix},$$

$$Y_s = \begin{bmatrix} 0.7680 & 6.5500 & 4.0374 \\ -4.9923 & -10.3429 & -5.1123 \end{bmatrix}.$$

According to (2.54), the desired saturated controller gain is designed by

$$K_s = Y_s P_s^{-1} = \begin{bmatrix} 0.6259 & -0.2472 & 1.0392 \\ -1.2999 & 0.3994 & -1.8571 \end{bmatrix}.$$

Simulation results are shown in Figs. 2.3–2.6, where the initial value is set as $x_0 = \begin{bmatrix} 0.1 & -0.5 & 0.8 \end{bmatrix}^T$. Fig. 2.3 plots the state trajectories of system without control inputs and Figs. 2.4–2.6 depict the state trajectories of system with the designed control inputs, where the above picture is for the case of quantized control while the one below is for the case of saturated control. It can be seen that, when the designed quantized/saturated sampled-data controllers are added into the system, the state trajectories converge to equilibrium point, which has confirmed that the designed sampled-data controller in Section 2.2 stabilizes the unstable system very well.

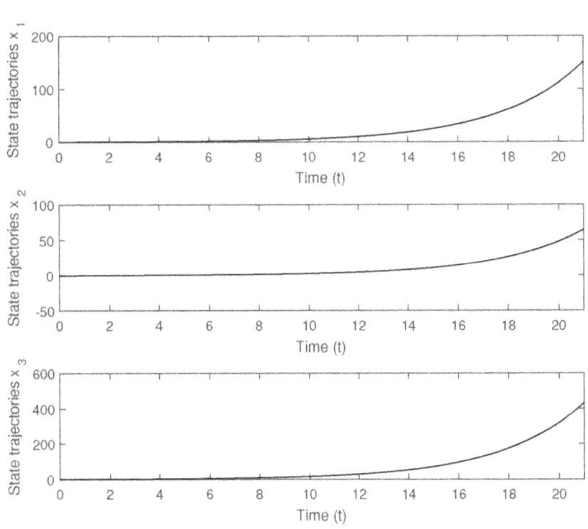

FIGURE 2.3
State trajectories without control inputs.

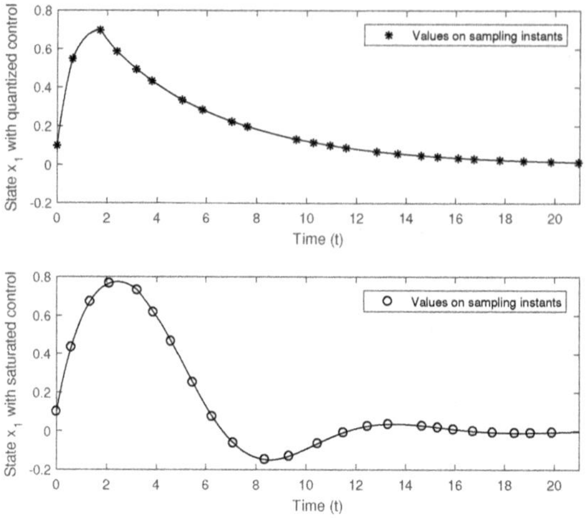

FIGURE 2.4
State trajectories x_1 with control inputs.

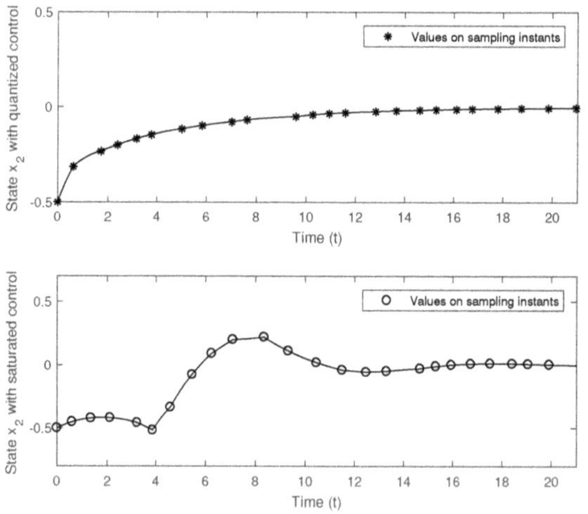

FIGURE 2.5
State trajectories x_2 with control inputs.

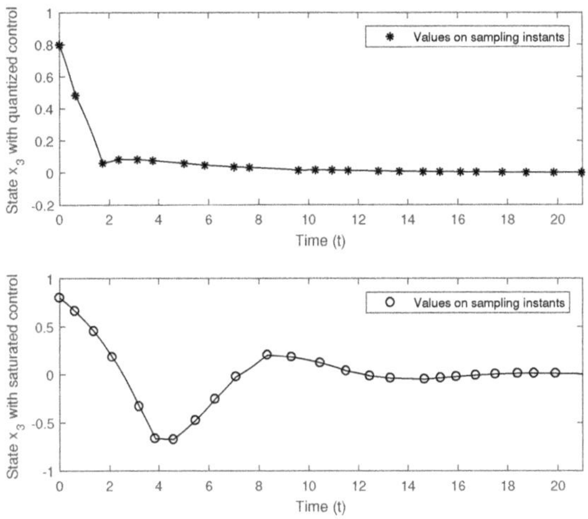

FIGURE 2.6
State trajectories x_3 with control inputs.

2.4 Summary

In this chapter, the stabilization problem has been studied for a class of sampled-data systems under noisy sampling intervals. We have converted the stochastic sampled-data control system into a discrete-time system. With the help of the matrix theories including matrix exponential computation, Vandermonde matrix, and Kronecker product operation, the sampled-data stabilization controller has been designed such that the stochastic sampled-data control system is stochastically stable. In addition, the stabilization problem has been discussed for a general class of stochastic sampled-data systems with multiple inputs under quantization and/or saturation effects, and a set of parallel results has been derived by using similar analysis techniques. Finally, some numerical simulation examples have been provided to demonstrate the effectiveness of the proposed design approach.

3

Distributed State Estimation with Nonuniform Samplings

This chapter deals with the distributed sampled-data H_∞ state estimation problem for a class of continuous-time nonlinear systems with infinite-distributed delays. To cater for possible implementation errors, the estimator gain is allowed to have certain bounded parameter variations. A sensor network is deployed to acquire the plant output by collaborating with their neighbors according to a given network topology. The individually sampled sensor measurement is transmitted to the corresponding estimator through a digital communication channel. By utilizing the input delay approach, the effect of the sampling intervals is transformed into an equivalent bounded TVD. A set of sampled-data distributed estimators is designed for the addressed nonlinear systems in order to meet the following three performance requirements: 1) the asymptotic convergence of the estimation error dynamics; 2) the H_∞ disturbance attenuation/rejection behaviuor against the exogenous disturbances; and 3) the resilience against possible gain variations. A Lyapunov functional approach is put forward to obtain the existence conditions for the desired estimators, which are then parameterized in light of the feasibility of some matrix inequalities. An illustrative numerical example is given to demonstrate the usefulness of the proposed estimator design algorithm.

3.1 Problem Formulation

Consider a target plant governed by the following continuous-time system:

$$\begin{cases} \dot{x}(t) = Ax(t) + f(x(t)) + \int_{-\infty}^{t} \kappa(t-r)g(x(r))dr + Bw(t) \\ z(t) = Mx(t) \end{cases} \tag{3.1}$$

with N sensors modeled by

$$y_i(t) = C_i x(t) + D_i v(t), \quad i = 1, 2, \ldots, N. \tag{3.2}$$

DOI: 10.1201/9781003307648-3

where $x(t) \in \mathbb{R}^{n_x}$ is the plant state; $z(t) \in \mathbb{R}^{n_z}$ is the controlled output that is to be estimated; $y_i(t) \in \mathbb{R}^{n_y}$ is the measured output received by sensor i; $w(t) \in \mathbb{R}^{n_w}$ and $v(t) \in \mathbb{R}^{n_v}$ are the exogenous disturbances belonging to $L_2[0, +\infty)$; and A, B, C_i, D_i and M are known constant matrices with appropriate dimensions.

Assumption 3.1 *The kernel* $\kappa(\cdot) : [0, +\infty) \rightarrow [0, +\infty)$ *for distributed time-delay is continuous, integrable and satisfies*

$$\breve{\kappa} = \int_0^{+\infty} \kappa(t)dt < +\infty, \qquad \int_0^{+\infty} t\kappa(t)dt < +\infty. \tag{3.3}$$

Assumption 3.2 $f(\cdot) : \mathbb{R}^{n_x} \rightarrow \mathbb{R}^{n_x}$ *and* $g(\cdot) : \mathbb{R}^{n_x} \rightarrow \mathbb{R}^{n_x}$ *are continuous nonlinear functions satisfying* $f(0) = 0$, $g(0) = 0$ *with the following constraints:*

$$\begin{aligned} &[f(x) - f(\varpi) - U_1(x - \varpi)]^T [f(x) - f(\varpi) - U_2(x - \varpi)] \leq 0, \\ &[g(x) - g(\varpi) - W_1(x - \varpi)]^T [g(x) - g(\varpi) - W_2(x - \varpi)] \leq 0 \end{aligned} \tag{3.4}$$

for all $x, \varpi \in \mathbb{R}^{n_x}$, *where* U_1, U_2, W_1, *and* W_2 *are constant matrices.*

In this chapter, the sensor network consists of N sensor nodes that are spatially distributed and the network topology is represented by a directed weighted graph $\mathcal{G} = (\mathcal{V}, \mathcal{Q}, \mathcal{W})$ of order N, where $\mathcal{V} = \{1, 2, \ldots, N\}$ and $\mathcal{Q} \subseteq \mathcal{V} \times \mathcal{V}$ are, respectively, the set of nodes and the set of edges, and $\mathcal{W} = [w_{ij}]_{N \times N}$ represents a weighted adjacency matrix with $w_{ij} > 0$ if $(i, j) \in \mathcal{Q}$. Moreover, assume $w_{ii} = 1$ for all $i \in \mathcal{V}$. $\mathcal{N}_i = \{j \in \mathcal{V} : (i, j) \in \mathcal{Q}\}$ denotes the set of neighbors of node $i \in \mathcal{V}$ plus the node itself.

In view of the given topology of the sensor network, the information received by sensor node i from both itself and its neighbours can be represented as follows:

$$\bar{y}_i(t) = \sum_{j \in \mathcal{N}_i} w_{ij}(y_j(t) - C_j \hat{x}_j(t)), \tag{3.5}$$

where $\hat{x}_i(t) \in \mathbb{R}^{n_x}$ is the estimate of state $x(t)$ on the node i.

For each $i \in \mathcal{V}$, the received signal is first sampled on an individual basis and then transmitted to the corresponding estimator. The sampled signal is generalized by a ZOH function, where the sequence of hold times is given by $0 = t_0^i < t_1^i < \cdots < \lim_{k \rightarrow +\infty} t_k^i = +\infty$. Then, we have

$$\hat{y}_i(t) = \bar{y}_i(t_k^i) = \bar{y}_i(t - (t - t_k^i)), \quad t \in [t_k^i, t_{k+1}^i), \tag{3.6}$$

where $\hat{y}_i(t)$ is the actual input to the estimator i and t_k^i is the sampling instants of node i. Moreover, the sampling interval in this chapter is assumed to satisfy $t_{k+1}^i - t_k^i \leq \tau_i$, where $\tau_i > 0$ is a known scalar for each $i \in \mathcal{V}$.

Based on the sampled signal, for each $i \in \mathcal{V}$, the distributed state estimators are constructed as follows:

$$\begin{cases} \dot{\hat{x}}_i(t) = A\hat{x}_i(t) + f(\hat{x}_i(t)) + \displaystyle\int_{-\infty}^{t} \kappa(t-r)g(\hat{x}_i(r))dr \\ \qquad + (K_i + \Delta K_i)\hat{y}_i(t) \\ \hat{z}_i(t) = M\hat{x}_i(t) \end{cases} \tag{3.7}$$

where $\hat{z}_i(t) \in \mathbb{R}^{n_z}$ is the estimate of output $z(t)$ on the node i. K_i is an appropriately dimensioned estimator parameter to be determined. ΔK_i represents the estimator parameter variation satisfying

$$\Delta K_i = K_i H_i F_i(t) T_i, \tag{3.8}$$

where H_i and T_i are matrices, which are known with appropriate dimensions, and the matrix $F_i(t)$ is unknown and satisfies $F_i^T(t)F_i(t) \leq I$.

Define $\tau_i(t) = t - t_k^i$ for $t \in [t_k^i, t_{k+1}^i)$, $k = 0, 1, 2, \ldots, \infty$. Then, (3.7) is rewritten as

$$\begin{cases} \dot{\hat{x}}_i(t) = A\hat{x}_i(t) + f(\hat{x}_i(t)) + \displaystyle\int_{-\infty}^{t} \kappa(t-r)g(\hat{x}_i(r))dr \\ \qquad + (K_i + \Delta K_i)\bar{y}_i(t - \tau_i(t)) \\ \hat{z}_i(t) = M\hat{x}_i(t). \end{cases} \tag{3.9}$$

It is easily seen that $\tau_i(t)$ satisfies $\tau_i(t) \in [0, \bar{\tau})$ with $\bar{\tau} = \max_{i \in \mathcal{V}}\{\tau_i\}$ and $\dot{\tau}_i(t) = 1$ for $t \neq t_k^i$, $k = 0, 1, 2, \ldots, \infty$.

Remark 3.1 *In this chapter, the sampling interval on each sensor is nonuniform and individually bounded. Therefore, at different sampling points and for different sensors, the sampling interval could be different, and so are the corresponding upper bounds. Such a sampling-induced issue renders traditional distributed filter structure unpractical in practice. To be specific, if we were to associate the sensor's neighbors with individual gains, the resulting filter design algorithm would be time-varying and difficult to be implemented. As such, to resolve the complications resulting from the nonuniform sampling, we have adopted a simplified filter structure as in (3.9), where the estimator gains related to the sensor's neighbors are assumed to be identical.*

Letting the state estimation error be $e_i(t) = x(t) - \hat{x}_i(t)$ and the output estimation error be $\tilde{z}_i(t) = z(t) - \hat{z}_i(t)$ for each $i \in \mathcal{V}$, we have

$$\begin{cases} \dot{e}_i(t) = Ae_i(t) + \displaystyle\int_{-\infty}^{t} \kappa(t-r)\tilde{g}(e_i(r))dr - (K_i + \Delta K_i)\sum_{j \in \mathcal{N}_i} w_{ij}C_j \\ \qquad \times e_j(t - \tau_i(t) + \tilde{f}(e_i(t)) - (K_i + \Delta K_i)\displaystyle\sum_{j \in \mathcal{N}_i} w_{ij}D_j \\ \qquad \times v(t - \tau_i(t)) + Bw(t) \\ \tilde{z}_i(t) = Me_i(t) \end{cases} \tag{3.10}$$

where $\tilde{f}(e_i(t)) = f(x(t)) - f(\hat{x}_i(t))$ and $\tilde{g}(e_i(t)) = g(x(t)) - g(\hat{x}_i(t))$.

Setting

$$\mathcal{K} = \text{diag}_N^i\{K_i\},\ \Delta\mathcal{K} = \text{diag}_N^i\{\Delta K_i\},\ \tilde{z}(t) = \text{col}_N^i\{\tilde{z}_i(t)\},\ \mathcal{M} = \text{diag}_N\{M\},$$
$$v_\tau(t) = \text{col}_N^i\{v(t-\tau_i(t))\},\ \bar{D} = \text{col}_N\{D_i\},\ I_i = \text{diag}\{\underbrace{0,\dots,0}_{i-1}, I, \underbrace{0,\dots,0}_{N-i}\},$$
$$\mathcal{C} = \text{diag}_N^i\{C_i\},\ \mathcal{B} = \text{col}_N\{B\},\ \mathcal{A} = \text{diag}_N\{A\},\ \mathcal{D} = \text{vec}_N^i\{I_i(\mathcal{W}\otimes I)\bar{D}\},$$
$$e(t) = \text{col}_N^i\{e_i(t)\},\ \mathcal{F}(e(t)) = \text{col}_N^i\{\tilde{f}(e_i(t)\},\ \mathcal{G}(e(t)) = \text{col}_N^i\{\tilde{g}(e_i(t)\},$$

the estimation error system can be further rewritten as the following compact form:

$$\begin{cases} \dot{e}(t) = \mathcal{A}e(t) + \mathcal{F}(e(t)) + \displaystyle\int_{-\infty}^{t} \kappa(t-r)\mathcal{G}(e(r))dr - (\mathcal{K}+\Delta\mathcal{K})\mathcal{D} \\ \qquad \times v_\tau(t) + \mathcal{B}w(t) - \displaystyle\sum_{i=1}^{N}(\mathcal{K}+\Delta\mathcal{K})I_i(\mathcal{W}\otimes I)\mathcal{C}e(t-\tau_i(t)) \\ \tilde{z}(t) = \mathcal{M}e(t). \end{cases} \tag{3.11}$$

The initial value of (3.11) is given as $e(\theta) = \phi(\theta) \in \mathcal{S}((-\infty,0],\mathbb{R}^{Nn_x})$, where $\mathcal{S}((-\infty,0],\mathbb{R}^{Nn_x})$ is the space of continuous functions $\phi(\cdot):(-\infty,0]\to\mathbb{R}^{Nn_x}$ with the norm satisfying $\sup_{-\infty<\theta\leq 0}\|\phi(\theta)\|^2 < \infty$.

Definition 3.1 *The estimation error system (3.11) with $w(t) = 0$ and $v_\tau(t) = 0$ is said to be asymptotically stable if*

$$\lim_{t\to\infty}\|e(t;\phi)\|^2 = 0. \tag{3.12}$$

In this chapter, we aim to design a set of distributed estimators such that the following two requirements are simultaneously satisfied:

a) The estimation error system with $w(t) = 0$ and $v_\tau(t) = 0$ is asymptotically stable.

b) With zero-initial condition, for all nonzero $w(t)$ and $v_\tau(t)$, the output estimation error $\tilde{z}(t)$ satisfies

$$\int_0^\infty \|\tilde{z}(t)\|^2 dt \leq \gamma^2 \int_0^\infty (\|w(t)\|^2 + \|v_\tau(t)\|^2)dt \tag{3.13}$$

where $\gamma > 0$ is a prescribed disturbance attenuation level.

3.2 Main Results

In this section, a sufficient condition is first derived to ensure the stability and H_∞ performance of the estimation error system (3.11). Then, based on the established analysis result, the desired distributed estimators are designed for the target plant (3.1). Before proceeding, we need the following lemmas in deriving our main results.

Lemma 3.1 *Under Assumption 3.2, the following inequalities are obtained:*

$$\begin{aligned}
&\begin{bmatrix} e(t) \\ \mathcal{F}(e(t)) \end{bmatrix}^T \begin{bmatrix} \tilde{U}_1 & \tilde{U}_2 \\ * & I \end{bmatrix} \begin{bmatrix} e(t) \\ \mathcal{F}(e(t)) \end{bmatrix} \leq 0, \\
&\begin{bmatrix} e(t) \\ \mathcal{G}(e(t)) \end{bmatrix}^T \begin{bmatrix} \tilde{W}_1 & \tilde{W}_2 \\ * & I \end{bmatrix} \begin{bmatrix} e(t) \\ \mathcal{G}(e(t)) \end{bmatrix} \leq 0,
\end{aligned} \tag{3.14}$$

where

$$\begin{aligned}
&\tilde{U}_1 = \frac{\hat{U}_1^T \hat{U}_2 + \hat{U}_2^T \hat{U}_1}{2}, \quad \tilde{U}_2 = -\frac{\hat{U}_1^T + \hat{U}_2^T}{2}, \quad \hat{U}_1 = \mathrm{diag}_N\{U_1\}, \\
&\tilde{W}_1 = \frac{\hat{W}_1^T \hat{W}_2 + \hat{W}_2^T \hat{W}_1}{2}, \quad \tilde{W}_2 = -\frac{\hat{W}_1^T + \hat{W}_2^T}{2}, \\
&\hat{U}_2 = \mathrm{diag}_N\{U_2\}, \quad \hat{W}_1 = \mathrm{diag}_N\{W_1\}, \quad \hat{W}_2 = \mathrm{diag}_N\{W_2\}.
\end{aligned}$$

Lemma 3.2 *[183] Let $Y = Y^T$, T, F and N be real matrices of appropriate dimensions. Then, for all F satisfying $FF^T \leq I$,*

$$Y + TFN + (TFN)^T < 0$$

holds if and only if there exists a positive scalar ε such that

$$Y + \varepsilon^{-1} TT^T + \varepsilon N^T N < 0.$$

Lemma 3.3 *[173] Let a positive semi-definite matrix M, a scalar function $\alpha(\cdot) : (-\infty, a] \to [0, +\infty)$ and a vector function $R(\cdot) : (-\infty, a] \to \mathbb{R}^n$ be all given. If the concerned integrations are well defined, then we have the following inequality:*

$$\begin{aligned}
\Big(\int_{-\infty}^{a} \alpha(s) R(s) ds\Big)^T M \Big(\int_{-\infty}^{a} \alpha(s) R(s) ds\Big) \\
\leq \int_{-\infty}^{a} \alpha(s) ds \Big(\int_{-\infty}^{a} \alpha(s) R^T(s) M R(s) ds\Big).
\end{aligned}$$

Furthermore, if we take

$$\alpha(t) = \begin{cases} 1: & b \leq t \leq a, \\ 0: & \text{otherwise}, \end{cases}$$

then, we obtain the well-known Jensen inequality as follows:

$$\Big(\int_{b}^{a} R(s) ds\Big)^T M \Big(\int_{b}^{a} R(s) ds\Big) \leq (a-b) \Big(\int_{b}^{a} R^T(s) M R(s) ds\Big).$$

Lemma 3.4 *[97] Let $z(t) \in W[a,b)$ and $z(a) = 0$, where $W[a,b)$ denotes the space of functions $\phi : [a,b] \to \mathbb{R}^n$, which are absolutely continuous on $[a,b)$ and have a finite $\lim_{\theta \to b^-} \phi(\theta)$ as well as the square integrable first order derivatives. Then, for any positive definite matrix $W \in \mathbb{R}^{n\times n}$, the following inequality holds:*

$$\int_a^b z^T(s)Wz(s)ds \leq \frac{4(b-a)^2}{\pi^2}\int_a^b \dot{z}^T(s)W\dot{z}(s)ds.$$

In the following theorem, a sufficient condition is derived under which the estimation error system is asymptotically stable and the H_∞ performance constraint is met.

Theorem 3.1 *Let the estimator parameters K_i ($i \in \mathcal{V}$) and the disturbance level $\gamma > 0$ be given. The estimation error dynamics governed by (3.11) is asymptotically stable (with $w(t) = 0$ and $v_\tau(t) = 0$) and achieves the H_∞ performance index (3.13) for all nonzero $w(t)$ and $v_\tau(t)$ (under the zero initial condition) if there exist matrices $P_1 > 0$, $P_2 > 0$, $Q > 0$, $Z_i > 0$ and $Y_i > 0$ ($i \in \mathcal{V}$), and scalars $\lambda_1 > 0$, $\lambda_2 > 0$ satisfying*

$$\bar{\Phi} = \begin{bmatrix} \bar{\Phi}_{11} & \Phi_{12} & 0 & \Phi_{14} & -\lambda_2\tilde{W}_2 & P_1 & P_1\mathcal{B} & \Phi_{18} & \mathcal{A}^T P_1 \\ * & \Phi_{22} & \bar{Z}^T & 0 & 0 & 0 & 0 & 0 & \Phi_{29} \\ * & * & \Phi_{33} & 0 & 0 & 0 & 0 & 0 & 0 \\ * & * & * & -\lambda_1 I & 0 & 0 & 0 & 0 & P_1 \\ * & * & * & * & \Phi_{55} & 0 & 0 & 0 & 0 \\ * & * & * & * & * & -\frac{1}{\breve{\kappa}}P_2 & 0 & 0 & P_1 \\ * & * & * & * & * & * & -\gamma^2 I & 0 & \mathcal{B}^T P_1 \\ * & * & * & * & * & * & * & -\gamma^2 I & \Phi_{89} \\ * & * & * & * & * & * & * & * & -P_1 R^{-1} P_1 \end{bmatrix}$$
$$< 0, \tag{3.15}$$

where

$$\Phi_{89} = -\mathcal{D}^T(\mathcal{K}+\Delta\mathcal{K})^T P_1,\ \Phi_{29} = -(\bar{\mathcal{K}}+\Delta\bar{\mathcal{K}})^T P_1,\ \bar{\mathcal{K}} = \mathrm{vec}_N^i\{\mathcal{K}I_i(\mathcal{W}\otimes I)\mathcal{C}\},$$

$$\bar{\Phi}_{11} = P_1\mathcal{A} + \mathcal{A}^T P_1 + Q - \sum_{i=1}^N \frac{\pi^2}{4}Y_i - \lambda_1\tilde{U}_1 - \lambda_2\tilde{W}_1 - \sum_{i=1}^N Z_i + \mathcal{M}^T\mathcal{M},$$

$$\Phi_{33} = -\sum_{i=1}^N Z_i - Q,\ \Phi_{55} = \breve{\kappa}P_2 - \lambda_2 I,\ \Delta\bar{\mathcal{K}} = \mathrm{vec}_N^i\{\Delta\mathcal{K}I_i(\mathcal{W}\otimes I)\mathcal{C}\},$$

$$\bar{Z} = \mathrm{vec}_N^i\{Z_i\},\ \hat{Z} = \mathrm{diag}_N^i\{Z_i\},\ \bar{Y} = \mathrm{vec}_N^i\{Y_i\},\ \hat{Y} = \mathrm{diag}_N^i\{Y_i\},$$

$$\Phi_{18} = -P_1(\mathcal{K}+\Delta\mathcal{K})\mathcal{D},\ \Phi_{12} = -P_1(\bar{\mathcal{K}}+\Delta\bar{\mathcal{K}}) + \bar{Z} + \frac{\pi^2}{4}\bar{Y},$$

$$\Phi_{22} = -2\hat{Z} - \frac{\pi^2}{4}\hat{Y},\ R = \sum_{i=1}^{N} \bar{\tau}^2 Y_i + \sum_{i=1}^{N} \bar{\tau}^2 Z_i,\ \Phi_{14} = P_1 - \lambda_1 \tilde{U}_2. \tag{3.16}$$

Proof *To prove the stability, we consider the following Lyapunov functional candidate:*

$$V(e(t)) = \sum_{i=1}^{4} V_i(e(t)) \tag{3.17}$$

where

$$\begin{aligned}
V_1(e(t)) &= e^T(t)P_1e(t) + \int_{t-\bar{\tau}}^{t} e^T(s)Qe(s)ds, \\
V_2(e(t)) &= \sum_{i=1}^{N} \int_{-\bar{\tau}}^{0} \int_{t+r}^{t} \bar{\tau}\dot{e}^T(s)Z_i\dot{e}(s)dsdr, \\
V_3(e(t)) &= \int_{0}^{+\infty} \kappa(s)ds \int_{t-s}^{t} \mathcal{G}^T(e(r))P_2\mathcal{G}(e(r))dr, \\
V_4(e(t)) &= \sum_{i=1}^{N} \int_{t-\tau_i(t)}^{t} \bar{\tau}^2\dot{e}^T(s)Y_i\dot{e}(s)ds - \sum_{i=1}^{N} \int_{t-\tau_i(t)}^{t} \frac{\pi^2}{4} \\
&\quad \times (e(s) - e(t-\tau_i(t)))^T Y_i (e(s) - e(t-\tau_i(t)))ds,
\end{aligned}$$

with $t \in [t_k^i, t_{k+1}^i)$, $i \in \mathcal{V}$, $k = 0, 1, 2, \ldots, \infty$.

Note that $V_4(e(t))$ *is discontinuous at the sampling instants* t_k^i, $i \in \mathcal{V}$, $k = 0, 1, 2, \ldots, \infty$. *From Lemma 3.4, it is easily known that* $V_4(e(t)) \geq 0$ *and we can further obtain* $\lim_{t \to (t_k^i)^-} V(e(t)) \geq V(e(t))|_{t=t_k^i}$ *since* $V_4(e(t)) = 0$ *at* $t = t_k^i$, $k = 0, 1, 2, \ldots, \infty$.

Calculating the time derivative of $V(e(t))$ *along the trajectory of system (3.11) with* $w(t) = 0$ *and* $v_\tau(t) = 0$ *yields*

$$\begin{aligned}
\dot{V}(e(t)) =& 2e^T(t)P_1\dot{e}(t) + e^T(t)Qe(t) - e^T(t-\bar{\tau})Qe(t-\bar{\tau}) \\
&+ \sum_{i=1}^{N} \bar{\tau}^2\dot{e}^T(t)Z_i\dot{e}(t) - \sum_{i=1}^{N} \int_{t-\bar{\tau}}^{t} \bar{\tau}\dot{e}^T(s)Z_i\dot{e}(s)ds \\
&+ \int_{0}^{+\infty} \kappa(s)\mathcal{G}^T(e(t))P_2\mathcal{G}(e(t))ds \\
&- \int_{0}^{+\infty} \kappa(s)\mathcal{G}^T(e(t-s))P_2\mathcal{G}(e(t-s))ds \\
&+ \sum_{i=1}^{N} \bar{\tau}^2\dot{e}^T(t)Y_i\dot{e}(t) - \sum_{i=1}^{N} \frac{\pi^2}{4}(e(t) - e(t-\tau_i(t)))^T \\
&\times Y_i(e(t) - e(t-\tau_i(t))).
\end{aligned} \tag{3.18}$$

From Lemma 3.3 and Assumption 3.1, it can be obtained that

$$\begin{aligned}
&-\sum_{i=1}^{N}\int_{t-\bar{\tau}}^{t}\bar{\tau}\dot{e}^T(s)Z_i\dot{e}(s)ds \\
&= -\sum_{i=1}^{N}\int_{t-\tau_i(t)}^{t}\bar{\tau}\dot{e}^T(s)Z_i\dot{e}(s)ds - \sum_{i=1}^{N}\int_{t-\bar{\tau}}^{t-\tau_i(t)}\bar{\tau}\dot{e}^T(s)Z_i\dot{e}(s)ds \\
&\leq -\sum_{i=1}^{N}(e(t)-e(t-\tau_i(t)))^T Z_i(e(t)-e(t-\tau_i(t))) \\
&\quad -\sum_{i=1}^{N}(e(t-\tau_i(t))-e(t-\bar{\tau}))^T Z_i(e(t-\tau_i(t))-e(t-\bar{\tau}))
\end{aligned} \tag{3.19}$$

and

$$\begin{aligned}
&-\int_{0}^{+\infty}\kappa(s)\mathcal{G}^T(e(t-s))P_2\mathcal{G}(e(t-s))ds \\
&= -\int_{-\infty}^{t}\kappa(t-r)\mathcal{G}^T(e(r))P_2\mathcal{G}(e(r))dr \\
&\leq -\frac{1}{\breve{\kappa}}(\int_{-\infty}^{t}\kappa(t-r)\mathcal{G}(e(r))dr)^T P_2(\int_{-\infty}^{t}\kappa(t-r)\mathcal{G}(e(r))dr).
\end{aligned} \tag{3.20}$$

Considering Lemma 3.1 and substituting (3.11), (3.19)–(3.20) into (3.18), we have

$$\begin{aligned}
\dot{V}(e(t)) \leq& 2e^T(t)P_1\Big(\mathcal{A}e(t)+\int_{-\infty}^{t}\kappa(t-r)\mathcal{G}(e(r))dr - \sum_{i=1}^{N}(\mathcal{K}+\Delta\mathcal{K}) \\
&\times I_i(\mathcal{W}\otimes I)\mathcal{C}e(t-\tau_i(t))\Big) - \sum_{i=1}^{N}(e(t)-e(t-\tau_i(t)))^T Z_i(e(t) \\
&- e(t-\tau_i(t))) - \lambda_1\begin{bmatrix} e(t) \\ \mathcal{F}(e(t))\end{bmatrix}^T\begin{bmatrix}\tilde{U}_1 & \tilde{U}_2 \\ * & I\end{bmatrix}\begin{bmatrix} e(t) \\ \mathcal{F}(e(t))\end{bmatrix} - \lambda_2 \\
&\times\begin{bmatrix} e(t) \\ \mathcal{G}(e(t))\end{bmatrix}^T\begin{bmatrix}\tilde{W}_1 & \tilde{W}_2 \\ * & I\end{bmatrix}\begin{bmatrix} e(t) \\ \mathcal{G}(e(t))\end{bmatrix} + \sum_{i=1}^{N}\bar{\tau}^2\dot{e}^T(t)(Y_i+Z_i)\dot{e}(t) \\
&-\sum_{i=1}^{N}\frac{\pi^2}{4}(e(t)-e(t-\tau_i(t)))^T Y_i(e(t)-e(t-\tau_i(t))) \\
&+ e^T(t)Qe(t) - e^T(t-\bar{\tau})Qe(t-\bar{\tau}) + \breve{\kappa}\mathcal{G}^T(e(t))P_2\mathcal{G}(e(t)) \\
&-\sum_{i=1}^{N}(e(t-\tau_i(t))-e(t-\bar{\tau}))^T Z_i(e(t-\tau_i(t))-e(t-\bar{\tau})) \\
&-\frac{1}{\breve{\kappa}}(\int_{-\infty}^{t}\kappa(t-r)\mathcal{G}(e(r))dr)^T P_2(\int_{-\infty}^{t}\kappa(t-r)\mathcal{G}(e(r))dr)
\end{aligned}$$

$$=\xi^T(t)\Phi\xi(t) + \dot{e}^T(t)R\dot{e}(t) \tag{3.21}$$

where

$$e_\tau(t) = \text{col}_N^i\{e(t-\tau_i(t))\},$$

$$\Phi_{11} = P_1\mathcal{A} + \mathcal{A}^T P_1 + Q - \sum_{i=1}^{N}\frac{\pi^2}{4}Y_i - \lambda_1\tilde{U}_1 - \lambda_2\tilde{W}_1 - \sum_{i=1}^{N} Z_i,$$

$$\Phi = \begin{bmatrix} \Phi_{11} & \Phi_{12} & 0 & \Phi_{14} & -\lambda_2\tilde{W}_2 & P_1 \\ * & \Phi_{22} & \bar{Z}^T & 0 & 0 & 0 \\ * & * & \Phi_{33} & 0 & 0 & 0 \\ * & * & * & -\lambda_1 I & 0 & 0 \\ * & * & * & * & \Phi_{55} & 0 \\ * & * & * & * & * & -\frac{1}{\breve{\kappa}}P_2 \end{bmatrix},$$

$$\xi(t) = \Big[e^T(t) \quad e_\tau^T(t) \quad e^T(t-\bar{\tau}) \quad \mathcal{F}^T(e(t)) \quad \mathcal{G}^T(e(t)) \quad \int_{-\infty}^{t}\kappa(t-r)\mathcal{G}^T(e(r))dr\Big]^T.$$

By using the Schur complement, it follows from (3.15) that $\dot{V}(e(t)) < 0$, which means that the estimation error system (3.11) with $w(t) = 0$ and $v_\tau(t) = 0$ is asymptotically stable.

Let us now move to the H_∞ performance analysis for the estimation error system (3.11). For all nonzero $w(t) \in L_2[0,\infty)$ and $v_\tau(t) \in L_2[0,\infty)$, it can be obtained from (3.11) and (3.21) that

$$\dot{V}(e(t)) + \|\tilde{z}(t)\|^2 - \gamma^2\|w(t)\|^2 - \gamma^2\|v_\tau(t)\|^2 \le \varsigma^T(t)\tilde{\Phi}\varsigma(t) + \dot{e}^T(t)R\dot{e}(t) \tag{3.22}$$

where $\varsigma(t) = \big[\xi^T(t) \quad w^T(t) \quad v_\tau^T(t)\big]^T$ and

$$\tilde{\Phi} = \begin{bmatrix} \bar{\Phi}_{11} & \Phi_{12} & 0 & \Phi_{14} & -\lambda_2\tilde{W}_2 & P_1 & P_1\mathcal{B} & \Phi_{18} \\ * & \Phi_{22} & \bar{Z}^T & 0 & 0 & 0 & 0 & 0 \\ * & * & \Phi_{33} & 0 & 0 & 0 & 0 & 0 \\ * & * & * & -\lambda_1 I & 0 & 0 & 0 & 0 \\ * & * & * & * & \Phi_{55} & 0 & 0 & 0 \\ * & * & * & * & * & -\frac{1}{\breve{\kappa}}P_2 & 0 & 0 \\ * & * & * & * & * & * & -\gamma^2 I & 0 \\ * & * & * & * & * & * & * & -\gamma^2 I \end{bmatrix}.$$

Furthermore, it can be derived from (3.15) that

$$\dot{V}(e(t)) + \|\tilde{z}(t)\|^2 - \gamma^2\|w(t)\|^2 - \gamma^2\|v_\tau(t)\|^2 < 0 \tag{3.23}$$

for all nonzero $w(t)$ and $v_\tau(t)$. Consequently, we obtain

$$\int_0^t (\|\tilde{z}(s)\|^2 - \gamma^2\|w(s)\|^2 - \gamma^2\|v_\tau(s)\|^2)ds$$

$$
\begin{aligned}
&= \int_0^t (\|\tilde{z}(s)\|^2 - \gamma^2\|w(s)\|^2 - \gamma^2\|v_\tau(s)\|^2 + \dot{V}(e(s)))ds \\
&\quad - V(e(t)) + V(e(0)) \\
&\leq \int_0^t (\|\tilde{z}(s)\|^2 - \gamma^2\|w(s)\|^2 - \gamma^2\|v_\tau(s)\|^2 + \dot{V}(e(s)))ds \\
&<0
\end{aligned}
\tag{3.24}
$$

under the zero-initial condition. By letting $t \to \infty$, the H_∞ performance constraint (3.13) is immediately satisfied and, therefore, the proof of the theorem is complete.

Remark 3.2 *In Theorem 3.1, with the help of Jensen inequality, a sufficient condition is obtained such that the estimation error dynamics is asymptotically stable with a given H_∞ disturbance attenuation level. It is worth noting that a similar analysis has been carried out in [212, 214]. Indeed, the use of Jensen inequality is likely to introduce a bit conservatism in the sufficient condition. For the sake of further reducing the conservatism, we could adopt other advance technologies to tackle time-delays such as Wirtinger-based inequality approach, which encompasses the Jensen one.*

In terms of the obtained results in Theorem 3.1, the algorithm for designing the estimator parameters is provided in the following theorem.

Theorem 3.2 *Let the disturbance rejection/attenuation level $\gamma > 0$ be given. The estimation error dynamics governed by system (3.11) is asymptotically stable (with $w(t) = 0$ and $v_\tau(t) = 0$) and satisfies the H_∞ performance requirement (3.13) for all nonzero $w(t)$ and $v_\tau(t)$ (under the zero initial condition) if there exist matrices $P_1 = \text{diag}_N\{P_{1i}\} > 0$, $Q > 0$, $P_2 > 0$, $X = \text{diag}_N\{X_i\}$, $Z_i > 0$, $Y_i > 0$ ($i \in \mathcal{V}$) and positive scalars λ_1, λ_2, ϵ_1, ϵ_2 satisfying*

$$
\Xi = \begin{bmatrix} \check{\Phi} & \Xi_{12} & \Xi_{13} \\ * & -\epsilon_1 I & 0 \\ * & * & -\epsilon_2 I \end{bmatrix} < 0 \tag{3.25}
$$

where

$$\Xi_{12} = \begin{bmatrix} -\mathcal{H}^T X^T & 0 & 0 & 0 & 0 & 0 & 0 & 0 & 0 \end{bmatrix}^T, \ \mathcal{T} = \text{diag}_N^i\{T_i\}, \ \mathcal{H} = \text{diag}_N^i\{H_i\},$$

$$\bar{X} = \text{vec}_N^i\{X I_i(\mathcal{W} \otimes I)\mathcal{C}\}, \ \bar{N} = \text{vec}_N^i\{\mathcal{T} I_i(\mathcal{W} \otimes I)\mathcal{C}\}, \ \bar{\Phi}_{28} = (\epsilon_1 + \epsilon_2)\bar{N}^T \mathcal{T}\mathcal{D},$$

$$\Xi_{13} = \begin{bmatrix} 0 & 0 & 0 & 0 & 0 & 0 & 0 & 0 & -\mathcal{H}^T X^T \end{bmatrix}^T, \ \bar{\Phi}_{12} = -\bar{X} + \bar{Z} + \frac{\pi^2}{4}\bar{Y},$$

$$\bar{\Phi}_{88} = -\gamma^2 I + (\epsilon_1 + \epsilon_2)\mathcal{D}^T \mathcal{T}^T \mathcal{T}\mathcal{D}, \ \bar{\Phi}_{22} = -2\hat{Z} - \frac{\pi^2}{4}\hat{Y} + (\epsilon_1 + \epsilon_2)\bar{N}^T \bar{N},$$

$$\check{\Phi} = \begin{bmatrix} \bar{\Phi}_{11} & \bar{\Phi}_{12} & 0 & \Phi_{14} & -\lambda_2 \tilde{W}_2 & P_1 & P_1\mathcal{B} & -X\mathcal{D} & \mathcal{A}^T P_1 \\ * & \bar{\Phi}_{22} & \bar{Z}^T & 0 & 0 & 0 & 0 & \bar{\Phi}_{28} & -\bar{X}^T \\ * & * & \Phi_{33} & 0 & 0 & 0 & 0 & 0 & 0 \\ * & * & * & -\lambda_1 I & 0 & 0 & 0 & 0 & P_1 \\ * & * & * & * & \Phi_{55} & 0 & 0 & 0 & 0 \\ * & * & * & * & * & -\frac{1}{\bar{\kappa}} P_2 & 0 & 0 & P_1 \\ * & * & * & * & * & * & -\gamma^2 I & 0 & \mathcal{B}^T P_1 \\ * & * & * & * & * & * & * & \bar{\Phi}_{88} & -\mathcal{D}^T X^T \\ * & * & * & * & * & * & * & * & -2P_1 + R \end{bmatrix},$$

and $\bar{Z}$, $\hat{Z}$, $\bar{Y}$ and $\hat{Y}$ are defined in (3.16). Furthermore, if (3.25) holds, then the estimator parameters are given by $K_i = P_{1i}^{-1} X_i$, $i = 1, 2, \ldots, N$.

Proof *Note that the matrix $\bar{\Phi}$ can be decomposed as $\hat{\Phi} + \Delta\hat{\Phi}$, where*

$$\bar{\mathcal{M}} = \begin{bmatrix} 0 & 0 & 0 & 0 & 0 & 0 & 0 & 0 & -\mathcal{H}^T \mathcal{K}^T P_1 \end{bmatrix}^T, \ \hat{\Phi}_{12} = -P_1 \bar{\mathcal{K}} + \bar{Z} + \frac{\pi^2}{4} \bar{Y},$$

$$\mathcal{M} = \begin{bmatrix} -\mathcal{H}^T \mathcal{K}^T P_1 & 0 & 0 & 0 & 0 & 0 & 0 & 0 & 0 \end{bmatrix}^T, \ \hat{\Phi}_{18} = -P_1 \mathcal{K} \mathcal{D},$$

$$\Xi_{14} = \begin{bmatrix} 0 & \bar{N} & 0 & 0 & 0 & 0 & 0 & \mathcal{T}\mathcal{D} & 0 \end{bmatrix}, \ F(t) = \mathrm{diag}_N^i \{F_i(t)\},$$

$$\Delta\hat{\Phi} = \mathcal{M}F(t)\Xi_{14} + (\mathcal{M}F(t)\Xi_{14})^T + \bar{\mathcal{M}}F(t)\Xi_{14} + (\bar{\mathcal{M}}F(t)\Xi_{14})^T,$$

$$\hat{\Phi} = \begin{bmatrix} \bar{\Phi}_{11} & \hat{\Phi}_{12} & 0 & \Phi_{14} & 0 & P_1 & -P_1\mathcal{B} & \hat{\Phi}_{18} & \mathcal{A}^T P_1 \\ * & \Phi_{22} & \hat{Z} & 0 & 0 & 0 & 0 & 0 & -\bar{\mathcal{K}}^T P_1 \\ * & * & \Phi_{33} & 0 & 0 & 0 & 0 & 0 & 0 \\ * & * & * & -\lambda_1 I & 0 & 0 & 0 & 0 & P_1 \\ * & * & * & * & \Phi_{55} & 0 & 0 & 0 & 0 \\ * & * & * & * & * & -\frac{1}{\bar{\kappa}} P_2 & 0 & 0 & P_1 \\ * & * & * & * & * & * & -\gamma^2 I & 0 & \mathcal{B}^T P_1 \\ * & * & * & * & * & * & * & -\gamma^2 I & -\mathcal{D}^T \mathcal{K}^T P_1 \\ * & * & * & * & * & * & * & * & -P_1 R^{-1} P_1 \end{bmatrix}.$$

By considering (3.15), one has

$$\bar{\Phi} = \hat{\Phi} + \mathcal{M}F(t)\Xi_{14} + (\mathcal{M}F(t)\Xi_{14})^T + \bar{\mathcal{M}}F(t)\Xi_{14} + (\bar{\mathcal{M}}F(t)\Xi_{14})^T < 0. \quad (3.26)$$

It follows from Lemma 3.2 that (3.26) is true if there exist positive scalars ϵ_1 and ϵ_2 such that

$$\hat{\Phi} + (\epsilon_1 + \epsilon_2)\Xi_{14}^T \Xi_{14} + \epsilon_1^{-1} \mathcal{M}\mathcal{M}^T + \epsilon_2^{-1} \bar{\mathcal{M}}\bar{\mathcal{M}}^T < 0. \quad (3.27)$$

On the other hand, noting $-P_1 R^{-1} P_1 \leq -2P_1 + R$ and $P_{1i} K_i = X_i$, we can conclude that inequality (3.27) can be satisfied if (3.25) holds. This completes the proof.

Remark 3.3 *In Theorems 3.1 and 3.2, the resilient distributed sampled-data H_∞ state estimation problem has been solved for a class of continuous time-delay systems over sensor networks and an algorithm for designing the desired distributed estimators has been given. Note that the established existence condition in Theorem 3.1 involves all the information from the target plant and the sensor network including the system parameters, the network topology, the bound of the distributed delay, the bounds on the nonlinear functions, the upper bounds of the sampling periods as well as the bounds on the parameter drifts of the estimator gains. Furthermore, the filter design algorithm provided in Theorem 3.2 is numerically tractable and can be implemented by using standard Matlab software package. Our main results exhibit the following distinct features: 1) multiple phenomena have been modeled that include nonlinearities, infinity-distributed delays and external disturbances; 2) multiple objectives have been considered that include stability, H_∞ constraints and resilience; and 3) multiple effects (from sampled-data, infinity-distributed delays as well as estimator parameter variations) are taken into account in the proposed algorithm.*

3.3 An Illustrative Example

In this section, a simulation example is provided to validate the proposed distributed state estimation approach.

Consider the target plant and the sensor measurement with the following parameters:

$$
\begin{aligned}
&A = \begin{bmatrix} 0.2 & 0 \\ 0.01 & -0.2 \end{bmatrix}, \quad B = \begin{bmatrix} -0.3 \\ 0.1 \end{bmatrix}, \quad M = \begin{bmatrix} 0.3 & -0.2 \end{bmatrix}, \\
&C_1 = \begin{bmatrix} 1.2 & 0.8 \end{bmatrix}, \quad C_2 = \begin{bmatrix} 1 & 0.9 \end{bmatrix}, \quad C_3 = \begin{bmatrix} 0.8 & -0.6 \end{bmatrix}, \\
&D_1 = -0.1, \quad D_2 = 0.2, \quad D_3 = 0.1.
\end{aligned}
$$

Let the nonlinear functions $f(x(t))$ and $g(x(t))$ be selected as

$$
\begin{aligned}
f(x(t)) &= 0.5((U_1 + U_2)x(t) + (U_2 - U_1)\sin(t)x(t)), \\
g(x(t)) &= 0.5((W_1 + W_2)x(t) + (W_2 - W_1)\cos(t)x(t)),
\end{aligned}
$$

where

$$U_1 = \begin{bmatrix} 0.2 & 0 \\ 0 & -0.15 \end{bmatrix}, \quad U_2 = \begin{bmatrix} 0.15 & 0 \\ 0 & -0.18 \end{bmatrix},$$
$$W_1 = \begin{bmatrix} 0.15 & 0 \\ 0 & -0.1 \end{bmatrix}, \quad W_2 = \begin{bmatrix} 0.12 & 0 \\ 0 & -0.15 \end{bmatrix},$$

and let the distributed time-delay kernel be chosen as $\kappa(s) = e^{-10s}$. It is easy to obtain that $\breve{\kappa} = 0.1$.

The topology of sensor networks is represented by a graph $\mathcal{G} = (\mathcal{V}, \mathcal{Q}, \mathcal{W})$ with the set of nodes $\mathcal{V} = \{1, 2, 3\}$, the set of edges $\mathcal{Q} = \{(1,1), (2,2), (2,3), (3,1), (3,3)\}$ and the following adjacency matrix:

$$\mathcal{W} = \begin{bmatrix} 1 & 0 & 0 \\ 0 & 1 & 1 \\ 1 & 0 & 1 \end{bmatrix}.$$

The parameters of the estimator gain variations are selected as follows:

$$H_1 = 0.2, \quad H_2 = -0.1, \quad H_3 = 0.1, \quad T_1 = -0.2,$$
$$T_2 = 0.3, \quad T_3 = 0.1, \quad F_1(t) = F_2(t) = F_3(t) = \sin(t).$$

The H_∞ performance attenuation level is set as $\gamma = 0.5$. The upper bounds of the sampling periods for each sensor node are taken as $\tau_1 = 0.12$, $\tau_2 = 0.15$, and $\tau_3 = 0.14$, respectively. Then, we have $\bar{\tau} = 0.15$. With the above parameters, using Matlab LMI Toolbox, the parameters of the desired state estimators are derived

$$K_1 = \begin{bmatrix} 1.7966 \\ -0.2572 \end{bmatrix}, \quad K_2 = \begin{bmatrix} 1.5690 \\ -0.1251 \end{bmatrix}, \quad K_3 = \begin{bmatrix} 1.8924 \\ -0.0292 \end{bmatrix}.$$

In the simulation, the exogenous disturbances are chosen as $w(t) = \sin(t)e^{-0.15t}$ and $v(t) = \cos(t)e^{-0.15t}$. The initial values of the state are taken randomly by following the uniform distribution over $[-1, 1]$.

Simulation results are shown in Figs. 3.1–3.5. Figs. 3.1–3.3 plot the measurements before and after being sampled on sensor nodes 1, 2, and 3, respectively, where the sampled measurements have been employed by the state estimators. Fig. 3.4 plots the output $z(t)$ and its estimates from estimator i ($i = 1, 2, 3$). The estimation errors $\tilde{z}_i(t)$ ($i = 1, 2, 3$) are depicted in Fig. 3.5. The simulation results show that the distributed estimation scheme presented in this chapter is indeed effective.

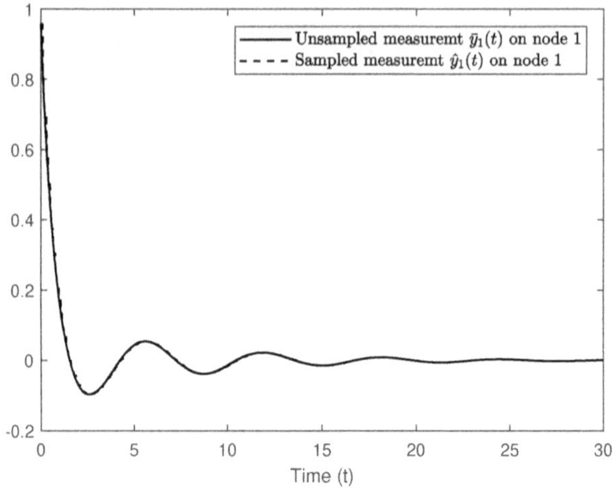

FIGURE 3.1
Unsampled measurement $\bar{y}_1(t)$ and sampled measurement $\hat{y}_1(t)$ on node 1.

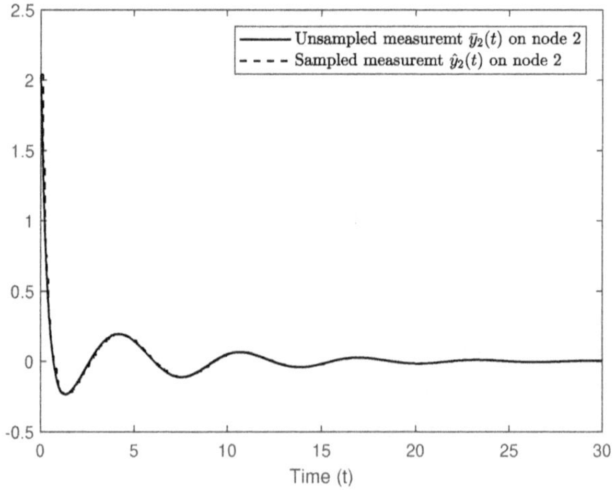

FIGURE 3.2
Unsampled measurement $\bar{y}_2(t)$ and sampled measurement $\hat{y}_2(t)$ on node 2.

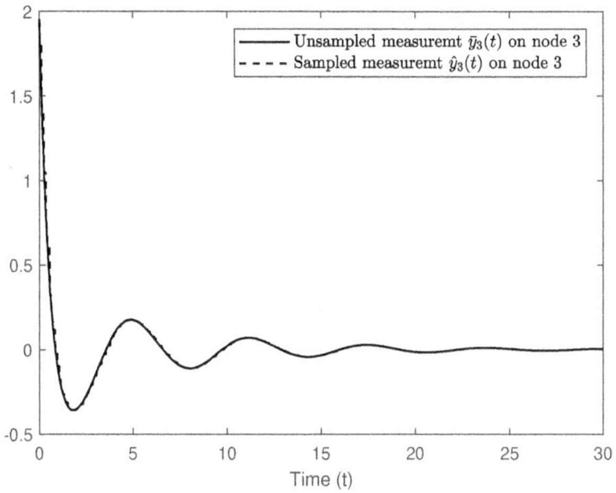

FIGURE 3.3
Unsampled measurement $\bar{y}_3(t)$ and sampled measurement $\hat{y}_3(t)$ on node 3.

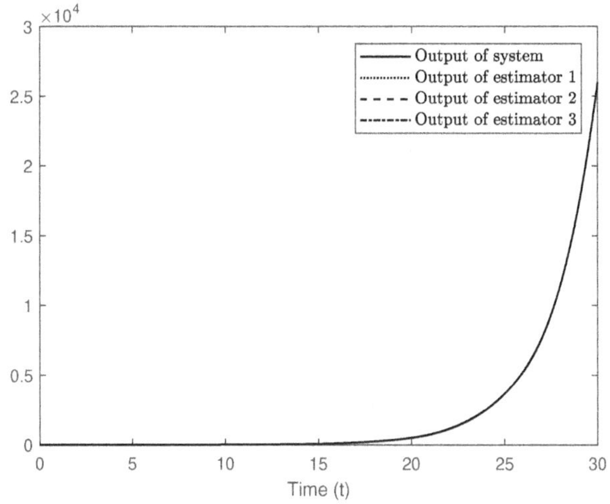

FIGURE 3.4
Output $z(t)$ and its estimates.

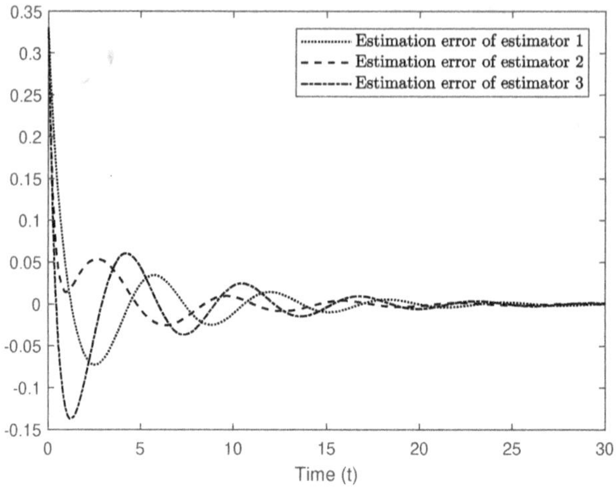

FIGURE 3.5
Estimation errors $\tilde{z}_i(t)$ $(i = 1, 2, 3)$.

3.4 Summary

In this chapter, the distributed sampled-data H_∞ state estimation problem has been discussed for a class of continuous-time systems with infinite-distributed delays and estimator parameter variations. A set of sensors has been deployed to measure the output of the plant and collaboratively share the measurement with their neighbors according to the given network topology. Then, the distributed state estimators have been designed based on the sampled measurement received by each sensor node. By taking advantage of the input delay approach, the effect of sampling intervals has been converted into an equivalent bounded TVD, and then a sufficient condition has been established to ensure both the asymptotical stability and the H_∞ performance requirement of the estimation error dynamics. By resorting to the LMI approach, the parameters of the desired distributed estimators have been obtained. Finally, we have used a numerical simulation example to illustrate the validity of the design approach proposed.

4

Event-Triggered Control for Switched Systems

This chapter is concerned with the ET control problem for switched systems. In order to reduce the communication burden, the ET strategy is adopted in the controller design of switched systems. We firstly consider a kind of switched DDSs with exogenous disturbances. The notions of boundedness and input-to-state practical stability are employed to characterize the control objectives in the presence of exogenous disturbances, switching signals and ET schemes. Some upper bounds on the system states are obtained and the Zeno phenomenon is shown to be excluded under the proposed ET scheme. The gain of the feedback controller, the parameters of the ET function and the switching signals are jointly designed for the underlying DDSs. Moreover, we also consider the ET control problem for the synchronization of switched stochastic complex networks with identical nodes. The notion of bounded synchronization in probability here is introduced to characterize the performance of the controlled dynamical networks in the presence of exogenous disturbances, switching rules and ET schemes. Upper bounds of the states of the switched stochastic complex networks are first provided, and then the controller gain, the ET parameters and the average dwell time are co-designed for switched subsystems. At last, some numerical examples are given to illustrate the effectiveness of our results.

4.1 Event-Triggered Control: The Input-to-State Stability

In this section, we endeavor to deal with the boundedness and the input-to-state stabilization problems for a kind of switched DDSs suffering from exogenous disturbances by using a purposely designed ET control strategy.

DOI: 10.1201/9781003307648-4

4.1.1 Problem Formulation

A: Description of the problem

Consider the following switched DDS with exogenous disturbances:

$$\begin{aligned}\dot{x}(t) =& A_{\sigma(t)}x(t) + f_{\sigma(t)}(t, x(t)) + B_{\sigma(t)}x(t-d)\\ &+ C_{\sigma(t)}u_{\sigma(t)}(t) + D_{\sigma(t)}v(t),\end{aligned} \tag{4.1}$$

where, for $t \in [0, \infty)$, $x(t) \in \mathbb{R}^n$ denotes the state variable, $d > 0$ is the delay of the system, $\sigma : [0, \infty) \to M = \{1, 2, \cdots, m\}$ is a switching signal, and m is the number of subsystems. $\sigma(\cdot)$ is a piecewise constant function of time such that $\sigma(t) = \sigma(\tau_k)$, for $t \in [\tau_k, \tau_{k+1})$, $k \in \mathbb{N}$, where time sequence $\{\tau_k\}_{k=0}^{\infty}$ satisfies $0 = \tau_0 < \tau_1 < \cdots < \tau_k < \cdots$ and $\tau_k \to \infty$ as $k \to \infty$ with τ_k denoting the moment of the k-th switching. For $i \in M$, $f_i : [0, \infty) \times \mathbb{R}^n \to \mathbb{R}^n$ are continuous and, for $\forall t \in \mathbb{R}_+$, $f(t, 0) = 0$. $u_i(t) \in \mathbb{R}^q$ denotes the control input, $v(t) \in \mathcal{L}_{\infty}^n$ represents an unknown exogenous disturbance, A_i, B_i, C_i and D_i are constant matrices with compatible dimensions. Here we assume that the solution of system (4.1) has global existence and is unique. In addition, we assume that the state of the system (4.1) does not jump at the switching instants, i.e., the solution $x(t)$ is everywhere continuous.

In this section, the state feedback control is adopted as follows:

$$u_{\sigma(t)}(t) = K_{\sigma(t)}x(t),$$

where K_i $(i \in M)$ are the feedback gain matrices which will be designed later.

In order to reduce the burden of the communications in the control loop, we introduce the ET mechanism into the execution of the control inputs. Let $\{t_k\}_{k=0}^{\infty}$ $(t_0 = 0)$ be the event-time sequence, which is defined iteratively as

$$t_{k+1} = \inf\{t : t > t_k,\ h(t) > 0\}, \tag{4.2}$$

where $h(t)$ is an event-generator function. According to the ZOH scheme, the control input $u_{\sigma(t)}(t)$ between successive updates remains a constant equal to the last control update, that is,

$$u_{\sigma(t)}(t) = K_{\sigma(t)}x(t_k), \quad t \in [t_k, t_{k+1}),\ k \in \mathbb{N}. \tag{4.3}$$

Under this ET control scheme, the sampled-data implementation of the closed-loop switched system in (4.1) is described as follows:

$$\begin{aligned}\dot{x}(t) =& A_{\sigma(t)}x(t) + f_{\sigma(t)}(t, x(t)) + B_{\sigma(t)}x(t-d)\\ &+ C_{\sigma(t)}K_{\sigma(t)}x(t_k) + D_{\sigma(t)}v(t)\end{aligned} \tag{4.4}$$

for $t \in [t_k, t_{k+1})$, $k \in \mathbb{N}$. Furthermore, by defining the measurement error as

$$\varepsilon(t) = x(t_k) - x(t), \tag{4.5}$$

the system (4.4) can be rewritten as follows:

$$\begin{aligned}\dot{x}(t) =&(A_{\sigma(t)} + C_{\sigma(t)}K_{\sigma(t)})x(t) + f_{\sigma(t)}(t, x(t)) \\ &+ B_{\sigma(t)}x(t-d) + C_{\sigma(t)}K_{\sigma(t)}\varepsilon(t) + D_{\sigma(t)}v(t).\end{aligned} \tag{4.6}$$

In this section, we develop an event-generator function $h(t)$ in the following form:

$$h(t) = \|\varepsilon(t)\|^2 - \zeta_1\|x(t_k)\|^2 - \zeta_2, \tag{4.7}$$

for $t \in [t_k, t_{k+1})$, $k \in \mathbb{N}$, where the positive scalars ζ_1 and ζ_2 are to be discussed in sequel.

Definition 4.1 *The closed-loop switched DDS (4.6) is said to be bounded with respect to the exogenous disturbance $v(t)$ if there are functions $\beta \in \mathcal{K}$, $\gamma \in \mathcal{K}$ and a nonnegative scalar θ such that the solution $x(t)$ satisfies*

$$\|x(t)\| < \beta(\|\xi\|_d) + \gamma(\|v\|_\infty) + \theta,$$

for the given event-generator function (4.7), the initial value $\xi \in \mathcal{C}([-d, 0]; \mathbb{R}^n)$ and the exogenous disturbance $v \in \mathcal{L}_\infty^n$.

Definition 4.2 *[78] The closed-loop switched DDS (4.6) is said to be input-to-state practically stable with respect to the exogenous disturbance $v(t)$ if there are functions $\beta \in \mathcal{KL}$, $\gamma \in \mathcal{K}$ and a nonnegative scalar θ such that the solution $x(t)$ satisfies*

$$\|x(t)\| < \beta(\|\xi\|_d, t) + \gamma(\|v\|_\infty) + \theta,$$

for the given event-generator function (4.7), the initial value $\xi \in \mathcal{C}([-d, 0]; \mathbb{R}^n)$ and the exogenous disturbance $v \in \mathcal{L}_\infty^n$.

Remark 4.1 *The term $\gamma(\|v\|_\infty) + \theta$ is used to deal with the effects from the switching signal, the exogenous disturbance and the ET control scheme on the dynamical performance of the closed-loop switched system. When $\theta = 0$ and $\|v\|_\infty = 0$, Definition 4.2 implies the asymptotical stability.*

Definition 4.3 *The switched DDS (4.1) is said to be input-to-state practically stabilizable if there exist feedback gain matrices K_i $(i \in M)$ and the ET parameters ζ_1, ζ_2 such that the closed-loop switched system (4.6) is input-to-state practically stable with respect to $v(t)$. Moreover, if $\theta = 0$, system (4.1) is said to be input-to-state stabilizable.*

B: Preliminaries results

Lemma 4.1 *For any $x, y \in \mathbb{R}^n$ and $Q > 0$ with appropriate dimension, the following inequality holds*

$$x^T y + y^T x \leq x^T Q x + y^T Q^{-1} y.$$

Lemma 4.2 *Let scalars α, β, γ, δ and $d > 0$ satisfy $\alpha+\beta > 0$, $\beta > 0$, $\gamma > 0$, $\delta > 0$, $d > 0$, and let $v(t) \in \mathcal{L}_\infty^n$. Furthermore, let $g : [t_0, \infty) \to \mathbb{R}_+$ satisfy the following delay differential inequality*

$$\dot{g}(t) \leq \alpha g(t) + \beta g(t-d) + \gamma + \delta \|v(t)\|^2, \tag{4.8}$$

for $t \in [t_0, \infty)$. Then, we have

$$g(t) \leq \Big[\sup_{\theta \in [t_0-d, t_0]} g(\theta) + \frac{\gamma + \delta \|v\|_\infty^2}{\alpha+\beta} \Big] e^{(\alpha+\beta)(t-t_0)} - \frac{\gamma + \delta \|v\|_\infty^2}{\alpha+\beta}, \tag{4.9}$$

for $t \in [t_0, \infty)$.

Proof *Let $\epsilon > 0$. Define the following continuous functions*

$$\begin{aligned} y_\epsilon(t) =& \int_{t_0}^{t} (\gamma + \delta \|v(s)\|^2) e^{-(\alpha+\beta)(s-t_0)} ds \\ &+ \sup_{\theta \in [t_0-d, t_0]} g(\theta) + \epsilon, \quad t \in [t_0, \infty), \end{aligned} \tag{4.10}$$

$$G(t) = e^{-(\alpha+\beta)(t-t_0)} g(t), \quad t \in [t_0 - d, \infty), \tag{4.11}$$

with the initial condition

$$G(t) \leq y_\epsilon(t), \quad t \in [t_0 - d, t_0].$$

Now, we claim that

$$G(t) \leq y_\epsilon(t), \quad t \in [t_0, \infty). \tag{4.12}$$

If (4.12) is not true, there must exist some $t > t_0$ such that $G(t) > y_\epsilon(t)$. Since $G(t)$ and $y_\epsilon(t)$ are continuous functions, we have

$$G(\bar{t}) = y_\epsilon(\bar{t}), \quad G(t) \leq y_\epsilon(t), \quad t \in [t_0 - d, \bar{t}), \tag{4.13}$$

where $\bar{t} \triangleq \inf\{t > t_0 | G(t) > y_\epsilon(t)\}$. We note that $G(t_0) = g(t_0) < \sup_{\theta \in [t_0-d, t_0]} g(\theta) + \epsilon = y_\epsilon(t_0)$, which yields $\bar{t} > t_0$. The continuities of $G(t)$ and $y_\epsilon(t)$ yield that there exists a sufficiently small positive scalar $\Delta t > 0$ such that

$$G(t) > y_\epsilon(t), \quad t \in (\bar{t}, \bar{t} + \Delta t).$$

Computing the upper right-hand Dini derivative of $G(t)$ at $\bar{t}$, we obtain

$$\begin{aligned} D^+ G(\bar{t}) =& \limsup_{h \to 0^+} \frac{G(\bar{t}+h) - G(\bar{t})}{h} \\ \geq& \limsup_{h \to 0^+} \frac{y_\epsilon(\bar{t}+h) - y_\epsilon(\bar{t})}{h} \end{aligned}$$

$$=(\gamma+\delta\|v(\bar{t})\|^2)e^{-(\alpha+\beta)(\bar{t}-t_0)}. \tag{4.14}$$

On the other hand, according to (4.8) and (4.11), we have

$$\begin{aligned}
&D^+G(\bar{t})\\
=&-(\alpha+\beta)e^{-(\alpha+\beta)(\bar{t}-t_0)}g(\bar{t})+e^{-(\alpha+\beta)(\bar{t}-t_0)}D^+g(\bar{t})\\
\le&\beta e^{-(\alpha+\beta)(\bar{t}-t_0)}[-g(\bar{t})+g(\bar{t}-d)]+(\gamma+\delta\|v(\bar{t})\|^2)e^{-(\alpha+\beta)(\bar{t}-t_0)}. \quad (4.15)
\end{aligned}$$

- *If $t_0 \le \bar{t}-d<\bar{t}$, then we have $G(\bar{t}-d)\le y_\epsilon(\bar{t}-d)<y_\epsilon(\bar{t})=G(\bar{t})$ since $y_\epsilon(t)$ is monotonically increasing on $[t_0,\infty)$. Then, we obtain*

$$g(\bar{t}-d)<e^{-(\alpha+\beta)d}g(\bar{t})<g(\bar{t}).$$

- *If $t_0-d<\bar{t}-d<t_0<\bar{t}$, then we have*

$$g(\bar{t}-d)\le \sup_{\theta\in[t_0-d,t_0]} g(\theta)<y_\epsilon(t_0)<y_\epsilon(\bar{t})=G(\bar{t})<g(\bar{t}).$$

It then follows from (4.15) that

$$D^+G(\bar{t})<(\gamma+\delta\|v(\bar{t})\|^2)e^{-(\alpha+\beta)(\bar{t}-t_0)},$$

which contradicts (4.14). Therefore, (4.12) is correct and then we have

$$\begin{aligned}
g(t)=&e^{(\alpha+\beta)(t-t_0)}G(t)\le e^{(\alpha+\beta)(t-t_0)}y_\epsilon(t)\\
=&e^{(\alpha+\beta)(t-t_0)}\int_{t_0}^{t}(\gamma+\delta\|v(s)\|^2)e^{-(\alpha+\beta)(s-t_0)}ds\\
&+(\sup_{\theta\in[t_0-d,t_0]} g(\theta)+\epsilon)e^{(\alpha+\beta)(t-t_0)}\\
\le&e^{(\alpha+\beta)(t-t_0)}\int_{t_0}^{t}(\gamma+\delta\|v\|_\infty^2)e^{-(\alpha+\beta)(s-t_0)}ds\\
&+(\sup_{\theta\in[t_0-d,t_0]} g(\theta)+\epsilon)e^{(\alpha+\beta)(t-t_0)}\\
=&\left[\sup_{\theta\in[t_0-d,t_0]} g(\theta)+\epsilon+\frac{\gamma+\delta\|v\|_\infty^2}{\alpha+\beta}\right]e^{(\alpha+\beta)(t-t_0)}\\
&-\frac{\gamma+\delta\|v\|_\infty^2}{\alpha+\beta}.
\end{aligned}$$

Let $\epsilon\to 0^+$, we get (4.9), which completes the proof.

Lemma 4.3 *For scalars $\alpha>\beta>0$, $\gamma>0$, $\delta>0$, $d>0$ and vector $v(t)\in\mathcal{L}_\infty^n$, let $g:[t_0,\infty)\to\mathbb{R}_+$ satisfy the following delay differential inequality*

$$\dot{g}(t)\le-\alpha g(t)+\beta g(t-d)+\gamma+\delta\|v(t)\|^2, \tag{4.16}$$

for $t \in [t_0, \infty)$. *Then, we have*

$$g(t) \leq \Big[\sup_{\theta \in [t_0 - d, t_0]} g(\theta) - \frac{\gamma + \delta\|v\|_\infty^2}{\kappa^*}\Big] e^{-\kappa^*(t-t_0)} + \frac{\gamma + \delta\|v\|_\infty^2}{\kappa^*}, \tag{4.17}$$

for $t \in [t_0, \infty)$, *where* κ^* *is the solution of* $\kappa = \alpha - \beta e^{\kappa d}$.

Proof *Define*

$$\begin{aligned} y_\epsilon(t) =& \int_{t_0}^{t} (\gamma + \delta\|v(s)\|^2) e^{\kappa^*(s-t_0)} ds, \\ &+ \sup_{\theta \in [t_0 - d, t_0]} g(\theta) + \epsilon, \quad t \in [t_0, \infty), \end{aligned} \tag{4.18}$$

$$G(t) = e^{\kappa^*(t-t_0)} g(t), \quad t \in [t_0 - d, \infty), \tag{4.19}$$

with the initial condition

$$G(t) \leq y_\epsilon(t), \quad t \in [t_0 - d, t_0].$$

The rest of the proof of (4.17) is similar to that of Lemma 4.2 and is therefore omitted.

4.1.2 Main Results

In this subsection, we present our main results on the boundedness and the input-to-state practical stability for the switched DDSs with the state feedback control (4.6) and the event-generator function (4.7). For doing so, we first assume that the functions $f_i(t, x)$ $(i \in M)$ satisfy the Lipschitz condition.

Assumption 4.1 *There exist positive constant* L_i $(i \in M)$ *such that, for any* x, $y \in \mathbb{R}^n$, *the following is true:*

$$\|f_i(t, x) - f_i(t, y)\| \leq L_i \|x - y\|, \quad i \in M. \tag{4.20}$$

Now, let us first discuss the boundedness in the following theorem.

Theorem 4.1 *Under Assumption 4.1 and with* $\zeta_0 \in (0, 1)$, *let the feedback gain matrices* $K_i \in \mathbb{R}^{q \times n}$ $(i \in M)$ *and the two ET parameters* $0 < \zeta_1 < \zeta_0 < 1$, $\zeta_2 > 0$ *be given. Assume that there exist positive definite matrices* Q_i, $S_i \in \mathbb{R}^{n \times n}$ $(i \in M)$, *constants* μ_i $(i \in M)$, $\tilde{d}_1 > d$ *and* $\tilde{d}_2 > d$ *such that*

$$\begin{aligned} &P_i(A_i + C_iK_i) + (A_i + C_iK_i)^T P_i + P_iB_iP_i^{-1}B_i^T P_i \\ &+ P_iC_iK_iS_i^{-1}K_i^T C_i^T P_i + P_iD_iQ_i^{-1}D_i^T P_i + P_i^2 \\ &+ L_i^2 I \leq \mu_i P_i, \quad i \in M. \end{aligned} \tag{4.21}$$

Furthermore, assume that the switching signal $\sigma(t)$ and the switching time sequence satisfy

$$(i)\ \text{for}\ \sigma(0) \in M_2\ \text{and for}\ k \in \mathbb{N}_+,\ \text{if}\ \sigma(\tau_k) \in M_1,\ \text{then}\ \sigma(\tau_{k+1}) \in M_2; \tag{4.22}$$

$$\begin{aligned}(ii)\ &\text{for}\ k \in \mathbb{N},\ \text{if}\ \sigma(\tau_k) \in M_1,\ \text{then}\ d \leq \tau_{k+1} - \tau_k \leq \tilde{d}_2;\ \text{and} \\ &\quad \text{if}\ \sigma(\tau_k) \in M_2,\ \text{then}\ d < \tilde{d}_1 \leq \tau_{k+1} - \tau_k;\end{aligned} \tag{4.23}$$

$$(iii)\ z^*(\tilde{d}_1 - d) - (b^* + 1)\tilde{d}_2 > 3 \ln \rho \tag{4.24}$$

where $\rho = \frac{\max_{i \in M} \lambda_{\max}(P_i)}{\min_{i \in M} \lambda_{\min}(P_i)}$, $b_i = \mu_i + \frac{\zeta_0 \zeta_1 \lambda_{\max}(S_i)}{(\zeta_0 - \zeta_1)(1 - \zeta_0)\lambda_{\min}(P_i)}$, $M_1 = \{i \in M | b_i + 1 > 0\}$, $M_2 = \{i \in M | b_i + 1 < 0\}$, $M_1 \cup M_2 = M$, $M_1 \cap M_2 = \varnothing$, $b^* = \max_{i \in M_1} b_i$, z_i^* *is the solution of* $z = -b_i - e^{dz}$ $(i \in M_2)$, *and* $z^* = \max_{i \in M_2} z_i^*$. *Then, the system (4.6) is bounded with respect to exogenous disturbance* $v(t)$.

Proof *Construct the following switched Lyapunov function*

$$V_{\sigma(t)}(x(t)) = x^T(t) P_{\sigma(t)} x(t), \tag{4.25}$$

where P_i $(i \in M)$ *are positive definite matrices. For simplicity, we denote* $V_{\sigma(t)}(t) \triangleq V_{\sigma(t)}(x(t))$ *and calculate the derivative of* $V_{\sigma(t)}(t)$ *along the trajectory of system (4.6) as follows:*

$$\begin{aligned}&\dot{V}_{\sigma(t)}(t) \\ =&x^T(t)\Big[(A_{\sigma(t)} + C_{\sigma(t)}K_{\sigma(t)})^T P_{\sigma(t)} + P_{\sigma(t)}(A_{\sigma(t)} \\ &+ C_{\sigma(t)}K_{\sigma(t)})\Big]x(t) + 2f^T_{\sigma(t)}(t, x(t))P_{\sigma(t)}x(t) \\ &+ 2x^T(t)P_{\sigma(t)}B_{\sigma(t)}x(t-d) + 2x^T P_{\sigma(t)}D_{\sigma(t)}v(t) \\ &+ 2x^T(t)P_{\sigma(t)}C_{\sigma(t)}K_{\sigma(t)}\varepsilon(t) \\ &\triangleq s_1 + s_2 + s_3 + s_4 + s_5.\end{aligned} \tag{4.26}$$

In view of (4.20), one has

$$s_2 \leq x^T(t)P_{\sigma(t)}P_{\sigma(t)}x(t) + L^2_{\sigma(t)}\|x(t)\|^2. \tag{4.27}$$

According to Lemma 4.1, we obtain

$$s_3 \leq x^T(t)P_{\sigma(t)}B_{\sigma(t)}P^{-1}_{\sigma(t)}B^T_{\sigma(t)}P_{\sigma(t)}x(t) + x^T(t-d)P_{\sigma(t)}x(t-d) \tag{4.28}$$

and, from Lemma 4.1, we further obtain

$$\begin{aligned}s_4 + s_5 \leq& x^T(t)P_{\sigma(t)}C_{\sigma(t)}K_{\sigma(t)}S^{-1}_{\sigma(t)}K^T_{\sigma(t)}C^T_{\sigma(t)}P_{\sigma(t)}x(t) \\ &+ x^T(t)P_{\sigma(t)}D_{\sigma(t)}Q^{-1}_{\sigma(t)}D^T_{\sigma(t)}P_{\sigma(t)}x(t) \\ &+ \varepsilon^T(t)S_{\sigma(t)}\varepsilon(t) + v^T(t)Q_{\sigma(t)}v(t)\end{aligned}$$

$$\begin{aligned}
\leq & x^T(t)P_{\sigma(t)}C_{\sigma(t)}K_{\sigma(t)}S_{\sigma(t)}^{-1}K_{\sigma(t)}^T C_{\sigma(t)}^T P_{\sigma(t)}x(t) \\
& + x^T(t)P_{\sigma(t)}D_{\sigma(t)}Q_{\sigma(t)}^{-1}D_{\sigma(t)}^T P_{\sigma(t)}x(t) \\
& + \lambda_{\max}(S_{\sigma(t)})\|\varepsilon(t)\|^2 + \lambda_{\max}(Q_{\sigma(t)})\|v(t)\|^2. \qquad (4.29)
\end{aligned}$$

Substituting (4.27)–(4.29) into (4.26) and using Assumption 4.1 together with (4.21), we arrive at

$$\begin{aligned}
& \dot{V}_{\sigma(t)}(t) \\
\leq & x^T(t)\big[P_{\sigma(t)}(A_{\sigma(t)} + C_{\sigma(t)}K_{\sigma(t)}) + (A_{\sigma(t)} \\
& + C_{\sigma(t)}K_{\sigma(t)})^T P_{\sigma(t)} + P_{\sigma(t)}B_{\sigma(t)}P_{\sigma(t)}^{-1}B_{\sigma(t)}^T P_{\sigma(t)} \\
& + P_{\sigma(t)}C_{\sigma(t)}K_{\sigma(t)}S_{\sigma(t)}^{-1}K_{\sigma(t)}^T C_{\sigma(t)}^T P_{\sigma(t)} \\
& + P_{\sigma(t)}D_{\sigma(t)}Q_{\sigma(t)}^{-1}D_{\sigma(t)}^T P_{\sigma(t)} + P_{\sigma(t)}^2 + L_{\sigma(t)}^2 I\big]x(t) \\
& + x^T(t-d)P_{\sigma(t)}x(t-d) + \lambda_{\max}(S_{\sigma(t)})\|\varepsilon(t)\|^2 \\
& + \lambda_{\max}(Q_{\sigma(t)})\|v(t)\|^2 \\
\leq & \mu_{\sigma(t)}x^T(t)P_{\sigma(t)}x(t) + x^T(t-d)P_{\sigma(t)}x(t-d) \\
& + \lambda_{\max}(S_{\sigma(t)})\|\varepsilon(t)\|^2 + \lambda_{\max}(Q_{\sigma(t)})\|v(t)\|^2. \qquad (4.30)
\end{aligned}$$

In view of (4.5), we note that $x(t_k) = \varepsilon(t) + x(t)$. *According to the triggered rule (4.2) with the event-generator function (4.7), we obtain*

$$\begin{aligned}
\|\varepsilon(t)\|^2 \leq & \zeta_1\|x(t_k)\|^2 + \zeta_2 = \zeta_1\|\varepsilon(t) + x(t)\|^2 + \zeta_2 \\
\leq & \frac{\zeta_1}{1-\zeta_0}\|x(t)\|^2 + \frac{\zeta_1}{\zeta_0}\|\varepsilon(t)\|^2 + \zeta_2,
\end{aligned}$$

which yields

$$\|\varepsilon(t)\|^2 \leq \frac{\zeta_0\zeta_1}{(\zeta_0-\zeta_1)(1-\zeta_0)}\|x(t)\|^2 + \frac{\zeta_0\zeta_2}{\zeta_0-\zeta_1}. \qquad (4.31)$$

Then, it follows from (4.30) and (4.31) that

$$\begin{aligned}
\dot{V}_{\sigma(t)}(t) \leq & b_{\sigma(t)}V_{\sigma(t)}(t) + V_{\sigma(t)}(t-d) + c_{\sigma(t)} \\
& + \lambda_{\max}(Q_{\sigma(t)})\|v(t)\|^2, \quad t \geq 0, \qquad (4.32)
\end{aligned}$$

where $c_i = \frac{\zeta_0\zeta_2\lambda_{\max}(S_i)}{\zeta_0-\zeta_1}$.

Using Lemmas 4.2 and 4.3, for $t \in [\tau_k, \tau_{k+1})$, *we obtain the following inequalities:*

$$\begin{aligned}
V_{\sigma(\tau_k)}(t) \leq & \Big[\sup_{\theta\in[\tau_k-d,\tau_k]} V_{\sigma(\tau_k)}(\theta) + r_{\sigma(\tau_k)}\Big]e^{(b^*+1)(t-\tau_k)} \\
& - r_{\sigma(\tau_k)}, \quad \text{when } \sigma(\tau_k) \in M_1, \qquad (4.33)
\end{aligned}$$

$$V_{\sigma(\tau_k)}(t) \leq \Big[\sup_{\theta\in[\tau_k-d,\tau_k]} V_{\sigma(\tau_k)}(\theta) - r_{\sigma(\tau_k)}\Big]e^{-z^*(t-\tau_k)} + r_{\sigma(\tau_k)}, \quad \text{when } \sigma(\tau_k) \in M_2, \tag{4.34}$$

where

$$r_{\sigma(\tau_k)} = \begin{cases} \dfrac{c_{\sigma(\tau_k)} + \lambda_{\max}(Q_{\sigma(\tau_k)})\|v\|_\infty^2}{b_{\sigma(\tau_k)}+1}, & \text{when } \sigma(\tau_k) \in M_1, \\ \dfrac{c_{\sigma(\tau_k)} + \lambda_{\max}(Q_{\sigma(\tau_k)})\|v\|_\infty^2}{z^*_{\sigma(\tau_k)}}, & \text{when } \sigma(\tau_k) \in M_2. \end{cases}$$

Defining $r = \max_{i\in M}\{r_i\}$, *it is derived from (4.33) and (4.34) that*

$$V_{\sigma(\tau_k)}(t) \leq \Big[\sup_{\theta\in[\tau_k-d,\tau_k]} V_{\sigma(\tau_k)}(\theta) + r\Big]e^{(b^*+1)(t-\tau_k)} - r, \quad \text{when } \sigma(\tau_k) \in M_1, \tag{4.35}$$

$$V_{\sigma(\tau_k)}(t) \leq \Big[\sup_{\theta\in[\tau_k-d,\tau_k]} V_{\sigma(\tau_k)}(\theta) - r\Big]e^{-z^*(t-\tau_k)} + r, \quad \text{when } \sigma(\tau_k) \in M_2. \tag{4.36}$$

Now, we claim that, for $k \in \mathbb{N}$, $t \in [\tau_k, \tau_{k+1})$, *the following is true:*

$$V_{\sigma(\tau_k)}(t) \leq \rho \min_{i\in M} \lambda_{\min}(P_i)\|\xi\|_d^2 + ar \tag{4.37}$$

where

$$a = \frac{(\rho^2+\rho)e^{(b^*+1)\tilde{d}_2} - \rho^2 e^{-z^*(\tilde{d}_1-d)+(b^*+1)\tilde{d}_2} - \rho}{1-\rho^3 e^{-z^*(\tilde{d}_1-d)+(b^*+1)\tilde{d}_2}} > \rho.$$

Let us prove (4.37) by mathematical induction. It is easy to know that

$$\sup_{\theta\in[-d,0]} V_{\sigma(0)}(\theta) \leq \max_{i\in M} \lambda_{\max}(P_i)\|\xi\|_d^2 = \rho \min_{i\in M} \lambda_{\min}(P_i)\|\xi\|_d^2.$$

In view of (4.24), we obtain $z^*\tilde{d}_1 > 3\ln\rho$. *Moreover, from (4.22) and (4.36), we have for* $t \in [0, \tau_1)$ *that*

$$\begin{aligned} V_{\sigma(0)}(t) \leq& \Big[\rho \min_{i\in M} \lambda_{\min}(P_i)\|\xi\|_d^2 - r\Big]e^{-z^*t} + r \\ <& \rho \min_{i\in M} \lambda_{\min}(P_i)\|\xi\|_d^2 + r, \\ V_{\sigma(\tau_1)}(\tau_1) \leq& \rho V_{\sigma(0)}(\tau_1) \\ \leq& \rho\Big[\rho \min_{i\in M} \lambda_{\min}(P_i)\|\xi\|_d^2 - r\Big]e^{-z^*\tau_1} + \rho r \end{aligned}$$

$$\leq \rho^2 \min_{i\in M} \lambda_{\min}(P_i)\|\xi\|_d^2 e^{-z^*\tilde{d}_1} + \rho r$$
$$\leq \frac{1}{\rho} \min_{i\in M} \lambda_{\min}(P_i)\|\xi\|_d^2 + \rho r,$$

which means that (4.37) is satisfied when $t \in [0, \tau_1]$.

If (4.37) is satisfied for $t \in [0, \tau_k]$ $(k \in \mathbb{N}_+)$*, then we will prove that (4.37) is correct for* $t \in [0, \tau_{k+1}]$. *According to (4.22), we consider three cases: (1)* $\sigma(\tau_{k-1}) \in M_2$, $\sigma(\tau_k) \in M_2$, *(2)* $\sigma(\tau_{k-1}) \in M_2$, $\sigma(\tau_k) \in M_1$ *and (3)* $\sigma(\tau_{k-1}) \in M_1$, $\sigma(\tau_k) \in M_2$.

From (4.37), we have

$$\begin{aligned}\sup_{\theta\in[\tau_{k-1}-d,\tau_{k-1}]} V_{\sigma(\tau_{k-1})}(\theta) \leq \rho \sup_{\theta\in[\tau_{k-1}-d,\tau_{k-1}]} V_{\sigma(\theta)}(\theta)\\ \leq \rho^2 \min_{i\in M} \lambda_{\min}(P_i)\|\xi\|_d^2 + \rho a r.\end{aligned} \tag{4.38}$$

Case 1: $\sigma(\tau_{k-1}) \in M_2$, $\sigma(\tau_k) \in M_2$.

From (4.23), (4.36), and (4.38), for $t \in [\tau_k - d, \tau_k) \subset [\tau_{k-1}, \tau_k)$*, we have*

$$\begin{aligned}&V_{\sigma(\tau_{k-1})}(t)\\ \leq&\Big[\sup_{\theta\in[\tau_{k-1}-d,\tau_{k-1}]} V_{\sigma(\tau_{k-1})}(\theta) - r\Big]e^{-z^*(t-\tau_{k-1})} + r\\ \leq&\big[\rho^2 \min_{i\in M} \lambda_{\min}(P_i)\|\xi\|_d^2 + \rho a r - r\big]e^{-z^*(\tau_k-\tau_{k-1}-d)} + r\\ \leq&\big[\rho^2 \min_{i\in M} \lambda_{\min}(P_i)\|\xi\|_d^2 + \rho a r - r\big]e^{-z^*(\tilde{d}_1-d)} + r.\end{aligned} \tag{4.39}$$

In view of (4.23), (4.24) and the definition of a*, it follows that*

$$V_{\sigma(\tau_{k-1})}(t) \leq \frac{1}{\rho} \min_{i\in M} \lambda_{\min}(P_i)\|\xi\|_d^2 + \frac{a}{\rho^2} r,$$

for $t \in [\tau_k - d, \tau_k)$*, which yields*

$$\begin{aligned}\sup_{\theta\in[\tau_k-d,\tau_k]} V_{\sigma(\tau_k)}(\theta) \leq& \rho \sup_{\theta\in[\tau_k-d,\tau_k]} V_{\sigma(\theta)}(\theta)\\ \leq& \min_{i\in M} \lambda_{\min}(P_i)\|\xi\|_d^2 + \frac{a}{\rho} r.\end{aligned}$$

Since $\sigma(\tau_k) \in M_2$*, for* $t \in [\tau_k, \tau_{k+1})$*, it follows that*

$$\begin{aligned}&V_{\sigma(\tau_k)}(t)\\ \leq&\Big[\sup_{\theta\in[\tau_k-d,\tau_k]} V_{\sigma(\tau_k)}(\theta) - r\Big]e^{-z^*(t-\tau_k)} + r\\ \leq&\big[\min_{i\in M} \lambda_{\min}(P_i)\|\xi\|_d^2 + \frac{a}{\rho} r - r\big]e^{-z^*(t-\tau_k)} + r\end{aligned}$$

$$\leq \min_{i\in M} \lambda_{\min}(P_i)\|\xi\|_d^2 + \frac{a}{\rho}r.$$

Moreover, from (4.24), we obtain

$$\begin{aligned}
&V_{\sigma(\tau_{k+1})}(\tau_{k+1})\\
\leq&\rho V_{\sigma(\tau_k)}(\tau_{k+1})\\
\leq&\rho\Big[\min_{i\in M}\lambda_{\min}(P_i)\|\xi\|_d^2 + \frac{a}{\rho}r - r\Big]e^{-z^*(\tau_{k+1}-\tau_k)} + \rho r\\
\leq&\rho\min_{i\in M}\lambda_{\min}(P_i)\|\xi\|_d^2 + ar.
\end{aligned}$$

Therefore, in Case 1, (4.37) is correct for $t \in [0, \tau_{k+1}]$.

Case 2: $\sigma(\tau_{k-1}) \in M_2$, $\sigma(\tau_k) \in M_1$.

We estimate the value of $V_{\sigma(\tau_k)}(t)$ $(t \in [\tau_k, \tau_{k+1}))$ *by using (4.23) and (4.24) as follows:*

$$\begin{aligned}
&V_{\sigma(\tau_k)}(t)\\
\leq&\Big[\sup_{\theta\in[\tau_k-d,\tau_k]} V_{\sigma(\tau_k)}(\theta) + r\Big]e^{(b^*+1)(t-\tau_k)} - r\\
\leq&\big[\rho\big(\rho^2\min_{i\in M}\lambda_{\min}(P_i)\|\xi\|_d^2 + \rho a r - r\big)e^{-z^*(\tau_k-\tau_{k-1}-d)}\\
&+\rho r + r\big]e^{(b^*+1)(\tau_{k+1}-\tau_k)} - r\\
\leq&\big[\rho^3\min_{i\in M}\lambda_{\min}(P_i)\|\xi\|_d^2 + (\rho^2 a - \rho)r\big]e^{-z^*(\widetilde{d}_1-d)+(b^*+1)\widetilde{d}_2}\\
&+(\rho+1)re^{(b^*+1)\widetilde{d}_2} - r\\
\leq&\min_{i\in M}\lambda_{\min}(P_i)\|\xi\|_d^2 + \frac{a}{\rho}r.
\end{aligned}$$

Then, we have

$$\begin{aligned}
V_{\sigma(\tau_{k+1})}(\tau_{k+1}) \leq&\rho V_{\sigma(\tau_k)}(\tau_{k+1})\\
\leq&\rho\min_{i\in M}\lambda_{\min}(P_i)\|\xi\|_d^2 + ar,
\end{aligned}$$

which means that (4.37) is correct for $t \in [0, \tau_{k+1}]$ *in Case 2.*

Case 3: $\sigma(\tau_{k-1}) \in M_1$, $\sigma(\tau_k) \in M_2$.

In this case, we claim that $\sigma(\tau_{k-2}) \in M_2$. *Otherwise, if* $\sigma(\tau_{k-2}) \in M_1$, *then* $\sigma(\tau_{k-1}) \in M_1$ *contradicts with (4.22). Similar to the proof in Case 2, we obtain*

$$V_{\sigma(\tau_{k-1})}(t) \leq \min_{i\in M}\lambda_{\min}(P_i)\|\xi\|_d^2 + \frac{a}{\rho}r, \quad t \in [\tau_k - d, \tau_k)$$

which yields

$$\sup_{\theta\in[\tau_k-d,\tau_k]} V_{\sigma(\tau_k)}(\theta) \leq\rho \sup_{\theta\in[\tau_k-d,\tau_k]} V_{\sigma(\theta)}(\theta)$$

$$\leq \rho \min_{i \in M} \lambda_{\min}(P_i) \|\xi\|_d^2 + ar.$$

Since $\sigma(\tau_k) \in M_2$, similar to the proof in Case 1, we obtain

$$\begin{aligned}
&V_{\sigma(\tau_k)}(t) \\
\leq &\Big[\sup_{\theta \in [\tau_k - d, \tau_k]} V_{\sigma(\tau_k)}(\theta) - r\Big] e^{-z^*(t-\tau_k)} + r \\
\leq &\rho \min_{i \in M} \lambda_{\min}(P_i) \|\xi\|_d^2 + ar, \qquad t \in [\tau_k, \tau_{k+1}), \\
&V_{\sigma(\tau_{k+1})}(\tau_{k+1}) \leq \rho V_{\sigma(\tau_k)}(\tau_{k+1}) \\
\leq &\rho\big[\rho \min_{i \in M} \lambda_{\min}(P_i) \|\xi\|_d^2 + ar - r\big] e^{-z^*(\tau_{k+1}-\tau_k)} + \rho r \\
\leq &\rho \min_{i \in M} \lambda_{\min}(P_i) \|\xi\|_d^2 + ar,
\end{aligned}$$

which means that (4.37) is satisfied for $t \in [0, \tau_{k+1}]$ in Case 3.

To this end, we have

$$V_{\sigma(t)}(t) \leq \rho \min_{i \in M} \lambda_{\min}(P_i) \|\xi\|_d^2 + ar, \quad t \in [0, \infty).$$

Setting

$$\begin{aligned}
\chi_1 &= \max\{\max_{i \in M_1} \frac{\lambda_{\max}(Q_i)}{b_i + 1}, \max_{i \in M_2} \frac{\lambda_{\max}(Q_i)}{z_i^*}\}, \\
\chi_2 &= \max\{\max_{i \in M_1} \frac{\zeta_0 \zeta_2 \lambda_{\max}(S_i)}{(\zeta_0 - \zeta_1)(b_i + 1)}, \max_{i \in M_2} \frac{\zeta_0 \zeta_2 \lambda_{\max}(S_i)}{(\zeta_0 - \zeta_1) z_i^*}\},
\end{aligned}$$

we then obtain

$$\begin{aligned}
&\|x(t)\| \\
\leq &\sqrt{\rho \|\xi\|_d^2 + \frac{ar}{\min_{i \in M} \lambda_{\min}(P_i)}} \\
\leq &\sqrt{\frac{a\chi_1}{\min_{i \in M} \lambda_{\min}(P_i)}} \|v\|_\infty + \sqrt{\frac{a\chi_2}{\min_{i \in M} \lambda_{\min}(P_i)}} \\
&+ \sqrt{\rho} \|\xi\|_d,
\end{aligned} \tag{4.40}$$

which implies that the switched DDS (4.6) is bounded with respect to the exogenous disturbance $v(t)$. The proof is complete.

Remark 4.2 *In Theorem 4.1, the boundedness issue is investigated for system (4.6) with switching signals, exogenous disturbances and ET control scheme. The control input is updated whenever the system switches or the measurement error $\varepsilon(t)$ exceeds the prescribed threshold relating to the latest sampled data. With the updated control signals, the state of system (4.6) is guaranteed to be driven into a bounded set. Note that (4.40) gives an upper bound of the norm of the system state.*

Remark 4.3 *In this section, we are concerned with a class of switched time-delayed systems with exogenous disturbances and the input-to-state stability has been adopted to describe the system stability. In order to handle the exogenous disturbances and establish the corresponding input-to-state stability criteria, we have proposed a new switching rule instead of the traditional dwell-time method. Comparing with the average dwell-time method, the advantages of our proposed method are that: 1) the proposed method is capable of dealing with the switched delayed systems with exogenous disturbances and ET mechanism, and thus the boundedness and input-to-state stabilization problem can be solved; and 2) in the proposed method, two cases of* $b_i > -1$ *and* $b_i < -1$ *are discussed separately, which can remove one assumption, which is usually required by using the traditional dwell-time method.*

We are now in a position to investigate the input-to-state practical stability of the switched DDS (4.6) with exogenous disturbance.

Theorem 4.2 *Under Assumption 4.1 and for* $\zeta_0 \in (0,1)$*, let the feedback gain matrices* $K_i \in \mathbb{R}^{q\times n}$ $(i \in M)$ *and the two ET parameters* $0 < \zeta_1 < \zeta_0 < 1$*,* $\zeta_2 > 0$ *be given. Assume that there exist positive definite matrices* Q_i*,* $S_i \in \mathbb{R}^{n\times n}$ $(i \in M)$ *and constants* $\mu_i < 0$ $(i \in M)$*,* $\omega > 0$ *and* $\tilde{d} > d$ *such that*

$$\begin{aligned} &P_i(A_i + C_iK_i) + (A_i + C_iK_i)^T P_i + P_iB_iP_i^{-1}B_i^T P_i \\ &+ P_iC_iK_iS_i^{-1}K_i^T C_i^T P_i + P_iD_iQ_i^{-1}D_i^T P_i + P_i^2 \\ &+ L_i^2 I \le \mu_i P_i, \quad i \in M. \end{aligned} \tag{4.41}$$

Furthermore, let the switching signal $\sigma(t)$ *and the switching time sequence satisfy that*

$$b_i = \mu_i + \frac{\zeta_0\zeta_1\lambda_{\max}(S_i)}{(\zeta_0 - \zeta_1)(1-\zeta_0)\lambda_{\min}(P_i)} < -1, \quad i \in M, \tag{4.42}$$

$$(z^* - \omega)(\tilde{d} - d) - 2\omega d \ge 2\omega + 2\ln\rho, \tag{4.43}$$

$$\tau_{k+1} - \tau_k \ge \tilde{d} > d, \quad k \in \mathbb{N}. \tag{4.44}$$

where $\rho = \frac{\max_{i\in M}\lambda_{\max}(P_i)}{\min_{i\in M}\lambda_{\min}(P_i)}$*,* $b_{\max} = \max_{i\in M} b_i$*,* z_i^* *is the solution of* $z = -b_i - e^{dz}$*, and* $z^* = \max_{i\in M} z_i^*$*. Then, the system (4.6) is input-to-state practically stable with respect to exogenous disturbances* $v(t)$*.*

Proof *Similar to the proof of Theorem 4.1, we consider the same Lyapunov function and obtain (4.32). Then, it follows that*

$$\begin{aligned} V_{\sigma(\tau_k)}(t) \le &\Big[\sup_{\theta\in[\tau_k-d,\tau_k]} V_{\sigma(\tau_k)}(\theta) - \tilde{r}_{\sigma(\tau_k)}\Big] e^{-z^*(t-\tau_k)} \\ &+ \tilde{r}_{\sigma(\tau_k)}, \quad t \in [\tau_k, \tau_{k+1}), \ \forall k \in \mathbb{N}, \end{aligned}$$

where

$$\tilde{r}_{\sigma(\tau_k)} = \frac{c_{\sigma(\tau_k)} + \lambda_{\max}(Q_{\sigma(\tau_k)})\|v\|_{\infty}^2}{z^*_{\sigma(\tau_k)}}.$$

Defining $\tilde{r} = \max_{i\in M}\{\tilde{r}_i\}$, *it can be derived that*

$$V_{\sigma(\tau_k)}(t) \leq \Big[\sup_{\theta\in[\tau_k-d,\tau_k]} V_{\sigma(\tau_k)}(\theta) - \tilde{r}\Big] e^{-z^*(t-\tau_k)} + \tilde{r}, \quad t \in [\tau_k, \tau_{k+1}),\ \forall k \in \mathbb{N}. \tag{4.45}$$

Now, we claim that for $\forall k \in \mathbb{N}$ *and* $t \in [\tau_k, \tau_{k+1})$, *the following is true:*

$$V_{\sigma(\tau_k)}(t) \leq \rho \min_{i\in M} \lambda_{\min}(P_i)\|\xi\|_d^2 e^{-\omega t} + \tilde{a}\tilde{r} \tag{4.46}$$

where

$$\tilde{a} = \frac{\rho - \rho e^{-z^*(\tilde{d}-d)}}{1-\rho^2 e^{-z^*(\tilde{d}-d)}} > \rho.$$

Let us prove (4.46) by mathematical induction. It is easy to know that

$$\sup_{\theta\in[-d,0]} V_{\sigma(0)}(\theta) \leq \max_{i\in M} \lambda_{\max}(P_i)\|\xi\|_d^2 = \rho \min_{i\in M} \lambda_{\min}(P_i)\|\xi\|_d^2.$$

From (4.43), we have $(z^* - \omega)\tau_1 \geq (z^* - \omega)\tilde{d} \geq 2\ln\rho$. *For* $t \in [0, \tau_1)$, *we have*

$$\begin{aligned}
V_{\sigma(0)}(t) \leq& \Big[\rho \min_{i\in M} \lambda_{\min}(P_i)\|\xi\|_d^2 - \tilde{r}\Big] e^{-z^* t} + r \\
<& \rho \min_{i\in M} \lambda_{\min}(P_i)\|\xi\|_d^2 e^{-z^* t} + \tilde{r} \\
\leq& \rho \min_{i\in M} \lambda_{\min}(P_i)\|\xi\|_d^2 e^{-\omega t} + \tilde{r}, \\
V_{\sigma(\tau_1)}(\tau_1) \leq& \rho V_{\sigma(0)}(\tau_1) \\
\leq& \rho\Big[\rho \min_{i\in M} \lambda_{\min}(P_i)\|\xi\|_d^2 - \tilde{r}\Big] e^{-z^*\tau_1} + \rho\tilde{r} \\
\leq& \rho^2 \min_{i\in M} \lambda_{\min}(P_i)\|\xi\|_d^2 e^{-\omega\tau_1 - 2\ln\rho} + \rho\tilde{r} \\
\leq& \min_{i\in M} \lambda_{\min}(P_i)\|\xi\|_d^2 e^{-\omega\tau_1} + \rho\tilde{r},
\end{aligned}$$

which means that (4.46) is satisfied when $t \in [0, \tau_1]$.

If (4.46) is satisfied for $t \in [0, \tau_k]$ $(k \in \mathbb{N}_+)$, *then we will prove that (4.46) is correct for* $t \in [0, \tau_{k+1}]$. *Note that*

$$\begin{aligned}
\sup_{\theta\in[\tau_{k-1}-d,\tau_{k-1}]} V_{\sigma(\tau_{k-1})}(\theta) \leq& \rho \sup_{\theta\in[\tau_{k-1}-d,\tau_{k-1}]} V_{\sigma(\theta)}(\theta) \\
&\leq \rho^2 \min_{i\in M} \lambda_{\min}(P_i)\|\xi\|_d^2 e^{-\omega(\tau_{k-1}-d)} + \rho\tilde{a}\tilde{r}.
\end{aligned} \tag{4.47}$$

For $t \in [\tau_k - d, \tau_k) \subset [\tau_{k-1}, \tau_k)$, from (4.43) and the definition of a, we have

$$\begin{aligned}
&V_{\sigma(\tau_{k-1})}(t)\\
\leq&\Big[\sup_{\theta\in[\tau_{k-1}-d,\tau_{k-1}]} V_{\sigma(\tau_{k-1})}(\theta) - \tilde{r}\Big]e^{-z^*(t-\tau_{k-1})} + \tilde{r}\\
\leq&\rho^2 \min_{i\in M}\lambda_{\min}(P_i)\|\xi\|_d^2 e^{-\omega(\tau_{k-1}-d)-z^*(t-\tau_{k-1})}\\
&+(\rho\tilde{a}-1)\tilde{r}e^{-z^*(\tau_k-\tau_{k-1}-d)} + \tilde{r}\\
\leq&\rho^2 \min_{i\in M}\lambda_{\min}(P_i)\|\xi\|_d^2 e^{(z^*-\omega)(\tau_{k-1}-t)+\omega d-\omega t}\\
&+(\rho\tilde{a}-1)\tilde{r}e^{-z^*(\tilde{d}-d)} + \tilde{r}\\
\leq&\rho^2 \min_{i\in M}\lambda_{\min}(P_i)\|\xi\|_d^2 e^{(z^*-\omega)(\tau_{k-1}-\tau_k+d)+\omega d-\omega t} + \frac{\tilde{a}}{\rho}\tilde{r}\\
\leq&\rho^2 \min_{i\in M}\lambda_{\min}(P_i)\|\xi\|_d^2 e^{-(z^*-\omega)(\tilde{d}-d)+\omega d-\omega t} + \frac{\tilde{a}}{\rho}\tilde{r}\\
\leq&\min_{i\in M}\lambda_{\min}(P_i)\|\xi\|_d^2 e^{-\omega t-\omega d-2\omega} + \frac{\tilde{a}}{\rho}\tilde{r},
\end{aligned} \tag{4.48}$$

which yields

$$\begin{aligned}
\sup_{\theta\in[\tau_k-d,\tau_k]} V_{\sigma(\tau_k)}(\theta) \leq&\rho \sup_{\theta\in[\tau_k-d,\tau_k]} V_{\sigma(\theta)}(\theta)\\
\leq&\rho \min_{i\in M}\lambda_{\min}(P_i)\|\xi\|_d^2 e^{-\omega\tau_k} + \tilde{a}\tilde{r}
\end{aligned}$$

and, for $t \in [\tau_k, \tau_{k+1})$, it follows that

$$\begin{aligned}
V_{\sigma(\tau_k)}(t) \leq&\Big[\sup_{\theta\in[\tau_k-d,\tau_k]} V_{\sigma(\tau_k)}(\theta) - \tilde{r}\Big]e^{-z^*(t-\tau_k)} + \tilde{r}\\
\leq&\rho \min_{i\in M}\lambda_{\min}(P_i)\|\xi\|_d^2 e^{-\omega\tau_k - z^*(t-\tau_k)}\\
&+(a-1)\tilde{r}e^{-z^*(t-\tau_k)} + \tilde{r}\\
\leq&\rho \min_{i\in M}\lambda_{\min}(P_i)\|\xi\|_d^2 e^{-\omega t} + \tilde{a}\tilde{r}.
\end{aligned}$$

Moreover, from (4.43), we obtain

$$\begin{aligned}
&V_{\sigma(\tau_{k+1})}(\tau_{k+1})\\
\leq&\rho V_{\sigma(\tau_k)}(\tau_{k+1})\\
\leq&\rho\big[\rho \min_{i\in M}\lambda_{\min}(P_i)\|\xi\|_d^2 e^{-\omega\tau_k} + \tilde{a}\tilde{r} - \tilde{r}\big]e^{-z^*(\tau_{k+1}-\tau_k)} + \rho\tilde{r}\\
\leq&\rho \min_{i\in M}\lambda_{\min}(P_i)\|\xi\|_d^2 e^{-\omega\tau_{k+1}} + \tilde{a}\tilde{r},
\end{aligned}$$

which means that (4.46) is correct for $t \in [0, \tau_{k+1}]$.

To this end, we conclude that

$$
\begin{aligned}
&\|x(t)\| \\
\leq &\sqrt{\rho\|\xi\|_d^2 e^{-\omega t} + \frac{\tilde{a}\tilde{r}}{\min_{i\in M}\lambda_{\min}(P_i)}} \\
\leq &\sqrt{\rho}\|\xi\|_d e^{-\frac{\omega}{2}t} + \sqrt{\frac{\tilde{a}\max_{i\in M}\frac{\lambda_{\max}(Q_i)}{z_i^*}}{\min_{i\in M}\lambda_{\min}(P_i)}}\|v\|_\infty \\
&+ \sqrt{\frac{\tilde{a}\max_{i\in M}\frac{\zeta_0\zeta_2\lambda_{\max}(S_i)}{(\zeta_0-\zeta_1)z_i^*}}{\min_{i\in M}\lambda_{\min}(P_i)}},
\end{aligned} \tag{4.49}
$$

which implies that the switched DDS (4.6) is input-to-state practically stable with respect to the exogenous disturbance $v(t)$. The proof is complete.

Corollary 4.1 *Let all the conditions in Theorem 4.2 be satisfied. When $\xi_2 = 0$ in (4.7), the switched DDS (4.6) is input-to-state stable with respect to the exogenous disturbance $v(t)$. When $v(t) = 0$ in (4.1) and $\xi_2 = 0$ in (4.7), then the switched DDS (4.6) is globally exponentially stable.*

Corollary 4.2 *If the system has only one subsystem, i.e. $m = 1$, then (4.6) is simplified into a DDS without switching. If we remove the condition (4.43) and replace (4.42) by*

$$
b + 1 = \mu + \frac{\zeta_0\zeta_1\lambda_{\max}(S)}{(\zeta_0 - \zeta_1)(1 - \zeta_0)\lambda_{\min}(P)} + 1 < 0,
$$

then, the system (4.6) without switching is input-to-state practically stable with respect to the exogenous disturbance $v(t)$, which is in accord with Theorem 1 in [78].

Remark 4.4 *In Theorem 4.2, the input-to-state practical stability is investigated for system (4.6) with switching signals, exogenous disturbances and ET control scheme. From (4.49), we note that $\|x(t)\|$ is upper bounded by a decreasing function. In fact, $\|x(t)\|$ is in the following bounded set:*

$$
\begin{aligned}
\mathcal{B} = \Bigg\{ x \in \mathbb{R}^n : \|x(t)\| \leq &\sqrt{\frac{\tilde{a}\max_{i\in M}\frac{\lambda_{\max}(Q_i)}{z_i^*}}{\min_{i\in M}\lambda_{\min}(P_i)}}\|v\|_\infty \\
&+ \sqrt{\frac{\tilde{a}\max_{i\in M}\frac{\zeta_0\zeta_2\lambda_{\max}(S_i)}{(\zeta_0-\zeta_1)z_i^*}}{\min_{i\in M}\lambda_{\min}(P_i)}} + \sqrt{\rho}\|\xi\|_d \Bigg\}.
\end{aligned}
$$

As is well known, for a given initial condition, if the updating times of the controller converge to a finite constant, the ET scheme induces undesired accumulation of event instants, which leads to the Zeno phenomenon. Now let us prove that the Zeno behaviours are actually excluded in our main results. In the following theorem, we will show that there exists a positive lower bound of the event intervals.

Theorem 4.3 *Let the event-generator function (4.7) be triggered at time* $\{t_k\}_{k=0}^{\infty}$. *If the conditions of Theorem 4.1 hold, then there exists a positive constant* T^* *such that* $t_{k+1} - t_k \geq T^*$ *for all* $k \in \mathbb{N}$.

Proof *Since* $\varepsilon(t) = x(t_k) - x(t)$, $t \in [t_k, t_{k+1})$, *we have from (4.6) and (4.20) that*

$$\begin{aligned}
&D^+\|\varepsilon(t)\|^2\\
=&-2\varepsilon^T(t)[(A_{\sigma(t)} + C_{\sigma(t)}K_{\sigma(t)})x(t) + f_{\sigma(t)}(t, x(t))\\
&+ B_{\sigma(t)}x(t-d) + C_{\sigma(t)}K_{\sigma(t)}\varepsilon(t) + D_{\sigma(t)}v(t)]\\
=&2\varepsilon^T(t)A_{\sigma(t)}\varepsilon(t) - 2\varepsilon^T(t)(A_{\sigma(t)} + C_{\sigma(t)}K_{\sigma(t)})x(t_k)\\
&- 2\varepsilon^T(t)f_{\sigma(t)}(t, x(t)) - 2\varepsilon^T(t)B_{\sigma(t)}x(t-d)\\
&- 2\varepsilon^T(t)D_{\sigma(t)}v(t).
\end{aligned}$$

In light of the elementary inequality $2a^T Hb \leq \|H\|(\|a\|^2 + \|b\|^2)$ *(for any* $a, b \in \mathbb{R}^n$ *and* $H \in \mathbb{R}^{n\times n}$*), we obtain that*

$$\begin{aligned}
&D^+\|\varepsilon(t)\|^2\\
\leq&[2\|A_{\sigma(t)}\| + \|B_{\sigma(t)}\| + \|A_{\sigma(t)} + C_{\sigma(t)}K_{\sigma(t)}\|\\
&+ \|D_{\sigma(t)}\| + 2L^2_{\sigma(t)} + 1]\|\varepsilon(t)\|^2 + \|D_{\sigma(t)}\|\|v(t)\|^2\\
&+ (\|A_{\sigma(t)} + C_{\sigma(t)}K_{\sigma(t)}\| + 2L^2_{\sigma(t)})\|x(t_k)\|^2\\
&+ \|B_{\sigma(t)}\|\|x(t-d)\|^2, \quad t \in [t_k, t_{k+1}).
\end{aligned} \tag{4.50}$$

For triggering time t_k, *there exists an* $l \in \mathbb{N}$ *such that* t_k *is in a switching interval* $\tau_l \leq t_k < \tau_{l+1}$.

Now we consider two cases:

<u>Case 1</u>: If there exists $s \in \mathbb{N}$ $(s \geq 2)$ *such that* $\tau_l \leq t_k < \tau_{l+1} < \tau_{l+s} \leq t_{k+1}$, *then it follows from the condition (4.23) that* $t_{k+1} - t_k \geq \tau_{l+s} - \tau_{l+1} \geq d$.

<u>Case 2</u>: If $\tau_l \leq t_k < t_{k+1} \leq \tau_{l+1}$, *then the event interval* $[t_k, t_{k+1})$ *is in one switching interval* $[\tau_l, \tau_{l+1})$. *According to (4.40), we have*

$$\|x(t_k)\|^2 \leq \rho\|\xi\|_d^2 + \frac{ar}{\min_{i\in M}\lambda_{\min}(P_i)} \tag{4.51}$$

$$\|x(t-d)\|^2 \leq \rho\|\xi\|_d^2 + \frac{ar}{\min_{i\in M}\lambda_{\min}(P_i)}. \tag{4.52}$$

Denote $\vartheta_{i1} = 2\|A_i\| + \|B_i\| + \|A_i + C_iK_i\| + \|D_i\| + 2L_i^2 + 1$, $\vartheta_{i2} = \|A_i + C_iK_i\| + 2L_i^2$, $\vartheta_{i3} = \|B_i\|$, $\vartheta_{i4} = \rho\|\xi\|_d^2 + \frac{ar}{\min_{i\in M}\lambda_{\min}(P_i)}$, *and* $\vartheta_{i5} = \|D_i\|$. *Then, for* $t \in [t_k, t_{k+1})$, *it follows from (4.50)–(4.52) that*

$$D^+\|\varepsilon(t)\|^2 \leq \vartheta_{\sigma(t)1}\|\varepsilon(t)\|^2 + \tilde{K}_{\sigma(t)} + \vartheta_{\sigma(t)5}\|v(t)\|^2, \tag{4.53}$$

where $\tilde{K}_i = \rho\|\xi\|_d^2(\vartheta_{i2} + \vartheta_{i3}) + (\vartheta_{i2} + \vartheta_{i3})\vartheta_{i4}$.

Multiplying $e^{-\vartheta_{\sigma(t)1}(t-t_k)}$ *on the both sides of (4.53), we have*

$$\begin{aligned} &D^+(e^{-\vartheta_{\sigma(t)1}(t-t_k)}\|\varepsilon(t)\|^2) \\ \leq&(\tilde{K}_{\sigma(t)} + \vartheta_{\sigma(t)5}\|v(t)\|^2)e^{-\vartheta_{\sigma(t)1}(t-t_k)}. \end{aligned} \tag{4.54}$$

Note that $\varepsilon(t_k) = 0$. *For any* $t \in [t_k, t_{k+1})$, *integrating from* t_k *to* t *on both sides of (4.54) leads to*

$$\begin{aligned} \|\varepsilon(t)\|^2 \leq& \int_{t_k}^{t} (\tilde{K}_{\sigma(\tau_r)} + \vartheta_{\sigma(\tau_r)5}\|v(s)\|^2)e^{\vartheta_{\sigma(\tau_r)1}(t-s)}ds \\ \leq& \int_{t_k}^{t} (\tilde{K}_{\sigma(\tau_r)} + \vartheta_{\sigma(\tau_r)5}\|v\|_\infty^2)e^{\vartheta_{\sigma(\tau_r)1}(t-s)}ds \\ =&\frac{\tilde{K}_{\sigma(\tau_r)} + \vartheta_{\sigma(\tau_r)5}\|v\|_\infty^2}{\vartheta_{\sigma(\tau_r)1}}(e^{\vartheta_{\sigma(\tau_r)1}(t-t_k)} - 1). \end{aligned}$$

In view of the triggering rule (4.2) and the event-generator function (4.7), we know that the system will not be triggered before $\|\varepsilon(t)\|^2 = \zeta_1\|x(t_k)\|^2 + \zeta_2$. *Letting* $\varsigma = \max_{i\in M}\frac{\tilde{K}_i + \vartheta_{i5}\|v\|_\infty^2}{\vartheta_{i1}}$, *the next triggering time* t_{k+1} *satisfies the following inequality*

$$\|\varepsilon(t)\|^2 = \zeta_1\|x(t_k)\|^2 + \zeta_2 \leq \varsigma(e^{\max_{i\in M}\vartheta_{i1}(t_{k+1}-t_k)} - 1),$$

which yields

$$\begin{aligned} t_{k+1} - t_k \geq& \frac{1}{\max_{i\in M}\vartheta_{i1}}\ln\left(1 + \frac{\zeta_1\|x(t_k)\|^2 + \zeta_2}{\varsigma}\right) \\ \geq&\frac{1}{\max_{i\in M}\vartheta_{i1}}\ln\left(1 + \frac{\zeta_2}{\varsigma}\right) > 0. \end{aligned}$$

Therefore

$$t_{k+1} - t_k \geq T^* = \min\{d, \frac{1}{\max_{i\in M}\vartheta_{i1}}\ln\left(1 + \frac{\zeta_2}{\varsigma}\right)\} \tag{4.55}$$

for $k \in \mathbb{N}$, *which means that the Zeno behaviuor is excluded. The proof is now complete.*

Theorem 4.4 *Let the event-generator function (4.7) be triggered at time* $\{t_k\}_{k=0}^{\infty}$. *If the conditions of Theorem 4.2 are satisfied, then there exists a positive constant* $\tilde{T}^*$ *such that*

$$t_{k+1} - t_k \geq \tilde{T}^* = \min\{d, \frac{1}{\max_{i\in M} \vartheta_{i1}} \ln\left(1 + \frac{\zeta_2}{\varsigma'}\right)\}$$

for $k \in \mathbb{N}$, *where* $\varsigma' = \max_{i\in M} \frac{\tilde{K}_i' + \vartheta_{i5}\|v\|_\infty^2}{\vartheta_{i1}}$, $\tilde{K}_i' = \rho\|\xi\|_d^2(\vartheta_{i2} + \vartheta_{i3}e^{\omega d}) + (\vartheta_{i2} + \vartheta_{i3})\vartheta_{i4}$, $i \in M$, ϑ_{i1}, ϑ_{i2}, ϑ_{i3}, ϑ_{i4}, *and* ϑ_{i5} *are the same as defined in Theorem 4.3.*

Proof *The proof is similar to that of Theorem 4.3 and is thus omitted.*

In the following theorem, we deal with the co-design problem for the feedback gain matrices K_i $(i \in M)$, the ET parameters ζ_1, ζ_2 for the controller (4.3) and the event-generator function (4.7), in order to guarantee the input-to-state practical stability of system (4.6).

Theorem 4.5 *Let Assumption 4.1 and (4.22)–(4.24) hold. For given* $\tilde{\mu}_i \in \mathbb{R}$ $(i \in M)$, *let* $\min_{i\in M} \tilde{\mu}_i < -1$. *Then, the switched DDS (4.6) is bounded if there exist positive definite matrices* $\tilde{P}_i \in \mathbb{R}^{n\times n}$, $\tilde{Q}_i \in \mathbb{R}^{n\times n}$, $\tilde{S}_i \in \mathbb{R}^{n\times n}$ *and constant matrices* $Y_i \in \mathbb{R}^{q\times n}$ $(i \in M)$ *such that*

$$\begin{bmatrix} \Pi_{11}^i & \Pi_{12}^i & \Pi_{13}^i & \Pi_{14}^i & \Pi_{15}^i \\ * & -\tilde{P}_i & 0 & 0 & 0 \\ * & * & -\tilde{Q}_i & 0 & 0 \\ * & * & * & -\tilde{S}_i & 0 \\ * & * & * & * & -I \end{bmatrix} < 0, \quad i \in M, \tag{4.56}$$

where $\Pi_{11}^i = A_i\tilde{P}_i + \tilde{P}_iA_i^T + C_iY_i + Y_i^TC_i^T + L_i^2I - \mu_i\tilde{P}_i$. $\Pi_{12}^i = B_i\tilde{P}_i$, $\Pi_{13}^i = D_i$, $\Pi_{14}^i = C_iY_i$, $\Pi_{15}^i = \tilde{P}_i$. *Moreover, the gain matrices* K_i *of the desired feedback controller (4.3) can be designed by*

$$K_i = Y_i\tilde{P}_i^{-1} \tag{4.57}$$

and the triggered parameters ζ_1 *and* ζ_2 *in (4.7) satisfy*

$$\zeta_2 > 0, \quad 0 < \zeta_1 < \min\left\{\frac{\zeta_0\iota_1}{\iota_1 - \iota_2}, \zeta_0\right\}. \tag{4.58}$$

where

$$\iota_1 = (1-\zeta_0)(\min_{i\in M} \tilde{\mu}_i + 1),\ \iota_2 = \zeta_0 \max_{i\in M} \frac{\lambda_{\max}(\tilde{S}_i)}{\lambda_{\min}(\tilde{P}_i^{-1})}.$$

Proof *We need to prove that (4.21) is true and the set* M_2 *is nonempty, that is, the following inequality holds:*

$$\min_{i\in M}\left\{\mu_i + \frac{\zeta_0\zeta_1\lambda_{\max}(S_i)}{(\zeta_0-\zeta_1)(1-\zeta_0)\lambda_{\min}(P_i)}\right\} < -1. \tag{4.59}$$

Pre- and post-multiplying (4.56) by $\mathrm{diag}\{\tilde{P}_i^{-1}, \tilde{P}_i^{-1}, I, \tilde{P}_i, I\}$, *we obtain*

$$\begin{bmatrix} \tilde{P}_i^{-1}\Pi_{11}^i\tilde{P}_i^{-1} & \tilde{P}_i^{-1}B_i^T & \tilde{P}_i^{-1}D_i^T & \tilde{P}_i^{-1}C_iY_i & \tilde{P}_i^{-1} \\ * & -\tilde{P}_i^{-1} & 0 & 0 & 0 \\ * & * & -\tilde{Q}_i & 0 & 0 \\ * & * & * & -\tilde{P}_i\tilde{S}_i\tilde{P}_i & 0 \\ * & * & * & * & -I \end{bmatrix} < 0,$$

which, by Schur complement, yields

$$\begin{aligned} &\tilde{P}_i^{-1}(A_i + C_iK_i) + (A_i + C_iK_i)^T\tilde{P}_i^{-1} + \tilde{P}_i^{-1}B_i\tilde{P}_iB_i^T\tilde{P}_i^{-1} \\ &+ \tilde{P}_i^{-1}C_iK_iS_i^{-1}K_i^TC_i^T\tilde{P}_i^{-1} + \tilde{P}_i^{-1}D_i\tilde{Q}_i^{-1}D_i^T\tilde{P}_i^{-1} \\ &+ \tilde{P}_i^{-1}\tilde{P}_i^{-1} + L_i^2I \leq \mu_i\tilde{P}_i^{-1}, \quad i \in M. \end{aligned}$$

Choose $\tilde{\mu}_i = \mu_i$, $\tilde{P}_i^{-1} = P_i$, $\tilde{Q}_i = Q_i$ *and* $\tilde{S}_i = S_i$. *Then, we obtain (4.21). Moreover, using the condition (4.58), we obtain*

$$\zeta_1 \in (0, \zeta_0), \quad \min_{i\in M}\mu_i + \frac{\zeta_0\zeta_1 \max_{i\in M}\frac{\lambda_{\max}(S_i)}{\lambda_{\min}(P_i)}}{(\zeta_0 - \zeta_1)(1-\zeta_0)} < -1,$$

which implies (4.59) holds. The proof is now complete.

Theorem 4.6 *Under Assumption 4.1 and let (4.42)–(4.43) hold. For given* $\tilde{\mu}_i \in \mathbb{R}$ $(i \in M)$, *let* $\max_{i\in M}\tilde{\mu}_i < -1$. *The switched DDS (4.6) is input-to-state practically stabilizable if there exist positive definite matrices* $\tilde{P}_i \in \mathbb{R}^{n\times n}$, $\tilde{Q}_i \in \mathbb{R}^{n\times n}$, $\tilde{S}_i \in \mathbb{R}^{n\times n}$ *and constant matrices* $Y_i \in \mathbb{R}^{q\times n}$ $(i \in M)$ *such that (4.56) holds. Moreover, the gain matrices* K_i *of the desired feedback controller (4.3) can be designed by (4.57), and the triggered parameters* ζ_1 *and* ζ_2 *in (4.7) satisfy*

$$\zeta_2 > 0, \quad 0 < \zeta_1 < \min\left\{\frac{\zeta_0\bar{\iota}_1}{\bar{\iota}_1 - \iota_2}, \zeta_0\right\}$$

where $\bar{\iota}_1 = (1-\zeta_0)(\max_{i\in M}\tilde{\mu}_i + 1)$.

Proof *The rest of the proof is similar to that of Theorem 4.5, which is thus omitted here.*

Remark 4.5 *In this section, we investigate both the boundedness and the input-to-state stability problems for a kind of switched DDSs with exogenous disturbances. An ET strategy is proposed in the feedback control loop in order to reduce the communication burden. We employ the notions of boundedness and input-to-state practical stability to characterize the control objectives in the simultaneous presence of exogenous disturbances, switching signals and*

ET schemes. An estimate of the upper bounds on the system states is obtained and Zeno phenomenon is proved to be excluded under our purposely designed ET scheme. It is worth mentioning that the gain of the feedback controller, the parameters of the event-triggering function and the switching signals can be co-designed to achieve a satisfactory trade-off between the system performance and the energy consumed for the addressed switched DDSs.

Remark 4.6 *So far, a unified framework has been established for the analysis problems of the boundedness and the input-to-state stabilization for a class of switched DDSs suffering from exogenous disturbances by using a purposely designed ET control strategy. The advantages of our proposed controller design algorithm can be highlighted as follows: 1) the system model under investigation is quite general, which addresses time-delays, exogenous disturbances, switching signals as well as ET control schemes; 2) the ET feedback controller and the switching signal can be jointly designed and this offers much flexibility for improving the system performance; and 3) the Zeno phenomenon is shown to be excluded with an explicit parameterization of the desired lower bound of the event intervals.*

4.2 Event-Triggered Pinning Synchronization Control

In this section, let us consider the pinning control problem for the synchronization of a class of nonlinear discrete-time switched stochastic complex networks with identical nodes under ET mechanisms.

4.2.1 Problem Formulation

Consider a class of switched stochastic complex dynamic network comprising N identical nodes as follows:

$$\begin{aligned} x_i(k+1) =& f_{\sigma(k)}(x_i(k)) + \sum_{j=1}^{N} l_{ij}\Gamma g_{\sigma(k)}(x_j(k)) \\ &+ h_{\sigma(k)}(x_i(k))\omega(k) + u_i(k) + \nu_i(k), \quad i = 1, 2, \cdots, N, \end{aligned} \tag{4.60}$$

where $x_i(k) = [x_{i1}(k), x_{i2}(k), \cdots, x_{in}(k)]^T \in \mathbb{R}^n$ is the state of node i, $u_i(k) = [u_{i1}(k), u_{i2}(k), \cdots, u_{in}(k)]^T \in \mathbb{R}^n$ represents the control input, $\nu_i(k) = [\nu_{i1}(k), \nu_{i2}(k), \cdots, \nu_{in}(k)]^T \in \mathbb{R}^n$ is the exogenous disturbance on the network that belongs to $\mathcal{L}_\infty^n$. The zero-mean random sequence $\omega(k) \in \mathbb{R}$ is defined on the probability space $(\Omega, \mathcal{F}, \text{Prob})$ with $\mathbb{E}(\omega^2(k)) = 1$. σ denotes the switching signal, which is a piecewise function of k and takes values in the finite set $\mathcal{I} = \{1, 2, \cdots, m\}$ with $m \geq 1$ being the number of the subsystems. $\{\tau_l\}_{l=0}^{\infty}$ denotes the switching time sequence of the system satisfying $0 = \tau_0 < \tau_1 < \cdots < \tau_l < \cdots$. For any $p \in \mathcal{I}$, f_p, g_p, $h_p : \mathbb{R}^n \to \mathbb{R}^n$ are nonlinear functions which denote the dynamics of, respectively, the nodes, the inner-couplings and the intensities of the stochastic noises. Γ stands for the inner-coupling matrix linking the subsystems and $L = (l_{ij})_{N \times N}$ is the coupled configuration matrix which governs the topological structure of the complex network. If there is a connection from node j to node i, then $l_{ij} > 0$, otherwise, $l_{ij} = 0$ for $i \neq j$.

For the unforced isolated node, its dynamics is given as follows:

$$s(k+1) = f_{\sigma(k)}(s(k)) + h_{\sigma(k)}(s(k))\omega(k). \tag{4.61}$$

The solution $s(k) = [s_1(k), s_2(k), \cdots, s_n(k)]^T$ is regarded as the target state to be synchronized by the switched stochastic complex dynamic network (4.60).

Due to the large size of a complex network, we adopt a pinning control scheme by which only a small part of the nodes are to be controlled in the network. In this section, we let the set of nodes $\mathcal{N}_0 \triangleq \{1, 2, \cdots, l\} \subseteqq \mathcal{N}$ be chosen to be pinned, and this will not cause the loss of any generality. Define the synchronization error to be

$$r_i(k) = x_i(k) - s(k), \quad i \in \mathcal{N}. \tag{4.62}$$

To lighten the communication burden in the channel from the controller to the actuator, we consider the ET scheme here. Let $\{k_s^i : s \in \mathbb{N}\}$ be the sequence of the triggering times that are determined in a iterative manner according to the following triggering rule:

$$k_{s+1}^i = \inf\left\{k \in \mathbb{N} : k > k_s^i, \quad \|\delta_i(k)\|^2 - \zeta\|r_i(k_s^i)\|^2 - \theta_i > 0\right\}, \tag{4.63}$$

where, for $k \in [k_s^i, k_{s+1}^i)$, $\delta_i(k) = r_i(k_s^i) - r_i(k)$ represents the measurement error. The nonnegative constants ζ and θ_i denote the parameters of the ET weight and the threshold, respectively, which are not both zero at the same time.

For the ith actuator, the corresponding control input is produced through a ZOH in the interval of the holding times as follows:

$$u_i(k) = K_{i\sigma(k)} r_i(k_s^i), \quad k \in [k_s^i, k_{s+1}^i),\ i \in \mathcal{N}_0. \tag{4.64}$$

It is easy to know that $r_i(k_s^i) = \delta_i(k) + r_i(k)$. Then, we obtain the pinning controller under the ET mechanism:

$$u_i(k) = \begin{cases} K_{i\sigma(k)}(\delta_i(k) + r_i(k)), & i \in \mathcal{N}_0, \\ 0, \quad i \in \mathcal{N} \setminus \mathcal{N}_0. \end{cases} \tag{4.65}$$

For any $p \in \mathcal{I}$, let $\tilde{f}_p(r_i(k)) = f_p(x_i(k)) - f_p(s(k))$, $\tilde{g}_p(r_i(k)) = g_p(x_i(k)) -$

$g_p(s(k))$ and $\tilde{h}_p(r_i(k)) = h_p(x_i(k)) - h_p(s(k))$. Then, the synchronization error system is obtained as follows:

$$\begin{cases} r_i(k+1) = \tilde{f}_{\sigma(k)}(r_i(k)) + \sum_{j=1}^{N} l_{ij}\Gamma\tilde{g}_{\sigma(k)}(r_j(k)) + \tilde{h}_{\sigma(k)}(r_i(k))\omega(k) \\ \qquad + K_{i\sigma(k)}(\delta_i(k) + r_i(k)) + \nu_i(k), \quad i \in \mathcal{N}_0, \\ r_i(k+1) = \tilde{f}_{\sigma(k)}(r_i(k)) + \sum_{j=1}^{N} l_{ij}\Gamma\tilde{g}_{\sigma(k)}(r_j(k)) + \tilde{h}_{\sigma(k)}(r_i(k))\omega(k) \\ \qquad + \nu_i(k), \quad i \in \mathcal{N}\backslash\mathcal{N}_0, \end{cases} \tag{4.66}$$

and (4.66) can be further written in the following form:

$$\begin{aligned} r(k+1) = & F_{\sigma(k)}(r(k)) + (L \otimes \Gamma)G_{\sigma(k)}(r(k)) + H_{\sigma(k)}(r(k))\omega(k) \\ & + K_{\sigma(k)}r(k) + K_{\sigma(k)}\delta(k) + \nu(k), \end{aligned} \tag{4.67}$$

where

$$\begin{aligned} & K_p = \mathrm{diag}\{K_{1p}, \cdots, K_{lp}, 0, \cdots, 0\},\ r(k) = [r_1^T(k) \cdots, r_N^T(k)]^T, \\ & \nu(k) = [\nu_1^T(k), \cdots, \nu_N^T(k)]^T,\ \delta(k) = [\delta_1^T(k), \cdots, \delta_l^T(k), 0, \cdots, 0]^T \\ & F_p(r(k)) = [\tilde{f}_p^T(r_1(k)), \tilde{f}_p^T(r_2(k)), \cdots, \tilde{f}_p^T(r_N(k))]^T, \\ & G_p(r(k)) = [\tilde{g}_p^T(r_1(k)), \tilde{g}_p^T(r_2(k)), \cdots, \tilde{g}_p^T(r_N(k))]^T, \\ & H_p(r(k)) = [\tilde{h}_p^T(r_1(k)), \tilde{h}_p^T(r_2(k)), \cdots, \tilde{h}_p^T(r_N(k))]^T, \quad p \in \mathcal{I}. \end{aligned}$$

The following assumptions are made to facilitate the investigation of the stochastic synchronization of the system (4.60).

Assumption 4.2 *For any $x,\ y \in \mathbb{R}^n$, the nonlinear vector-valued functions f_p, g_p ($p \in \mathcal{I}$) satisfy*

$$\begin{aligned} & (f_p(x) - f_p(y) - U_{1p}(x-y))^T(f_p(x) - f_p(y) - U_{2p}(x-y)) \leq 0, \\ & (g_p(x) - g_p(y) - W_{1p}(x-y))^T(g_p(x) - g_p(y) - W_{2p}(x-y)) \leq 0, \end{aligned} \tag{4.68}$$

for $\forall p \in \mathcal{I}$, where U_{1p}, U_{2p}, W_{1p} and $W_{2p} \in \mathbb{R}^{n\times n}$ are known real matrices.

Assumption 4.3 *For any $x,\ y \in \mathbb{R}^n$, the noise intensity function h_p ($p \in \mathcal{I}$) satisfies the following inequalities*

$$(h_p(x) - h_p(y))^T(h_p(x) - h_p(y)) \leq \kappa_p(\|x-y\|^2), \quad \forall p \in \mathcal{I}, \tag{4.69}$$

where $\kappa_p \in \mathcal{K}_\infty$ are concave functions and there exist positive constants $\lambda_p^* > 0$ such that $\lambda_p^* Id - \mathcal{N}\kappa_p \in \mathcal{K}$, $p \in \mathcal{I}$.

Remark 4.7 *In Assumption 4.2, (4.68)) is referred to as the sector-bounded condition that is more general than the commonly used Lipschitz condition. In Assumption 4.3, when κ_p ($p \in \mathcal{I}$) are chosen as $\kappa_p(u) = c_p u$ ($c_p > 0$), (4.69)) reduces to the global Lipschitz condition.*

The following definitions and lemma will be used in this section.

Definition 4.4 *[94] For any $k > s \geq k_0$, let $N_\sigma(s,k)$ be the switching number of σ over $[s,k)$. If $N_\sigma(s,k) \leq N_0 + (k-s)/\tau_a$ holds for any given $\tau_a > 0$ and $N_0 \geq 0$, then τ_a and N_0 are called average dwell time and chatter bound, respectively. We often choose $N_0 = 0$ for simplicity.*

Definition 4.5 *[79] For a given positive constant $\varepsilon \in (0,1)$, the synchronization error system is ultimately bounded in probability $1-\varepsilon$ if there exists a compact set $\mathcal{M} = \{x \in \mathbb{R}^{nN} : \|x\| \leq r\}$ such that, for any initial value $r(0) \in \mathbb{R}^{nN}$, the solution $r(k)$ of system satisfies*

$$\text{Prob}\{\lim_{k\to+\infty} \textit{dist}(r(k), \mathcal{M}) = 0\} > 1 - \varepsilon,$$

where $\textit{dist}(x,\mathcal{M}) \triangleq \inf_{y\in\mathcal{M}} \|x-y\|$ denotes the distance from a point $x \in \mathbb{R}^{nN}$ to the compact set $\mathcal{M}$. In this case, the positive constant r is called the error bound in the probability.

Definition 4.6 *[79] The switched stochastic complex network (4.60) is said to be quasi-synchronized to the isolate node (4.61) in probability $1-\varepsilon$ if the synchronization error dynamical system (4.67) is ultimately bounded in probability $1-\varepsilon$. In particular, the switched stochastic complex network (4.60) is said to be completely synchronized to the dynamical system (4.61) in probability $1-\varepsilon$ if the error bound r satisfies $r = 0$.*

Lemma 4.4 *[66] For any $\alpha \in \mathcal{K}_\infty$, there is a $\hat{\alpha} \in \mathcal{K}_\infty$ satisfying 1) $\hat{\alpha}(s) \leq \alpha(s)$ for any $s \in \mathbb{R}_+$; and 2) $Id - \hat{\alpha} \in \mathcal{K}$.*

The main objective of this section is to design the ET pinning controller (4.64) such that the switched stochastic complex network (4.60) is quasi- (and completely) synchronized to the isolate node (4.61) in probability $1-\varepsilon$ with a given positive constant $\varepsilon \in (0,1)$.

4.2.2 Main Results

Theorem 4.7 *For system (4.67), let the positive constant $\varepsilon \in (0,1)$ and the controller (4.65) be given. Assume that there are functions $V_p : \mathbb{R}^{nN} \to \mathbb{R}^+$, a, b, α_p, $\beta \in \mathcal{K}_\infty$, $\eta_p \in \mathcal{K}$ and positive constants $\rho_p > 0$, $\lambda \geq 1$, $\beta_0, \gamma_p \in (0,1)$, $\forall p \in \mathcal{I}$ such that*

(A1) $a(\|r\|) \leq V_p(r) \leq b(\|r\|), \quad \forall p \in \mathcal{I}$.
(A2) $\alpha_p - (1-\gamma_p)Id \in \mathcal{K}_\infty$, $\forall p \in \mathcal{I}$, $\beta(s) \leq \min_{p\in\mathcal{I}}[\alpha_p(s) - (1-\gamma_p)s]$, *for any* $s \in \mathbb{R}_+$, *and* $Id - \frac{\beta_0}{\max_{p\in\mathcal{I}}\gamma_p}\beta \in \mathcal{K}$, *where* $\gamma = \max_{p\in\mathcal{I}}\gamma_p$.
(A3) $\mathbb{E}\{V_p(r(k+1)) - V_p(r(k))\} \leq -\alpha_p(\mathbb{E}(V_p(r(k)))) + \gamma_p(\eta_p(\|\nu(k)\|) + \rho_p))$, $\forall p \in \mathcal{I}$.
(A4) $V_p(r) \leq \lambda V_q(r), \quad \forall p, q \in \mathcal{I}$.
(A5) The average dwell time $\tau_a > \tau_a^* = -\ln\lambda / \ln\gamma$.
Then, the synchronization error dynamics (4.67) is ultimately bounded in probability $1-\varepsilon$, *and there exists a function* $\tilde{\eta} \in \mathcal{K}$ *and a scalar* $\tilde{\rho} \geq 0$ *such that the error bound is equal to* $\tilde{\eta}(\|\nu\|_\infty) + \tilde{\rho}$.

Proof *For any initial condition* $r_0 = r(0) \in \mathbb{R}^{nN}$, $r(k)$ *is the solution of the system (4.67). Let* $\eta(u) = \max_{p\in\mathcal{I}}\eta_p(u)$ *and* $\rho = \max_{p\in\mathcal{I}}\rho_p$. *According to the condition (A2), we have*

$$\hat{\alpha} \triangleq \frac{\beta_0}{\gamma}\beta \in \mathcal{K}_\infty, \quad \text{and} \quad Id - \hat{\alpha} \in \mathcal{K}. \tag{4.70}$$

Then, for any $p \in \mathcal{I}$, *it follows from the condition (A3) that*

$$\begin{aligned}\mathbb{E}V_p(r(k+1)) \leq& \gamma_p\mathbb{E}V_p(r(k)) - [\alpha_p - (1-\gamma_p)Id](\mathbb{E}(V_p(r(k))))\\ &+ \gamma_p(\eta_p(\|\nu(k)\|) + \rho_p))\\ \leq& \gamma\mathbb{E}V_p(r(k)) - \beta(\mathbb{E}(V_p(r(k)))) + \gamma(\eta(\|\nu\|_\infty) + \rho)\\ \leq& \gamma(Id - \hat{\alpha})(\mathbb{E}(V_p(r(k)))) + \gamma(\eta(\|\nu\|_\infty) + \rho).\end{aligned} \tag{4.71}$$

For any $\mu > 1$, *denote the set*

$$\mathcal{B} \triangleq \{r \in \mathbb{R}^{nN} : V_{\sigma(0)}(r) \leq \lambda\hat{\alpha}^{-1}(\mu\eta(\|\nu\|_\infty) + \mu\rho)\}. \tag{4.72}$$

Now, let us consider two cases: (1) $r_0 \in \mathcal{B}$ *and (2)* $r_0 \in \mathbb{R}^{nN} \setminus \mathcal{B}$.
Case 1: $r_0 \in \mathcal{B}$. *In this case, we have*

$$\mathbb{E}V_{\sigma(0)}(r(0)) \leq \lambda\hat{\alpha}^{-1}(\mu\eta(\|\nu\|_\infty) + \mu\rho). \tag{4.73}$$

If the times k *and* $k+1$ *are both in the first interval between two consecutive switching moments, that is,* $0 = \tau_0 \leq k < k+1 < \tau_1$, *we let* $\tilde{\tau}_1 = \min\{\tau_1 - 1, \tau_a^*\}$ *and proceed to prove the following inequalities by mathematical induction:*

$$\mathbb{E}V_{\sigma(k)}(r(k)) \leq \lambda\gamma^{\min\{k,\tilde{\tau}_1\}}\hat{\alpha}^{-1}(\mu\eta(\|\nu\|_\infty) + \mu\rho), \quad k \in [0, \tau_1). \tag{4.74}$$

It is obvious that (4.74) holds for $k = 0$. *For* $k \in [0, \tilde{\tau}_1]$, *according to the condition (A5), we have*

$$\lambda\gamma^{\tau_a^*} = \lambda\gamma^{-\frac{\ln\lambda}{\ln\gamma}} = e^{\ln\lambda}(e^{\ln\gamma})^{-\frac{\ln\lambda}{\ln\gamma}} = e^{\ln\lambda}e^{-\ln\lambda} = 1, \tag{4.75}$$

which implies $\lambda\gamma^k \geq 1$. *If (4.74) is satisfied for* $k \in [0, \tilde{\tau}_1]$, *we have from (4.71) that*

$$\begin{aligned}\mathbb{E}V_{\sigma(k+1)}(r(k+1)) \leq& \gamma(Id - \hat{\alpha})(\mathbb{E}(V_{\sigma(k)}(r(k)))) + \gamma(\eta(\|\nu\|_\infty) + \rho)\\ \leq& \gamma(Id - \hat{\alpha})(\lambda\gamma^k\hat{\alpha}^{-1}(\mu\eta(\|\nu\|_\infty) + \mu\rho)) + \gamma(\eta(\|\nu\|_\infty) + \rho)\\ \leq& \lambda\gamma^{k+1}\hat{\alpha}^{-1}(\mu\eta(\|\nu\|_\infty) + \mu\rho) - \gamma\hat{\alpha}(\hat{\alpha}^{-1}(\mu\eta(\|\nu\|_\infty) + \mu\rho))\\ &+ \gamma(\eta(\|\nu\|_\infty) + \rho)\\ \leq& \lambda\gamma^{k+1}\hat{\alpha}^{-1}(\mu\eta(\|\nu\|_\infty) + \mu\rho),\end{aligned}$$

which means that (4.74) is satisfied for $k+1 \in [0, \tilde{\tau}_1+1] \cap [0, \tau_1)$ *and (4.74) has been proven when* $k \in [0, \tilde{\tau}_1]$. *It is easy to obtain that* $\mathbb{E}V_{\sigma(\tilde{\tau}_1+1)}(r(\tilde{\tau}_1 + 1)) \leq \lambda\gamma^{\tilde{\tau}_1+1}\hat{\alpha}^{-1}(\mu\eta(\|\nu\|_\infty) + \mu\rho) \leq \lambda\gamma^{\tilde{\tau}_1}\hat{\alpha}^{-1}(\mu\eta(\|\nu\|_\infty) + \mu\rho)$. *If (4.74) is true for* $k \in [\tilde{\tau}_1, \tau_1 - 1)$, *then we have*

$$\begin{aligned}\mathbb{E}V_{\sigma(k+1)}(r(k+1)) \leq& \gamma(Id - \hat{\alpha})(\mathbb{E}(V_{\sigma(k)}(r(k)))) + \gamma(\eta(\|\nu\|_\infty) + \rho)\\ \leq& \gamma(Id - \hat{\alpha})(\lambda\gamma^{\tilde{\tau}_1}\hat{\alpha}^{-1}(\mu\eta(\|\nu\|_\infty) + \mu\rho)) + \gamma(\eta(\|\nu\|_\infty) + \rho)\\ \leq& \lambda\gamma^{\tilde{\tau}_1+1}\hat{\alpha}^{-1}(\mu\eta(\|\nu\|_\infty) + \mu\rho) - \gamma\hat{\alpha}(\hat{\alpha}^{-1}(\mu\eta(\|\nu\|_\infty) + \mu\rho))\\ &+ \gamma(\eta(\|\nu\|_\infty) + \rho)\\ \leq& \lambda\gamma^{\tilde{\tau}_1}\hat{\alpha}^{-1}(\mu\eta(\|\nu\|_\infty) + \mu\rho),\end{aligned}$$

which means that (4.74) is satisfied for time $k + 1 \in [\tilde{\tau}_1 + 1, \tau_1)$. *Therefore, (4.74) is correct by induction.*

Now we consider $k = \tau_1$. *It follows from the condition (A4), (4.71), and (4.74) that*

$$\begin{aligned}\mathbb{E}V_{\sigma(\tau_1)}(r(\tau_1)) \leq& \lambda\mathbb{E}V_{\sigma(\tau_1-1)}(r(\tau_1))\\ \leq& \lambda\gamma(Id - \hat{\alpha})\big(\mathbb{E}(V_{\sigma(\tau_1-1)}(r(\tau_1 - 1)))\big) + \lambda\gamma(\eta(\|\nu\|_\infty) + \rho)\\ \leq& \lambda\gamma(Id - \hat{\alpha})\big(\lambda\gamma^{\tilde{\tau}_1}\hat{\alpha}^{-1}(\mu\eta(\|\nu\|_\infty) + \mu\rho)\big) + \lambda\gamma(\eta(\|\nu\|_\infty) + \rho)\\ \leq& \lambda\gamma^{\tilde{\tau}_1+1}\lambda\hat{\alpha}^{-1}(\mu\eta(\|\nu\|_\infty) + \mu\rho) - \lambda\gamma(\mu\eta(\|\nu\|_\infty) + \mu\rho)\\ &+ \lambda\gamma(\eta(\|\nu\|_\infty) + \rho)\\ \leq& \lambda\gamma^{\tilde{\tau}_1+1}\lambda\hat{\alpha}^{-1}(\mu\eta(\|\nu\|_\infty) + \mu\rho).\end{aligned}$$

According to the Definition 4.4, we note that $N_\sigma(0, k) \leq k/\tau_a$. *Define* $\tilde{\tau}_k = \min\{\tau_k - 1, \tau_a^* + \tau_{k-1}\}$, $k \geq 1$. *When* $k \in [\tau_l, \tau_{l+1})$ $(l \in \mathbb{N})$, *we have by induction that*

$$\begin{aligned}\mathbb{E}V_{\sigma(\tau_k)}(r(\tau_k)) \leq& \lambda^{N_\sigma(0,k)}\gamma^{\sum_{i=0}^{l-1}(\tilde{\tau}_{i+1}-\tau_i+1)+\min\{k,\tilde{\tau}_{l+1}\}-\tau_l}\lambda\hat{\alpha}^{-1}(\mu\eta(\|\nu\|_\infty) + \mu\rho)\\ \leq& \lambda^{k/\tau_a}\gamma^k\lambda\hat{\alpha}^{-1}(\mu\eta(\|\nu\|_\infty) + \mu\rho).\end{aligned}$$

Therefore, in view of the condition (A5), we obtain

$$\mathbb{E}V_{\sigma(k)}(r(k)) \leq \lambda\hat{\alpha}^{-1}(\mu\eta(\|\nu\|_\infty) + \mu\rho)\}, \quad k \in \mathbb{N}. \tag{4.76}$$

Applying Chebyshev's inequality yields that

$$\mathrm{Prob}\Big\{V_{\sigma(k)}(r(k)) \geq \frac{1}{\varepsilon}\lambda\hat{\alpha}^{-1}(\mu\eta(\|\nu\|_\infty)+\mu\rho)\Big\} \leq \frac{\varepsilon\mathbb{E}V_{\sigma(k)}(r(k))}{\lambda\hat{\alpha}^{-1}(\mu\eta(\|\nu\|_\infty)+\mu\rho)} \leq \varepsilon. \tag{4.77}$$

Using the condition (A1), we obtain

$$\begin{aligned}&\mathrm{Prob}\Big\{a(\|r(k)\|) \geq \frac{\lambda}{\varepsilon}\hat{\alpha}^{-1}(\mu\eta(\|\nu\|_\infty)+\mu\rho)\Big\}\\ \leq&\mathrm{Prob}\Big\{V_{\sigma(k)}(r(k)) \geq \frac{\lambda}{\varepsilon}\hat{\alpha}^{-1}(\mu\eta(\|\nu\|_\infty)+\mu\rho)\Big\},\end{aligned}$$

which implies that

$$\mathrm{Prob}\Big\{\|r(k)\| \geq a^{-1}\diamond\frac{\lambda\hat{\alpha}^{-1}}{\varepsilon}(\mu\eta(\|\nu\|_\infty)+\mu\rho)\Big\} \leq \varepsilon$$

or

$$\mathrm{Prob}\Big\{\|r(k)\| < a^{-1}\diamond\frac{\lambda\hat{\alpha}^{-1}}{\varepsilon}(\mu\eta(\|\nu\|_\infty)+\mu\rho)\Big\} > 1-\varepsilon. \tag{4.78}$$

Case 2: $r_0 \in \mathbb{R}^{nN}\setminus\mathcal{B}$. In this case, we notice that $\mathbb{E}V_{\sigma(0)}(r(0)) > \lambda\hat{\alpha}^{-1}(\mu\eta(\|\nu\|_\infty)+\mu\rho)$. If $\mathbb{E}V_{\sigma(k)}(r(k)) > \lambda\hat{\alpha}^{-1}(\mu\eta(\|\nu\|_\infty)+\mu\rho)$ for all $k\in\mathbb{N}$, then we have

$$\eta(\|\nu\|_\infty)+\rho < \frac{1}{\mu}\hat{\alpha}(\frac{1}{\lambda}\mathbb{E}V_{\sigma(k)}(r(k))). \tag{4.79}$$

- *If the times k and $k+1$ are both in the same interval between two consecutive switching moments, that is, there exists $r\in\mathbb{N}$ such that $\tau_r \leq k < k+1 < \tau_{r+1}$, then it follows from (4.71) and (4.79) that*

$$\begin{aligned}\mathbb{E}V_{\sigma(k+1)}(r(k+1)) \leq&\gamma\mathbb{E}V_{\sigma(k)}(r(k)) - (1-\frac{1}{\mu})\gamma\hat{\alpha}(\frac{1}{\lambda}\mathbb{E}V_{\sigma(k)}(r(k)))\\ \leq&\gamma\mathbb{E}V_{\sigma(k)}(r(k)).\end{aligned} \tag{4.80}$$

- *If the times k and $k+1$ are not in the same interval between two consecutive switching moments, that is, there exists $r\in\mathbb{N}$ such that $\tau_r \leq k < k+1 = \tau_{r+1}$, then we have from (4.71) and (4.79) that*

$$\begin{aligned}\mathbb{E}V_{\sigma(\tau_{r+1})}(r(\tau_{r+1})) \leq&\lambda\gamma\mathbb{E}V_{\sigma(\tau_r)}(r(\tau_{r+1}-1)) - \lambda\gamma(1-\frac{1}{\mu})\\ &\times\hat{\alpha}(\mathbb{E}V_{\sigma(\tau_r)}(r(\tau_{r+1}-1)))\\ \leq&\lambda\gamma\mathbb{E}V_{\sigma(\tau_r)}(r(\tau_{r+1}-1)).\end{aligned} \tag{4.81}$$

From (4.80) and (4.81), we conclude that

$$\mathbb{E}V_{\sigma(k}(r(k)) \leq \lambda^{N_\sigma(0,k)}\gamma^k\mathbb{E}V_{\sigma(0)}(r(0)) \leq (\lambda^{\frac{1}{\tau_a}}\gamma)^k\mathbb{E}V_{\sigma(0)}(r(0)),$$

and thus there exists a function $\phi \in \mathcal{KL}$ *such that*

$$\mathbb{E}V_{\sigma(k)}(r(k)) \leq \phi(V_{\sigma(0)}(r(0)), k), \quad k \in \mathbb{N}. \tag{4.82}$$

Using Chebyshev's inequality again yields

$$\text{Prob}\Big\{V_{\sigma(k)}(r(k)) \geq \frac{1}{\varepsilon}\phi(V_{\sigma(0)}(r(0)), k)\Big\} \leq \frac{\varepsilon\mathbb{E}V_{\sigma(k)}(r(k))}{\phi(V_{\sigma(0)}(r(0)), k)} \leq \varepsilon.$$

Then, together with the condition (A1), we derive that

$$\begin{aligned}&\text{Prob}\Big\{a(\|r(k)\|) \geq \frac{1}{\varepsilon}\phi(b(\|r(0)\|), k)\Big\}\\ \leq&\text{Prob}\Big\{a(\|r(k)\|) \geq \frac{1}{\varepsilon}\phi(V_{\sigma(0)}(r(0)), k)\Big\} \leq \varepsilon,\end{aligned}$$

which indicates

$$\text{Prob}\Big\{\|r(k)\| \geq a^{-1} \diamond \frac{\phi}{\varepsilon}(b(\|r(0)\|), k)\Big\} \leq \varepsilon$$

or

$$\text{Prob}\Big\{\|r(k)\| < a^{-1} \diamond \frac{\phi}{\varepsilon}(b(\|r(0)\|), k)\Big\} > 1-\varepsilon, \quad k \in \mathbb{N}. \tag{4.83}$$

On the other hand, if $\mathbb{E}V_{\sigma(0)}(r(0)) > \lambda\hat{\alpha}^{-1}(\mu\eta(\|\nu\|_\infty) + \mu\rho)$ *but there is some* $k > 0$ *such that* $\mathbb{E}V_{\sigma(k)}(r(k)) \leq \lambda\hat{\alpha}^{-1}(\mu\eta(\|\nu\|_\infty) + \mu\rho)$*, we define* $\bar{k} > 0$ *as follows:*

$$\bar{k} = \min\big\{k : \mathbb{E}V_{\sigma(k)}(r(k)) \leq \lambda\hat{\alpha}^{-1}(\mu\eta(\|\nu\|_\infty) + \mu\rho)\big\}. \tag{4.84}$$

When $k \in [0, \bar{k}-1]$*, it follows that*

$$\mathbb{E}V_{\sigma(k)}(r(k)) > \lambda\hat{\alpha}^{-1}(\mu\eta(\|\nu\|_\infty) + \mu\rho). \tag{4.85}$$

Similarly, for $k \in [0, \bar{k}-1]$*, we know that (4.82) is true and then obtain*

$$\text{Prob}\Big\{\|r(k)\| < a^{-1} \diamond \frac{\phi}{\varepsilon}(b(\|r(0)\|), k)\Big\} > 1-\varepsilon, \quad k \in [0, \bar{k}-1]. \tag{4.86}$$

Along the similar line of the proof in Case 1, we have

$$\mathbb{E}V_{\sigma(k)}(r(k)) \leq \lambda\hat{\alpha}^{-1}(\mu\eta(\|\nu\|_\infty) + \mu\rho)\}, \quad k \geq \bar{k} \tag{4.87}$$

and eventually

$$\text{Prob}\Big\{\sup_{k\geq\bar{k}}\|r(k)\| < a^{-1} \diamond \frac{\lambda\hat{\alpha}^{-1}}{\varepsilon}(\mu\eta(\|\nu\|_\infty) + \mu\rho)\Big\} > 1-\varepsilon. \tag{4.88}$$

Combining (4.86) and (4.88) gives rise to

$$\text{Prob}\Big\{\|r(k)\| < a^{-1} \diamond \frac{\phi}{\varepsilon}(b(\|r(0)\|), k) + a^{-1} \diamond \frac{\lambda\hat{\alpha}^{-1}}{\varepsilon}(\mu\eta(\|\nu\|_\infty) + \mu\rho)\Big\} > 1-\varepsilon.$$

Since $\phi \in \mathcal{KL}$, we have $\lim_{k\to+\infty}[a^{-1} \diamond \frac{\phi}{\varepsilon}(b(\|r(0)\|), k)] = 0$, which implies that

$$\text{Prob}\Big\{ \limsup_{k\to+\infty} \|r(k)\| < a^{-1} \diamond \frac{\lambda\hat{\alpha}^{-1}}{\varepsilon}(\mu\eta(\|\nu\|_\infty) + \mu\rho)\Big\} > 1 - \varepsilon.$$

Letting $\mu \to 1^+$, we have

$$\text{Prob}\Big\{ \limsup_{k\to+\infty} \|r(k)\| < a^{-1} \diamond \frac{\lambda\hat{\alpha}^{-1}}{\varepsilon}(\eta(\|\nu\|_\infty) + \rho)\Big\} > 1 - \varepsilon. \tag{4.89}$$

Using the properties of $\mathcal{K}$-function, it follows that

$$\text{Prob}\Big\{ \limsup_{k\to+\infty} \|r(k)\| < a^{-1} \diamond \frac{\lambda\hat{\alpha}^{-1}}{\varepsilon}(2\eta(\|\nu\|_\infty)) + a^{-1} \diamond \frac{\lambda\hat{\alpha}^{-1}}{\varepsilon}(2\rho)\Big\} > 1 - \varepsilon.$$

Define $\tilde{\eta} \triangleq a^{-1} \diamond \frac{\lambda\hat{\alpha}^{-1}}{\varepsilon} \diamond (2\eta)$, $\tilde{\rho} \triangleq a^{-1} \diamond \frac{\lambda\hat{\alpha}^{-1}}{\varepsilon}(2\rho)$ and a set $\mathcal{M} \triangleq \{x \in \mathbb{R}^{nN} : \|x\| \leq \tilde{\eta}(\|\nu\|_\infty) + \tilde{\rho}\}$. To this end, we conclude that

$$\text{Prob}\{ \lim_{k\to+\infty} dist(r(k), \mathcal{M}) = 0\} > 1 - \varepsilon, \tag{4.90}$$

which completes the proof.

Remark 4.8 *For switched systems, there are essentially two approaches to analyze stability by using Lyapunov direct method. One involves finding a common Lyapunov function for all the subsystems, and the other utilizes multiple Lyapunov functions. Since a common Lyapunov function may not exist for a switched system, we use multiple Lyapunov function approach in this section. In Theorem 4.7, the condition (A4) is a common restriction on the Lyapunov functions when dealing with average dwell-time switching signals [14, 15]. For example, quadratic Lyapunov functions satisfy this hypothesis. In view of interchangeability of p and q in (A4), it requires that $\lambda \geq 1$. The trivial case $\lambda = 1$ implies the existence of a common Lyapunov function.*

Now, we are ready to investigate the problem of stochastic synchronization between the switched stochastic complex network (4.60) and the isolate node (4.61).

Theorem 4.8 *Under Assumptions 4.2 and 4.3, let the constants $\varepsilon \in (0, 1)$, $0 \leq \zeta < \zeta_0 < 1$, $\theta_i \geq 0$ ($i \in \mathcal{N}_0$) and feedback gain matrices K_{ip} ($i \in \mathcal{N}_0$, $p \in \mathcal{I}$) be given, and let τ_a denote the average dwell time. The synchronization error system (4.67) is ultimately bounded in probability $1 - \varepsilon$ if there exist positive constants $\lambda_{1p} > 0$, $\lambda_{2p} > 0$, $\lambda_{3p} > 0$, $\varpi_{1p} > 0$, $\varpi_{2p} > 0$, $\gamma_p \in (0, 1)$, $d_p > 1$ and positive definite matrices $Q_{ip} \in \mathbb{R}^{n\times n}$ ($i \in \mathcal{N}$, $p \in \mathcal{I}$) such that*

$$\frac{\zeta_0\zeta}{(\zeta_0 - \zeta)(1 - \zeta_0)}\lambda_{2p} < \lambda_{1p}, \quad p \in \mathcal{I}. \tag{4.91}$$

$$\gamma_p = 1 - \frac{1}{d_p \lambda_{\max}(Q_p)} \big[\lambda_{1p} - \frac{\zeta_0 \zeta \lambda_{2p}}{(\zeta_0 - \zeta)(1 - \zeta_0)}\big], \tag{4.92}$$

$$\begin{bmatrix} \Pi_p & S_p^T \\ S_p & -Q_p^{-1} \end{bmatrix} < 0, \quad p \in \mathcal{I}, \tag{4.93}$$

$$\tau_a > \tau_a^* = -\ln \lambda / \ln \gamma, \quad \lambda = \frac{\max_{p \in \mathcal{I}} \lambda_{\max}(Q_p)}{\min_{p \in \mathcal{I}} \lambda_{\min}(Q_p)}, \ \gamma = \max_{p \in \mathcal{I}} \gamma_p, \tag{4.94}$$

where $Q_p = \text{diag}\{Q_{1p}, Q_{2p}, \cdots, Q_{Np}\}$, $S_p = [K_p, I_{nN}, L \otimes \Gamma, K_p, I_{nN}]$, $\bar{U}_p = U_{1p}^T U_{2p} + U_{2p}^T U_{1p}$, $\bar{W}_p = W_{1p}^T W_{2p} + W_{2p}^T W_{1p}$, $\tilde{U}_p = U_{1p} + U_{2p}$, $\tilde{W}_p = W_{1p} + W_{2p}$, $\Pi_{11}^p = (\lambda_{\max}(Q_p)\lambda_p^* + \lambda_{1p}) I_{nN} - \varpi_{1p} I_N \otimes \bar{U}_p - \varpi_2 I_N \otimes \bar{W}_p - Q_p$, $\Pi_{12}^p = \varpi_{1p} I_N \otimes \tilde{U}_p^T$, $\Pi_{13}^p = \varpi_{2p} I_N \otimes \tilde{W}_p^T$, *and*

$$\Pi_p = \begin{bmatrix} \Pi_{11}^p & \Pi_{12}^p & \Pi_{13}^p & 0 & 0 \\ * & -2\varpi_{1p} I_{nN} & 0 & 0 & 0 \\ * & * & -2\varpi_{2p} I_{nN} & 0 & 0 \\ * & * & * & -\lambda_{2p} I_{nN} & 0 \\ * & * & * & * & -\lambda_{3p} I_{nN} \end{bmatrix}, \quad p \in \mathcal{I}.$$

Proof *According to Assumption 4.2, for* $i \in \mathcal{N}$, $p \in \mathcal{I}$, *we have*

$$\begin{aligned} &2\varpi_1 [\tilde{f}_p^T(r_i(k)) \tilde{f}_p(r_i(k)) - \tilde{f}_p^T(r_i(k)) U_{2p} r_i(k) \\ &\quad - r_i^T(k) U_{1p}^T \tilde{f}_p(r_i(k)) + r_i^T(k) U_{1p}^T U_{2p} r_i(k)] \le 0, \\ &2\varpi_2 [\tilde{g}_p^T(r_i(k)) \tilde{g}_p(r_i(k)) - \tilde{g}_p^T(r_i(k)) W_{2p} r_i(k) \\ &\quad - r_i^T(k) W_{1p}^T \tilde{g}_p(r_i(k)) + r_i^T(k) W_{1p}^T W_{2p} r_i(k)] \le 0, \end{aligned}$$

which yield

$$\begin{aligned} &2\varpi_1 \tilde{f}_p^T(r_i(k)) \tilde{f}_p(r_i(k)) + \varpi_1 r_i^T(k) \bar{U}_p r_i(k) - \varpi_1 [\tilde{f}_p^T(r_i(k)) \tilde{U}_p r_i(k) \\ &\quad + r_i^T(k) \tilde{U}_p^T \tilde{f}_p(r_i(k))] \le 0, \quad \forall p \in \mathcal{I}, \qquad (4.95) \\ &2\varpi_2 \tilde{g}_p^T(r_i(k)) \tilde{g}_p(r_i(k)) + \varpi_2 r_i^T(k) \bar{W}_p r_i(k) - \varpi_2 [\tilde{g}_p^T(r_i(k)) \tilde{W}_p r_i(k) \\ &\quad + r_i^T(k) \tilde{W}_p^T \tilde{g}_p(r_i(k))] \le 0 \quad \forall p \in \mathcal{I}. \qquad (4.96) \end{aligned}$$

From (4.95) and (4.96), it follows that

$$\begin{aligned} \mathcal{F}_p(k) &\triangleq 2\varpi_1 F_p^T(r(k)) F_p(r(k)) + \varpi_1 r^T(k)(I_N \otimes \bar{U}_p) r(k) - \varpi_1 \\ &\times [F_p^T(r(k))(I_N \otimes \tilde{U}_p) r(k) + r^T(k)(I_N \otimes \tilde{U}_p^T) F_p(r(k))] \le 0, \quad \forall p \in \mathcal{I}, \end{aligned} \tag{4.97}$$

$$\begin{aligned} \mathcal{G}_p(k) &\triangleq 2\varpi_2 G_p^T(r(k)) G_p(r(k)) + \varpi_2^T r(k)(I_N \otimes \bar{W}_p) r(k) - \varpi_2 \\ &\times [G_p^T(r(k))(I_N \otimes \tilde{W}_p) r(k) + r^T(k)(I_N \otimes \tilde{W}_p^T) G_p(r(k))] \le 0, \quad \forall p \in \mathcal{I}. \end{aligned} \tag{4.98}$$

To study the ultimately bounded synchronization in probability $1-\varepsilon$ for error system (4.67), we choose the following Lyapunov function

$$V_{\sigma(k)}(r) = r^T(k)Q_{\sigma(k)}r(k) = \sum_{i=1}^{\mathcal{N}} r_i^T(k)Q_{i\sigma(k)}r_i(k). \tag{4.99}$$

The mathematical expectation of the difference of $V_p(r(k))$ ($p \in \mathcal{I}$) is calculated as follows:

$$\begin{aligned}\mathbb{E}\Delta V_p(r(k)) =& \mathbb{E}V_p(r(k+1)) - \mathbb{E}V_p(r(k))\\ =& \mathbb{E}\{[K_p r(k) + F_p(r(k)) + (L\otimes\Gamma)G_p(r(k)) + K_p\delta(k) + \nu(k)]^T Q_p\\ & \times [K_p r(k) + F_p(r(k)) + (L\otimes\Gamma)G_p(r(k)) + K_p\delta(k) + \nu(k)]\}\\ & + \mathbb{E}\{H_p^T(r(k))Q_p H_p(r(k))\} - \mathbb{E}\{r^T(k)Q_p r(k)\}.\end{aligned} \tag{4.100}$$

In view of Assumption 4.3, we derive that

$$\begin{aligned}\mathbb{E}\{H_p^T(r(k))Q_pH_p(r(k)\} =& \mathbb{E}\{\sum_{i=1}^{\mathcal{N}} \tilde{h}_p^T(r_i(k))Q_{ip}\tilde{h}_p(r_i(k))\}\\ \le& \mathbb{E}\{\lambda_{\max}(Q_p)\sum_{i=1}^{\mathcal{N}} \tilde{h}_p^T(r_i(k))\tilde{h}_p(r_i(k))\}\\ \le& \mathbb{E}\{\lambda_{\max}(Q_p)\sum_{i=1}^{\mathcal{N}} \kappa_p(\|r_i(k)\|^2)\}\\ \le& \mathbb{E}\{\mathcal{N}\lambda_{\max}(Q_p)\kappa_p(\|r(k)\|^2)\}\\ \le& \mathbb{E}\{\lambda_{\max}(Q_p)[\lambda_p^*\|r(k)\|^2 - (\lambda_p^* Id - \mathcal{N}\kappa_p)(\|r(k)\|^2)]\}\\ \le& -\mathbb{E}\{\lambda_{\max}(Q_p)(\lambda_p^* Id - \mathcal{N}\kappa_p)(\frac{1}{\lambda_{\max}(Q_p)}V_p(r(k)))\}\\ & + \mathbb{E}\{\lambda_{\max}(Q_p)\lambda_p^*\|r(k)\|^2\}.\end{aligned}$$

Since $\kappa_p \in \mathcal{K}$ ($p\in\mathcal{I}$) are concave functions, we have $a\kappa_p(u) \le \kappa_p(au)$ for $a\in\mathbb{R}_+$ and $\mathbb{E}\{\kappa_p(u)\} \le \kappa_p(\mathbb{E}\{u\})$. Subsequently, we obtain

$$\begin{aligned}&\mathbb{E}\{H_p^T(r(k))Q_pH_p(r(k)\}\\ \le& \mathbb{E}\{\lambda_{\max}(Q_p)\lambda_p^*\|r(k)\|^2\} - \mathbb{E}\{(\lambda_p^* Id - \mathcal{N}\kappa_p)(V_p(r(k)))\}\\ \le& \mathbb{E}\{\lambda_{\max}(Q_p)\lambda_p^*\|r(k)\|^2\} - (\lambda_p^* Id - \mathcal{N}\kappa_p)(\mathbb{E}\{V_p(r(k))\}).\end{aligned}$$

Letting $z_p^T(k) \triangleq [e^T(k), F_p^T(r(k)), G_p^T(r(k)), \delta^T(k), \nu^T(k)]^T$ ($p\in\mathcal{I}$), we have

$$\mathbb{E}\{\Delta V_p(r(k)) + \lambda_{1p}\|r(k)\|^2 - \lambda_{2p}\|\delta(k)\|^2 - \lambda_{3p}\|\nu(k)\|^2 - \mathcal{F}_p(k) - \mathcal{G}_p(k)\}$$

$$\leq \mathbb{E}\{z_p^T(k)S_p^TQ_pS_pz_p(k) + z_p^T(k)\Pi_pz_p(k)\} - (\lambda_p^* Id - \mathcal{N}\kappa_p)(\mathbb{E}\{V_p(r(k))\}). \tag{4.101}$$

According to the condition (4.93) and the Schur Complement Lemma, we know that

$$\mathbb{E}\{z_p^T(k)S_p^TQ_pS_pz_p(k) + z_p^T(k)\Pi_pz_p(k)\} \leq 0$$

and it then follows from (4.97), (4.98), and (4.101) that

$$\begin{aligned}\mathbb{E}\{\Delta V_p(r(k))\} \leq & -\lambda_{1p}\mathbb{E}\{\|r(k)\|^2\} - (\lambda_p^* Id - \mathcal{N}\kappa_p)(\mathbb{E}\{V_p(r(k))\}) \\ & + \lambda_{2p}\mathbb{E}\{\|\delta(k)\|^2\} + \lambda_{3p}\|\nu(k)\|^2. \end{aligned} \tag{4.102}$$

In view of the event trigger rule (4.63), the measurement error satisfies that $\|\delta_i(k)\|^2 \leq \zeta\|r_i(k_s^i)\|^2 + \theta_i$. *Then, we have*

$$\|\delta_i(k)\|^2 \leq \zeta\|r_i(k) + \delta_i(k)\|^2 + \theta_i \leq \frac{\zeta}{1-\zeta_0}\|r_i(k)\|^2 + \frac{\zeta}{\zeta_0}\|\delta_i(k)\|^2 + \theta_i,$$

which implies that

$$\mathbb{E}\{\|\delta_i(k)\|^2\} \leq \frac{\zeta_0\zeta}{(\zeta_0-\zeta)(1-\zeta_0)}\mathbb{E}\{\|r_i(k)\|^2\} + \frac{\zeta_0\theta_i}{\zeta_0-\zeta}, \quad i \in \mathcal{N}_p$$

and

$$\begin{aligned}\mathbb{E}\{\|\delta(k)\|^2\} = & \sum_{i=1}^{l}\mathbb{E}\{\|\delta_i(k)\|^2\} \\ \leq & \frac{\zeta_0\zeta}{(\zeta_0-\zeta)(1-\zeta_0)}\mathbb{E}\{\|r(k)\|^2\} + \frac{\zeta_0}{\zeta_0-\zeta}\sum_{i=1}^{l}\theta_i. \end{aligned} \tag{4.103}$$

Substituting (4.103) into (4.102) leads to

$$\begin{aligned}\mathbb{E}\{\Delta V_p(r(k))\} \leq & -\Big[\frac{1}{\lambda_{\max}(Q_p)}(\lambda_{1p} - \frac{\zeta_0\zeta\lambda_{2p}}{(\zeta_0-\zeta)(1-\zeta_0)})Id + \lambda_p^* Id - \mathcal{N}\kappa_p\Big] \\ & \times (\mathbb{E}\{V_p(r(k))\}) + \lambda_{3p}\|\nu(k)\|^2 + \frac{\zeta_0\lambda_{2p}}{\zeta_0-\zeta}\sum_{i=1}^{l}\theta_i. \end{aligned} \tag{4.104}$$

Let $\alpha_p(u) \triangleq \frac{1}{\lambda_{\max}(Q_p)}(\lambda_{1p} - \frac{\zeta_0\zeta\lambda_{2p}}{(\zeta_0-\zeta)(1-\zeta_0)})u + \lambda_p^* u - \mathcal{N}\kappa_p(u)$. *From Assumption 4.3 and (4.91), we know that* $\alpha_p \in \mathcal{K}_\infty$, *and then it follows that*

$$\alpha_p - (1-\gamma_p)Id = (1-\frac{1}{d})\frac{1}{\lambda_{\max}(Q_p)}\big[\lambda_{1p} - \frac{\zeta_0\zeta\lambda_{2p}}{(\zeta_0-\zeta)(1-\zeta_0)}\big]Id + \lambda_p^* Id - \mathcal{N}\kappa_p,$$

which implies $\alpha_p - (1-\gamma_p)Id \in \mathcal{K}_\infty$.

Let $\hat{\alpha}(u) \triangleq \min_{p\in\mathcal{I}}\frac{\beta_0}{\gamma}[\alpha_p(u) - (1-\gamma_p)u]$. *We know that* $\hat{\alpha} \in \mathcal{K}_\infty$

and $Id - \hat{\alpha} \in \mathcal{K}$*. Define* $\eta_p(u) \triangleq \frac{\lambda_{3p}}{\gamma_p}u^2$*,* $\rho_p \triangleq \frac{1}{\gamma_p}\frac{\zeta_0\lambda_{2p}}{\zeta_0-\zeta}\sum_{i=1}^{l}\theta_i$*,* $a(u) \triangleq \min_{p\in\mathcal{I}}\lambda_{\min}(Q_p)u^2$ *and* $b(u) \triangleq \max_{p\in\mathcal{I}}\lambda_{\max}(Q_p)u^2$*. Then, it follows from Theorem 4.7 that the synchronization error dynamics (4.67) is ultimately bounded in probability* $1-\varepsilon$ *for given* $\varepsilon \in (0,1)$*. Furthermore, we choose*

$$\tilde{\eta}(u) \triangleq a^{-1} \diamond \frac{\lambda\hat{\alpha}^{-1}}{\varepsilon} \diamond (2\max_{p\in\mathcal{I}}\eta_p)(u), \quad \tilde{\rho} \triangleq a^{-1} \diamond \frac{\lambda\hat{\alpha}^{-1}}{\varepsilon}(2\max_{p\in\mathcal{I}}\rho_p). \quad (4.105)$$

Then, the synchronization error bound is $\tilde{\eta}(\|\nu\|_\infty) + \tilde{\rho}$ *and the proof is complete.*

Remark 4.9 *In Theorem 4.8, a sufficient criterion is provided for the dynamic analysis of the synchronization error system (4.67) when the control gain matrices* K_{ip}*, the switching signal* $\sigma(k)$ *and the ET parameters* ζ*,* θ_i *are given in advance. Furthermore, Theorem 4.8 gives a quantitative estimation of the upper bound of the synchronization error in (4.105).*

Theorem 4.9 *Under Assumptions 4.2 and 4.3, let the constants* $\varepsilon \in (0,1)$*,* $\theta_i \geq 0$ $(i \in \mathcal{N}_0)$ *be given. The synchronization error system (4.67) is ultimately bounded in probability* $1-\varepsilon$ *if there exist positive constants* $\lambda_{1p} > 0$*,* $\lambda_{2p} > 0$*,* $\lambda_{3p} > 0$*,* $\varpi_{1p} > 0$*,* $\varpi_{2p} > 0$*, constant matrices* $Y_i \in \mathbb{R}^{n\times n}$ $(i \in \mathcal{N}_0)$ *and positive definite matrices* $Q_{ip} \in \mathbb{R}^{n\times n}$ $(i \in \mathcal{N},\ p \in \mathcal{I})$ *such that*

$$\begin{bmatrix} \Pi_p & \tilde{S}_p^T \\ \tilde{S}_p & -Q_p \end{bmatrix} < 0, \quad p \in \mathcal{I}, \quad (4.106)$$

where $Q_p = \mathrm{diag}\{Q_{1p}, Q_{2p}, \cdots, Q_{Np}\}$*,* $Y = \mathrm{diag}\{Y_1, Y_2, \cdots, Y_l, 0, \cdots, 0\}$*,* $\tilde{S}_p = [Y, Q_p, Q_p(L\otimes\Gamma), Y, Q_p]$*.* γ_p*,* Π_p *and the average dwell time* τ_a *have been defined in Theorem 4.7. Furthermore, the control gain matrices are designed by*

$$K_{ip} = Q_{ip}^{-1}Y_i, \quad i \in \mathcal{N}_p,\ p \in \mathcal{I} \quad (4.107)$$

with the ET parameters satisfying

$$0 \leq \zeta < \min_{p\in\mathcal{I}} \frac{\zeta_0\lambda_{1p}}{\frac{\zeta_0\lambda_{2p}}{1-\zeta_0} + \lambda_{1p}} \quad \text{and} \quad \theta_i \geq 0,\ i \in \mathcal{N}_p, \quad (4.108)$$

and the synchronization error bound is equal to $\tilde{\eta}(\|\nu\|_\infty) + \tilde{\rho}$*, where* $\tilde{\eta}$ *and* $\tilde{\rho}$ *are defined in (4.105).*

Proof *It is easy to obtain (4.91) from (4.108). Consider*

$$\begin{bmatrix} I_{5nN} & 0 \\ 0 & Q_p^{-1} \end{bmatrix}\begin{bmatrix} \Pi_p & \tilde{S}_p^T \\ \tilde{S}_p & -Q_p \end{bmatrix}\begin{bmatrix} I_{5nN} & 0 \\ 0 & Q_p^{-1} \end{bmatrix} = \begin{bmatrix} \Pi & \tilde{S}_p^T Q_p^{-1} \\ Q_p^{-1}\tilde{S}_p & -Q_p^{-1} \end{bmatrix}.$$

According to the definition of control gain matrices (4.107), we obtain

$$Q_p^{-1}\tilde{S}_p = Q_p^{-1}[Y, Q_p, Q_p(L \otimes \Gamma), Y, Q_p] = [K_p, I_{nN}, L \otimes \Gamma, K_p, I_{nN}] = S_p,$$

which shows that (4.93) holds if (4.106) is true. Therefore, the proof is complete.

Remark 4.10 *Theorem 4.9 shows how to design proper pinning control gain matrices and the ET parameters to guarantee the desired synchronization of switched stochastic complex networks (4.60) in probability sense. According to (4.106)–(4.108), the pinning gain matrices K_i and the parameters of the event-generator function ζ, θ_i can be obtained by using LMI toolbox.*

4.3 Illustrative Examples

In this section, two simulation examples are given to demonstrate the ET control approaches presented in this chapter.

4.3.1 Example 1

Consider a switched DDS (4.1) with $m = 2$ whose parameters are given as follows:

$$A_1 = \begin{bmatrix} -5 & 0.6 \\ -0.5 & -1.3 \end{bmatrix}, B_1 = \begin{bmatrix} 0.2 & 0.1 \\ -0.2 & 0.3 \end{bmatrix}, C_1 = \begin{bmatrix} -1 \\ 1.3 \end{bmatrix}, C_2 = \begin{bmatrix} -1.1 \\ 2 \end{bmatrix},$$
$$D_1 = \begin{bmatrix} -1.5 & 0.2 \\ -0.2 & 0.1 \end{bmatrix}, A_2 = \begin{bmatrix} -8 & 1 \\ -1 & 0.2 \end{bmatrix}, B_2 = \begin{bmatrix} 0.1 & 0.2 \\ -0.2 & 0.1 \end{bmatrix},$$
$$D_2 = \begin{bmatrix} 0.1 & 0.2 \\ -0.2 & 0.1 \end{bmatrix}, f_1(t,x) = \sin(0.5x),\ f_2(t,x) = \sin(0.3x).$$

We select two scalars $\mu_1 = 0.01$ and $\mu_2 = -10.5$. It is easy to verify that the two functions $f_1(t,x)$ and $f_2(t,x)$ satisfy the Lipschtiz condition. Since $\min_{i \in M} \mu_i = \mu_2 < -1$, we can apply the results of Theorem 4.5 to solve the corresponding LMI (4.56). A set of feasible solution (4.56) is obtained by using Matlab toolbox as follows:

$$\tilde{P}_1 = \begin{bmatrix} 0.3670 & 0.0730 \\ 0.0730 & 1.0798 \end{bmatrix}, \tilde{Q}_1 = \begin{bmatrix} 3.0158 & 0 \\ 0 & 3.0158 \end{bmatrix},$$
$$\tilde{S}_1 = \begin{bmatrix} 3.0158 & 0 \\ 0 & 3.0158 \end{bmatrix}, Y_1 = \begin{bmatrix} -0.0675 & -0.0808 \end{bmatrix},$$
$$\tilde{P}_2 = \begin{bmatrix} 1.1354 & 0.0444 \\ 0.0444 & 0.3006 \end{bmatrix}, \tilde{S}_2 = \begin{bmatrix} 4.9768 & -1.6144 \\ -1.6144 & 7.0242 \end{bmatrix},$$

$$\tilde{Q}_2 = \begin{bmatrix} 3.3097 & 0.2751 \\ 0.2751 & 3.3988 \end{bmatrix}, \ Y_2 = \begin{bmatrix} -0.0675 & -0.0808 \end{bmatrix}.$$

Furthermore, according to (4.58), we calculate the upper bound of the triggered weight parameter:

$$0 < \zeta_1 < \min\left\{\frac{9.5\zeta_0(1-\zeta_0)}{9.5 - 0.4799\zeta_0}, \zeta_0\right\}.$$

Here, we choose that $\zeta_0 = 0.1$, which yields $0 < \zeta_1 < 0.0905$. Then, according to Theorem 4.5, the feedback gain matrices K_i $(i = 1, 2)$ are designed as follows:

$$K_1 = Y_1\tilde{P}_1^{-1} = \begin{bmatrix} -0.1714 & -0.0632 \end{bmatrix},$$
$$K_2 = Y_2\tilde{P}_2^{-1} = \begin{bmatrix} -0.1202 & -5.3043 \end{bmatrix}.$$

For simulation purpose, let the time interval be $[0, 30s]$ and the step be $0.05s$. The unknown exogenous disturbance function $v(t) = [v_1(t), v_2(t)]^T$ is assumed as $v_1(t) = v_0 \sin(t)$ and $v_2(t) = v_0 \cos(t)$, where v_0 is a random number obeying the uniform distribution over the interval $(0, 1)$. The delay $d = 0.1$ and the initial values are $x_1(t) = 0.5 + t$ and $x_2(0) = 0.5 + t$ for $t \in [-0.1, 0]$. In this simulation, we choose the ET parameters $\zeta_1 = 0.09$ and $\zeta_2 = 0.01$.

The simulation results for the controlled switched system are shown in Figs. 4.1–4.3. Fig. 4.1 shows the dynamic evolutions of $x_1(t)$ and $x_2(t)$, which enter into a bounded domain under the ET control, the switching signal and the nonzero exogenous disturbances. The ET instants are depicted in Fig. 4.2, from which one can find that the frequency of control updating is greatly reduced. According to Theorem 4.3, we compute the inter-execution interval satisfying

$$\begin{aligned} t_{k+1} - t_k \geq T^* &= \min\left\{d, \frac{1}{\max_{i\in M}\vartheta_{i1}} \ln\left(1 + \frac{\zeta_2}{\varsigma}\right)\right\} \\ &\geq 7.768 \times 10^{-8}. \end{aligned}$$

Remark 4.11 *In Theorem 4.3, it has been shown that there exists a positive lower bound of the event intervals, i.e., T^*. In order to derive such a lower bound, some enlarging technologies have been employed and the derived lower bound T^* might be conservative. Actually, the real infimum of all the event intervals would be larger than this lower bound T^* and the Zeno phenomenon could be expected to excluded in the practical applications.*

The switching signal $\sigma(t)$ here is assumed to take the form

$$\sigma(t) = \begin{cases} 1, & t \in \big[10.1s, 10.1s + 10\big), \\ 2, & t \in \big[10.1s + 10, 10.1(s+1)\big), \end{cases} \quad s \in \mathbb{N}.$$

It can be verified that this switching signal satisfies (4.22)–(4.24), as can be seen in Fig. 4.3.

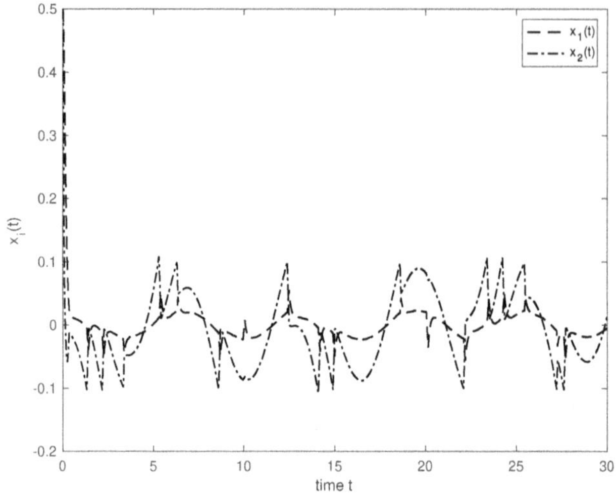

FIGURE 4.1
Boundedness of switched DDSs with exogenous disturbances.

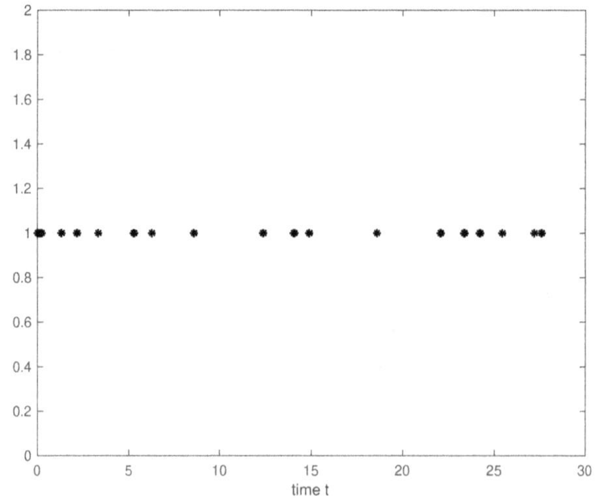

FIGURE 4.2
The event-triggering instants.

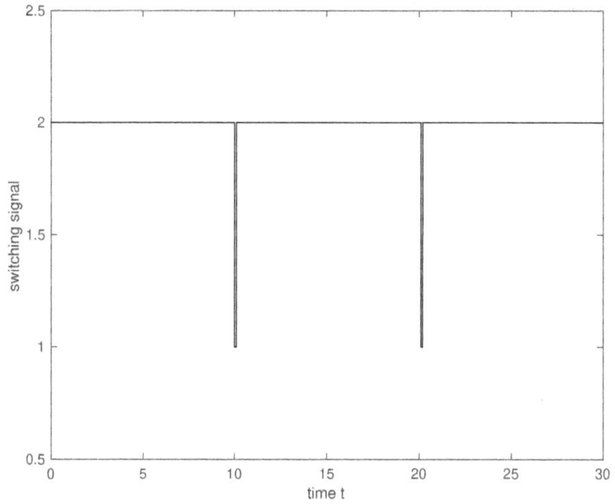

FIGURE 4.3
The switching signal.

4.3.2 Example 2

Consider a discrete switched stochastic complex network (4.60) with $N = 3$, $n = 2$, and $m = 2$. The nonlinear dynamical functions are

$$f_1(x_i(k)) = \begin{bmatrix} 0.5x_1 + 0.7x_2 + \tanh(0.2x_1 + 0.2x_2) \\ 0.45x_1 + 0.8x_2 - \tanh(0.2x_1 + 0.25x_2) \end{bmatrix},$$

$$f_2(x_i(k)) = \begin{bmatrix} 0.45x_1 + 0.7x_2 + \tanh(0.25x_1) \\ 0.45x_1 + 0.75x_2 - \tanh(0.2x_1 + 0.15x_2) \end{bmatrix},$$

and the inner-coupling functions are

$$g_1(x_i(k)) = \begin{bmatrix} -0.2x_1 + 0.2x_2 + \tanh(0.1x_2) \\ 0.3x_2 - \tanh(0.3x_1 + 0.1x_2) \end{bmatrix},$$

$$g_2(x_i(k)) = \begin{bmatrix} -0.25x_1 + 0.25x_2 + \tanh(0.05x_1 + 0.05x_2) \\ 0.2x_2 - \tanh(0.3x_1) \end{bmatrix}.$$

It is clear that f_1, f_2, g_1, and g_2 satisfy Assumption 4.2 with

$$U_{11} = \begin{bmatrix} 0.5 & 0.7 \\ 0.25 & 0.55 \end{bmatrix}, U_{21} = \begin{bmatrix} 0.7 & 0.9 \\ 0.45 & 0.8 \end{bmatrix}, U_{12} = \begin{bmatrix} 0.45 & 0.7 \\ 0.25 & 0.6 \end{bmatrix},$$

$$U_{22} = \begin{bmatrix} 0.7 & 0.7 \\ 0.45 & 0.75 \end{bmatrix}, W_{11} = \begin{bmatrix} -0.2 & 0.2 \\ -0.3 & 0.2 \end{bmatrix}, W_{21} = \begin{bmatrix} -0.2 & 0.3 \\ 0 & 0.3 \end{bmatrix},$$

$$W_{12} = \begin{bmatrix} -0.25 & 0.25 \\ -0.3 & 0.2 \end{bmatrix}, \; W_{21} = \begin{bmatrix} -0.2 & 0.3 \\ 0 & 0.2 \end{bmatrix}.$$

Let the noise intensity functions be

$$h_1(x_i(k)) = \begin{bmatrix} 0.1 & 0 \\ 0 & 0.2 \end{bmatrix} x_i(k), \quad h_2(x_i(k)) = \begin{bmatrix} 0.2 & 0 \\ 0 & 0.3 \end{bmatrix} x_i(k),$$

which satisfy Assumption 4.3 with $\kappa_1(u) = 0.04u$ and $\kappa_2(u) = 0.09u$. Furthermore, we assume that the coupled configuration matrix, the exogenous disturbance and the inner-coupling matrix are as follows:

$$L = \begin{bmatrix} -0.6 & 0.3 & 0.3 \\ 0.5 & -0.5 & 0 \\ 0.2 & 0.6 & -0.8 \end{bmatrix}, \; \nu_i(k) = \begin{bmatrix} 0.01\sin(k) \\ 0.01\cos(k) \end{bmatrix} \quad \text{and} \quad \Gamma = I_2.$$

It should be pointed out that the isolate node is an unstable dynamical system, which is illustrated in Fig. 4.6. Moreover, the switched stochastic complex network with nonlinear coupling structures cannot tract the target motion of the isolate node without control input, and therefore we need to control all the nodes. A feasible solution is obtained by using Matlab toolbox as follows:

$$\lambda_{11} = 1.5772, \quad \lambda_{21} = 91.5290, \quad \lambda_{31} = 102.2384, \quad \lambda_1^* = 0.2598,$$
$$\varpi_{11} = 32.1096, \quad \varpi_{21} = 39.1709, \quad \lambda_{12} = 2.2549, \quad \lambda_{22} = 112.1441,$$
$$\lambda_{32} = 130.2788, \quad \lambda_2^* = 0.3703, \quad \varpi_{11} = 42.3274, \quad \varpi_{21} = 50.5449,$$

$$Q_{11} = \begin{bmatrix} 15.1125 & -2.2853 \\ -2.2853 & 18.5825 \end{bmatrix}, \quad Q_{21} = \begin{bmatrix} 14.8447 & -2.4484 \\ -2.4484 & 18.5287 \end{bmatrix},$$

$$Q_{31} = \begin{bmatrix} 13.1530 & -2.2065 \\ -2.2065 & 15.8572 \end{bmatrix}, \quad Q_{12} = \begin{bmatrix} 20.9224 & -3.5714 \\ -3.5714 & 23.0668 \end{bmatrix},$$

$$Q_{22} = \begin{bmatrix} 20.5699 & -3.7792 \\ -3.7792 & 23.0550 \end{bmatrix}, \quad Q_{32} = \begin{bmatrix} 18.0516 & -3.4244 \\ -3.4244 & 19.8278 \end{bmatrix},$$

$$Y_{11} = \begin{bmatrix} -9.2833 & -7.0396 \\ -7.0396 & -7.3778 \end{bmatrix}, \quad Y_{21} = \begin{bmatrix} -8.9360 & -6.8771 \\ -6.8771 & -7.5091 \end{bmatrix},$$

$$Y_{31} = \begin{bmatrix} -8.4276 & -6.0150 \\ -6.0150 & -5.9573 \end{bmatrix}, \quad Y_{12} = \begin{bmatrix} -12.3561 & -7.6741 \\ -7.6741 & -9.6194 \end{bmatrix},$$

$$Y_{22} = \begin{bmatrix} -11.8668 & -7.5233 \\ -7.5233 & -9.7518 \end{bmatrix}, \quad Y_{32} = \begin{bmatrix} -11.1432 & -6.4749 \\ -6.4749 & -7.9301 \end{bmatrix},$$

and the control feedback gain matrices are calculated as follows:

$$K_{11} = \begin{bmatrix} -0.6843 & -0.5358 \\ -0.4630 & -0.4629 \end{bmatrix}, \quad K_{21} = \begin{bmatrix} -0.6780 & -0.5419 \\ -0.4607 & -0.4769 \end{bmatrix},$$

$$K_{31} = \begin{bmatrix} -0.7212 & -0.5328 \\ -0.4797 & -0.4498] \end{bmatrix} \quad K_{12} = \begin{bmatrix} -0.6649 & -0.4499 \\ -0.4356 & -0.4867 \end{bmatrix},$$

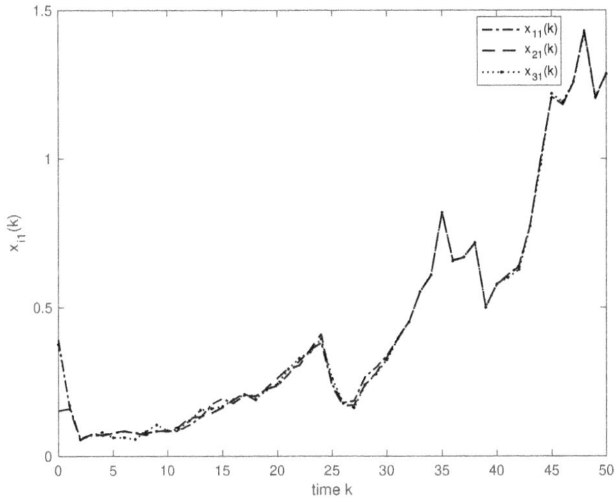

FIGURE 4.4
The state trajectories $x_{i1}(k)$ $(i = 1, 2, 3)$.

$$K_{22} = \begin{bmatrix} -0.6566 & -0.4572 \\ -0.4340 & -0.4979 \end{bmatrix}, \quad K_{32} = \begin{bmatrix} -0.7023 & -0.4493 \\ -0.4478 & -0.4775 \end{bmatrix}.$$

Then, we know $\max_{p\in\mathcal{I}} \lambda_{\max}(Q_p) = 0.0839$, $\min_{p\in\mathcal{I}} \lambda_{\min}(Q_p) = 0.0388$, $\lambda = 2.1624$. Here, we choose $\varepsilon = 0.25$, $\zeta_0 = 0.5$, $\zeta = 0.001$, $\theta_1 = \theta_2 = 0.001$, $d_1 = d_1 = 100$, $\gamma_1 = 0.8339$, $\gamma_2 = 0.6872$, which satisfy (4.91) and (4.92). Moreover, $\tau_a^* = 4.2458$ and the switching sequence $\{\tau_k\}_{k=0}^{\infty}$ is assumed to be $\{5k\}_{k=0}^{\infty}$ that satisfies (4.94). Thus, the switching signal is effective as can be seen in Fig. 4.7. It follows that $\alpha_1(u) = 16.666u$, $\alpha_1(u) = 31.2493u$, $\beta_0 = 20$, $\hat{\alpha}(u) = 0.9993u$, $\eta_1 = 122.6027$, $\eta_2 = 189.5792$. Then, according to (4.105), we conclude that switched stochastic complex network is the ultimately bounded in probability 0.75 and the upper bound of synchronization errors is calculated as 14.9871. Figs. 4.4 and 4.5 show the state trajectories of the closed-loop switched stochastic complex dynamical network with nonlinear coupling, which enters a bounded domain under the ET control and the switching signal even though the states suffer with nonzero exogenous disturbances. The ET instants are depicted in Fig. 4.8, from which one can see that the frequency of control updating is greatly reduced.

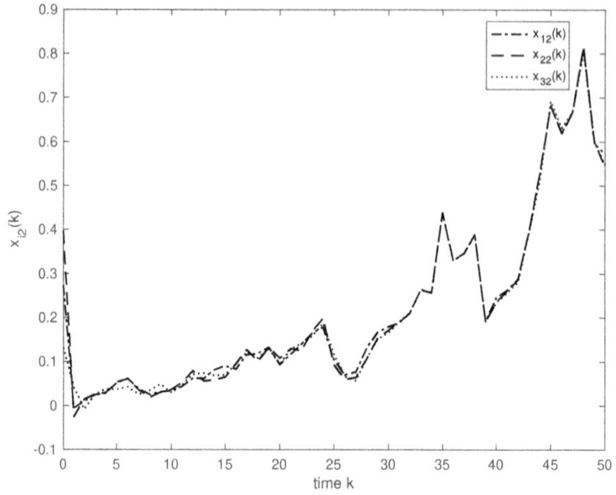

FIGURE 4.5
The state trajectories $x_{i2}(k)$ $(i = 1, 2, 3)$.

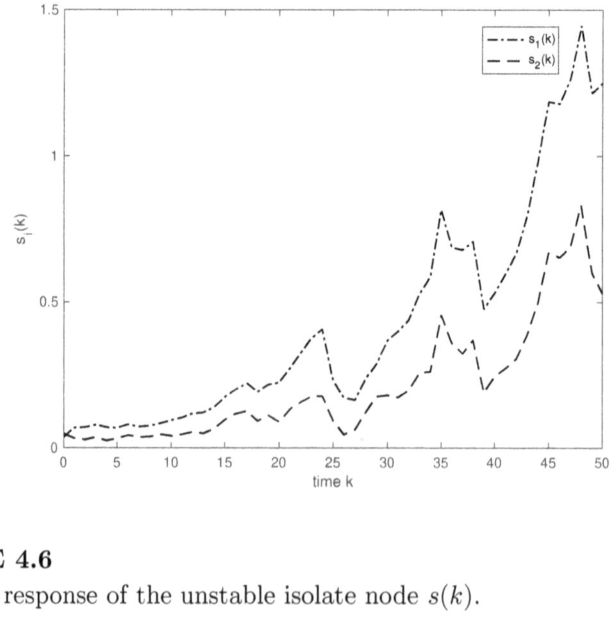

FIGURE 4.6
The state response of the unstable isolate node $s(k)$.

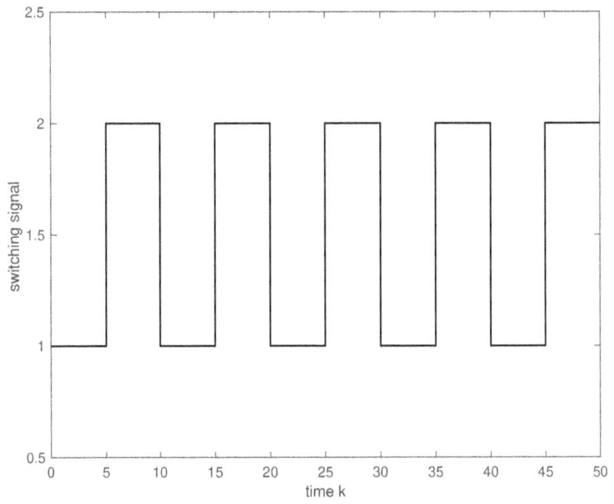

FIGURE 4.7
The switching signal.

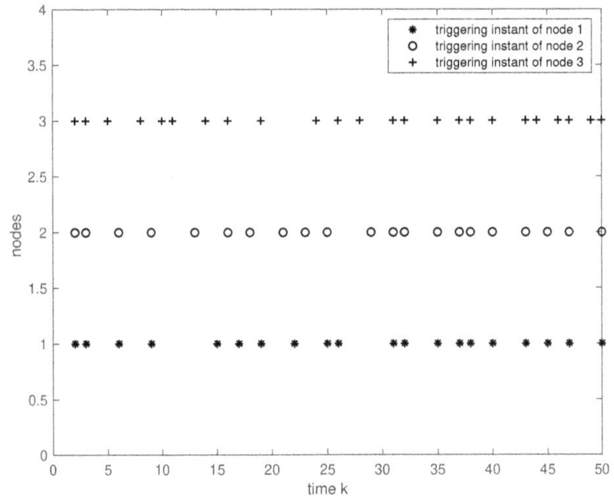

FIGURE 4.8
The triggering instants of the pinned nodes.

4.4 Summary

In this chapter, we have dealt with the ET control problem for two classes of switched systems with exogenous disturbances. In order to save the communication resource, the ET control strategy has been proposed by which the control signal is updated only when a certain condition is violated. The Lyapunov functional approach has been conducted to derive sufficient conditions for the existence of the desired controllers whose gain matrices have been obtained by solving two sets of LMIs. In the end, the results of this chapter have been demonstrated by some simulation examples.

5

Event-Triggered H_∞ State Estimation for State-Saturated Systems

This chapter addresses the ET H_∞ state estimation problem for state-saturated systems. In this system under consideration, a saturation function is introduced to constrain the state variables to stay within a bounded set. Firstly, the ET distributed H_∞ state estimation problem is investigated for a class of state-saturated systems with ROMDs over sensor networks. The mixed delays, which comprise both discrete and distributed delays, are allowed to occur in a random manner governed by two mutually independent Bernoulli distributed random variables. Moreover, the ET H_∞ state estimation problem is also studied for a class of state-saturated complex networks subject to quantization effects as well as randomly occurring distributed delays. The main purpose is to design ET state estimators such that the error dynamics of state estimation is exponentially mean-square stable with a prescribed H_∞ performance index. Sufficient conditions are derived via intensive stochastic analysis to guarantee the existence of the desired estimators, and the parameters of the desired estimators are then obtained in light of the feasibility of certain sets of matrix inequalities. Finally, some numerical examples are employed to illustrate the usefulness of the proposed ET estimation algorithms.

5.1 Distributed Event-Triggered H_∞ State Estimation in Sensor Networks

5.1.1 Problem Formulation

Consider a target plant of the following form:

$$\begin{cases} x(k+1) =\sigma\Big(Ax(k) + \alpha(k)Bx(k-\tau(k)) + \beta(k)E\sum_{s=1}^{+\infty}\kappa_s x(k-s) \\ \qquad\qquad + Lw(k)\Big) \\ \qquad z(k) =Mx(k) \end{cases} \tag{5.1}$$

DOI: 10.1201/9781003307648-5

where $x(k) \in \mathbb{R}^{n_x}$ is the plant vector, $z(k) \in \mathbb{R}^{n_z}$ is the signal to be estimated, and $w(k) \in \mathbb{R}^{n_w}$ represents the exogenous disturbance belonging to $l_2[0, +\infty)$. A, B, E, L and M are known matrices with appropriate dimensions. $\tau(k)$ is the TVD satisfying $0 < \tau_m \leq \tau(k) \leq \tau_M$. The initial values are set as $x(-l) = 0$ for $l = 1, 2, \ldots$ and $x(0) = \bar{x}$. $\alpha(k) \in \mathbb{R}$ and $\beta(k) \in \mathbb{R}$ are two mutually independent Bernoulli distributed random variables govern, respectively, the phenomena of randomly occurring discrete delays and distributed delays with

$$\begin{aligned} &\text{Prob}\{\alpha(k) = 1\} = \alpha, \quad \text{Prob}\{\alpha(k) = 0\} = 1 - \alpha, \\ &\text{Prob}\{\beta(k) = 1\} = \beta, \quad \text{Prob}\{\beta(k) = 0\} = 1 - \beta, \end{aligned} \tag{5.2}$$

with $\alpha \in [0, 1]$ and $\beta \in [0, 1]$ being known constants. The non-negative constants κ_s $(s = 1, 2, \ldots, +\infty)$ satisfy

$$\bar{\kappa} = \sum_{s=1}^{+\infty} \kappa_s < +\infty \quad \text{and} \quad \sum_{s=1}^{+\infty} s\kappa_s < +\infty. \tag{5.3}$$

The saturated function $\sigma : \mathbb{R}^{n_x} \to \mathbb{R}^{n_x}$ in (5.1) is defined as follows:

$$\sigma(x) = \begin{bmatrix} \sigma_1(x_1) & \sigma_2(x_2) & \cdots & \sigma_{n_x}(x_{n_x}) \end{bmatrix}^T \tag{5.4}$$

with $\sigma_i(x_i) = \text{sign}(x_i) \min\{\delta_i, |x_i|\}$, $i = 1, 2, \ldots, n_x$, where "sign" means the signum function, x_i is the ith element of the vector x and δ_i stands for the saturation level. For convenience of later analysis, denote $\aleph = \text{diag}\{\delta_1, \ldots, \delta_{n_x}\}$.

The ith $(i = 1, 2, \ldots, N)$ sensor node in the sensor network is described by

$$y_i(k) = C_i x(k) + D_i v(k), \tag{5.5}$$

where $y_i(k) \in \mathbb{R}^{n_y}$ is the measured output received by the ith node, $v(k) \in \mathbb{R}^{n_v}$ denotes the measurement noise belonging to $l_2[0, +\infty)$, and C_i and D_i are known constant matrices.

The topology of the sensor network is represented by a directed weighted graph $\mathcal{G} = (\mathcal{V}, \mathcal{Q}, \mathcal{W})$ of order N, where $\mathcal{V} = \{1, 2, \ldots, N\}$ is the set of nodes, $\mathcal{Q} \subseteq \mathcal{V} \times \mathcal{V}$ is the set of edges, and $\mathcal{W} = [w_{ij}]_{N \times N}$ describes a weighted adjacency matrix with nonnegative elements w_{ij} satisfying $w_{ij} > 0 \iff (i, j) \in \mathcal{Q}$, which implies that sensor node i can obtain the information from node j. $\mathcal{N}_i = \{j \in \mathcal{V} : (i, j) \in \mathcal{Q}\}$ denotes the set of neighbors of node $i \in \mathcal{V}$.

In this section, for the energy-saving purpose, an ET mechanism is adopted to decide when each sensor node should transmit certain information to its neighbors. For each $i \in \mathcal{V}$, the triggering instants can be represented as $0 \leq t_0^i < t_1^i < \cdots < t_l^i < \cdots$, and t_{l+1}^i satisfies

$$t_{l+1}^i = \min\left\{k | k > t_l^i,\ \sigma_i r_i^T(k) r_i(k) < (r_i(k) - r_i(t_l^i))^T (r_i(k) - r_i(t_l^i))\right\} \tag{5.6}$$

where σ_i is a given positive scalar, $r_i(t_l^i)$ is the innovation transmitted at the latest event time, and $r_i(k) = y_i(k) - C_i\hat{x}_i(k)$ with $\hat{x}_i(k)$ being the estimate of state $x(k)$ for node i.

In terms of the ET information, for each $i \in \mathcal{V}$, the following distributed state estimator is constructed:

$$\begin{cases} \hat{x}_i(k+1) = A\hat{x}_i(k) + \alpha B\hat{x}_i(k-\tau(k)) + \beta E\sum_{s=1}^{+\infty}\kappa_s\hat{x}_i(k-s) \\ \qquad\qquad + K_i r_i(k) + \sum_{j\in\mathcal{N}_i} K_{ij}w_{ij}r_j(t^j_{l_j(k)}) \\ \hat{z}_i(k) = M\hat{x}_i(k) \end{cases} \tag{5.7}$$

where $l_j(k) = \arg\min_{m\in\mathbb{Z}_+:k\geq t_m^j}\{k - t_m^j\}$ $(j \in \mathcal{N}_i)$, $\hat{z}_i(k)$ is the estimate of output $z(k)$ on the node i; and K_i and K_{ij} are the estimator parameters to be designed.

Defining $\varepsilon_i(k) = r_i(k) - r_i(t_l^i)$ for $k \in [t_l^i, t_{l+1}^i)$ $(l = 0, 1, 2, \ldots, \infty)$, (5.7) can be rewritten as follows:

$$\begin{cases} \hat{x}_i(k+1) = A\hat{x}_i(k) + \alpha B\hat{x}_i(k-\tau(k)) + \beta E\sum_{s=1}^{+\infty}\kappa_s\hat{x}_i(k-s) \\ \qquad\qquad + \sum_{j\in\mathcal{N}_i} K_{ij}w_{ij}\big(C_jx(k) + D_jv(k) - C_j\hat{x}_j(k) - \varepsilon_j(k)\big) \\ \qquad\qquad + K_i(C_ix(k) + D_iv(k) - C_i\hat{x}_i(k)) \\ \hat{z}_i(k) = M\hat{x}_i(k). \end{cases} \tag{5.8}$$

Setting $\tilde{z}_i(k) = z(k) - \hat{z}_i(k)$ and

$$\bar{M} = \text{diag}\{\underbrace{M, M, \ldots, M}_{N}\}, \quad \hat{x}(k) = \begin{bmatrix}\hat{x}_1^T(k) & \hat{x}_2^T(k) & \cdots & \hat{x}_N^T(k)\end{bmatrix}^T,$$

$$\bar{E} = \text{diag}\{\underbrace{E, E, \ldots, E}_{N}\}, \quad \varepsilon(k) = \begin{bmatrix}\varepsilon_1^T(k) & \varepsilon_2^T(k) & \cdots & \varepsilon_N^T(k)\end{bmatrix}^T,$$

$$\bar{B} = \text{diag}\{\underbrace{B, B, \ldots, B}_{N}\}, \quad \tilde{z}(k) = \begin{bmatrix}\tilde{z}_1^T(k) & \tilde{z}_2^T(k) & \cdots & \tilde{z}_N^T(k)\end{bmatrix}^T,$$

$$\bar{K} = \text{diag}\{K_1, K_2, \ldots, K_N\}, \quad \tilde{M} = \underbrace{\begin{bmatrix}M^T & M^T & \ldots & M^T\end{bmatrix}^T}_{N},$$

$$\bar{C} = \text{diag}\{C_1, C_2, \ldots, C_N\}, \quad D = \begin{bmatrix}D_1^T & D_2^T & \cdots & D_N^T\end{bmatrix}^T,$$

$$\bar{A} = \text{diag}\{\underbrace{A, A, \ldots, A}_{N}\}, \quad C = \begin{bmatrix}C_1^T & C_2^T & \cdots & C_N^T\end{bmatrix}^T,$$

we obtain

$$\begin{cases} \hat{x}(k+1) = \bar{A}\hat{x}(k) - (K+\bar{K})\bar{C}\hat{x}(k) + \alpha\bar{B}\hat{x}(k-\tau(k)) + \beta\bar{E}\sum_{s=1}^{+\infty}\kappa_s \\ \qquad \times \hat{x}(k-s) + (K+\bar{K})Cx(k) + (K+\bar{K})Dv(k) - K\varepsilon(k) \\ \vec{z}(k) = \tilde{M}x(k) - \bar{M}\hat{x}(k) \end{cases} \tag{5.9}$$

where $K = [\tilde{K}_{ij}]_{N\times N}$ with

$$\tilde{K}_{ij} = \begin{cases} K_{ij}w_{ij}, & i \in \mathcal{V},\ j \in \mathcal{N}_i, \\ 0, & i \in \mathcal{V},\ j \neq \mathcal{N}_i. \end{cases} \tag{5.10}$$

Obviously, K is a sparse matrix that satisfies $K \in \mathcal{L}_{n_x \times n_y}$ with

$$\mathcal{L}_{p\times q} = \left\{ \bar{T} = [T_{ij}] \in \mathbb{R}^{Np\times Nq} | T_{ij} \in \mathbb{R}^{p\times q},\ T_{ij} = 0 \text{ if } j \neq \mathcal{N}_i \right\}. \tag{5.11}$$

Letting $\vec{x}(k) = \begin{bmatrix} x^T(k) & \hat{x}^T(k) \end{bmatrix}^T$, the following augmented system is obtained:

$$\begin{cases} \vec{x}(k+1) = \mathcal{K}_1\vec{x}(k) + \mathcal{B}\vec{x}(k-\tau(k)) + \mathcal{E}\sum_{s=1}^{+\infty}\kappa_s\vec{x}(k-s) \\ \qquad - \mathcal{K}_2\varepsilon(k) + \mathcal{K}_3 v(k) + \mathcal{H}(k) \\ \vec{z}(k) = \mathcal{M}\vec{x}(k) \end{cases} \tag{5.12}$$

where

$$\mathcal{M} = \begin{bmatrix} \tilde{M} & -\bar{M} \end{bmatrix},\ \mathcal{B} = \text{diag}\{0, \alpha\bar{B}\},$$

$$\mathcal{E} = \text{diag}\{0, \beta\bar{E}\},\ S = \begin{bmatrix} I & 0 \end{bmatrix},\ \mathcal{H}(k) = \begin{bmatrix} h(k) \\ 0 \end{bmatrix},$$

$$h(k) = \sigma\Big(AS\vec{x}(k) + \alpha(k)BS\vec{x}(k-\tau(k)) + \beta(k)E\sum_{s=1}^{+\infty}\kappa_s S\vec{x}(k-s) + Lw(k)\Big),$$

$$\mathcal{K}_1 = \begin{bmatrix} 0 & 0 \\ (K+\bar{K})C & \bar{A} - (K+\bar{K})\bar{C} \end{bmatrix},\ \mathcal{K}_3 = \begin{bmatrix} 0 \\ (K+\bar{K})D \end{bmatrix},\ \mathcal{K}_2 = \begin{bmatrix} 0 \\ K \end{bmatrix}.$$

Definition 5.1 *The augmented system (5.12) is said to be exponentially mean-square stable in case of $w(k) = 0$ and $v(k) = 0$ if there exist two constants $a > 0$ and $0 < b < 1$ such that*

$$\mathbb{E}\{\|\vec{x}(k)\|^2\} \leq ab^k \sup_{\theta\in\mathbb{Z}_-} \mathbb{E}\{\|\phi(\theta)\|^2\}, \quad \forall k \geq 0 \tag{5.13}$$

where $\phi(\theta)$ is the initial value of (5.12) defined as $\vec{x}(\theta) = \phi(\theta)$, $\theta \in \mathbb{Z}_-$.

The aim of the section is to design a set of ET distributed estimators to achieve the following two requirements simultaneously.

a) The augmented system (5.12) with $w(k) = 0$ and $v(k) = 0$ is exponentially stable in the mean square.

b) Under the zero-initial condition, the output estimation error $\vec{z}(k)$ satisfies

$$\mathbb{E}\left\{\sum_{k=0}^{+\infty}\|\vec{z}(k)\|^2\right\} < \gamma^2\sum_{k=0}^{+\infty}\|v(k)\|^2 + \gamma^2\sum_{k=0}^{+\infty}\|w(k)\|^2 \tag{5.14}$$

for all nonzero $w(k)$ and $v(k)$, where $\gamma > 0$ is a given disturbance attenuation level.

5.1.2 Main Results

In this section, sufficient conditions are first given to guarantee both the exponentially mean-square stability and H_∞ performance of the augmented system (5.12). Then, on the basis of such established conditions, the desired distributed estimators are designed for the target plant (5.1).

Before stating our main results, the following lemmas are introduced to facilitate later development.

Lemma 5.1 *[65] Let $\mathcal{Y}_{n_x}$ be the set of $n_x \times n_x$ diagonal matrices whose diagonal elements are either* 1 *or* 0. *a) There are 2^{n_x} elements in $\mathcal{Y}_{n_x}$, where its ith element is denoted as Y_i, $i \in [1, 2^{n_x}]$. b) By defining $Y_i^- = I - Y_i$ and letting $G \in \mathbb{R}^{n_x \times n_x}$ satisfy $\|G\|_\infty \leq 1$, for any vector $\nu \in \mathbb{R}^{n_x}$, we have*

$$\aleph^{-1}\sigma(Ax(k)+\nu) \in co\left\{Y_i\aleph^{-1}(Ax(k)+\nu) + Y_i^- G\aleph^{-1}x(k), i \in [1, 2^{n_x}]\right\},$$

where co$\{\ldots\}$ *is the convex hull of a set.*

Lemma 5.2 *[65] For any symmetric positive definite matrix P, the map $x \mapsto x^T P x$ is convex.*

Lemma 5.3 *[106] For the given positive semi-definite matrix $W \in \mathbb{R}^{m\times m}$, vector $X_i \in \mathbb{R}^m$ and a set of scalars $a_i > 0$ ($i = 1, 2, \ldots$), if the series concerned is convergent, then one has*

$$\left(\sum_{i=1}^{+\infty} a_i X_i\right)^T W\left(\sum_{i=1}^{+\infty} a_i X_i\right) \leq \left(\sum_{i=1}^{+\infty} a_i\right)\left(\sum_{i=1}^{+\infty} a_i X_i^T W X_i\right).$$

Lemma 5.4 *[137] Let $P = \mathrm{diag}\{P_1, P_2, \ldots, P_N\}$, where $P_i \in \mathbb{R}^{n_x\times n_x}$ ($1 \leq i \leq N$) are invertible matrices. If $X = PK$ for $K \in \mathbb{R}^{Nn_x\times Nn_y}$, then one has $K \in \mathcal{L}_{n_x\times n_y} \Longleftrightarrow X \in \mathcal{L}_{n_x\times n_y}$.*

In the following theorem, sufficient conditions are obtained to ensure that the augmented system (5.12) is exponentially stable in the mean square and the H_∞ performance constraint (5.14) is satisfied.

Theorem 5.1 *Let the estimator gains K, $\bar{K}$ and the disturbance level $\gamma > 0$ be given. Assume that there exist matrices $P = \mathrm{diag}\{P_1, P_2\} > 0$, $Q > 0$, $R > 0$ and G, and a scalar $\lambda > 0$ satisfying*

$$\|G\|_\infty \leq 1, \tag{5.15}$$

$$\Phi_j = \begin{bmatrix} -P^{-1} & 0 & \bar{\Pi}_1 \\ * & -P_1^{-1} & \bar{\Pi}_{2j} \\ * & * & \hat{\Phi}_{22} \end{bmatrix} < 0, \quad j \in [1, 2^{n_x}] \tag{5.16}$$

where

$$\hat{\Phi}_{22} = \begin{bmatrix} \mathcal{Z} & 0 & 0 & 0 & 0 & \lambda \mathcal{C}^T \Lambda D \\ * & \mathcal{R} & 0 & 0 & 0 & 0 \\ * & * & \mathcal{Q} & 0 & 0 & 0 \\ * & * & * & -\lambda I & 0 & 0 \\ * & * & * & * & -\gamma^2 I & 0 \\ * & * & * & * & * & \mathcal{Z}_6 \end{bmatrix}, \quad \mathcal{C} = \begin{bmatrix} C & -\bar{C} \end{bmatrix},$$

$$\begin{aligned}
&\mathcal{Q} = -\frac{1}{\bar{\kappa}} Q + \beta(1-\beta) S^T E^T T_j P_1 T_j E S, \quad \mathcal{Z}_6 = -\gamma^2 I + \lambda D^T \Lambda D, \\
&\bar{\Pi}_1 = \begin{bmatrix} \mathcal{K}_1 & \mathcal{B} & \mathcal{E} & -\mathcal{K}_2 & 0 & \mathcal{K}_3 \end{bmatrix}, \quad \Lambda = \mathrm{diag}\{\sigma_1 I, \sigma_2 I, \ldots, \sigma_N I\}, \\
&\bar{\Pi}_{2j} = \begin{bmatrix} T_j A S + \aleph T_j^- G \aleph^{-1} S & \alpha T_j B S & \beta T_j E S & 0 & T_j L & 0 \end{bmatrix}, \\
&\mathcal{Z} = -P + \bar{\kappa} Q + (1 + \tau_M - \tau_m) R + \lambda \mathcal{C}^T \Lambda \mathcal{C} + \mathcal{M}^T \mathcal{M}, \\
&\mathcal{R} = -R + \alpha(1-\alpha) S^T B^T T_j P_1 T_j B S.
\end{aligned} \tag{5.17}$$

Then, we have that 1) the augmented system (5.12) (with $w(k) = 0$ and $v(k) = 0$) is exponentially stable in the mean square, and 2) under zero-initial condition, the H_∞ performance constraint (5.14) is satisfied for all nonzero $w(k)$ and $v(k)$.

Proof *To prove the stability of the augmented system (5.12) (with $w(k) = 0$ and $v(k) = 0$), we consider the following Lyapunov functional candidate:*

$$V(k) = V_1(k) + V_2(k) + V_3(k) \tag{5.18}$$

where

$$V_1(k) = \vec{x}^T(k) P \vec{x}(k), \quad V_2(k) = \sum_{s=1}^{+\infty} \kappa_s \sum_{i=k-s}^{k-1} \vec{x}^T(i) Q \vec{x}(i),$$

$$V_3(k) = \sum_{i=k-\tau(k)}^{k-1} \vec{x}^T(i) R \vec{x}(i) + \sum_{j=k-\tau_M+1}^{k-\tau_m} \sum_{i=j}^{k-1} \vec{x}^T(i) R \vec{x}(i).$$

Let $\mathcal{H}_1(k) = \Big[\sigma^T(AS\vec{x}(k) + \alpha(k)BS\vec{x}(k-\tau(k)) + \beta(k)ES\sum_{s=1}^{+\infty}\kappa_s\vec{x}(k-s)) \quad 0\Big]^T$. *From (5.12), one has*

$$
\begin{aligned}
&\mathbb{E}\{\Delta V_1(k)\}\\
&= \mathbb{E}\{V_1(k+1) - V_1(k)\}\\
&= \mathbb{E}\Big\{\Big[\mathcal{K}_1\vec{x}(k) + \mathcal{B}\vec{x}(k-\tau(k)) + \mathcal{E}\sum_{s=1}^{+\infty}\kappa_s\vec{x}(k-s) + \mathcal{H}_1(k) - \mathcal{K}_2\varepsilon(k)\Big]^T P\\
&\quad \times\Big[\mathcal{K}_1\vec{x}(k) + \mathcal{B}\vec{x}(k-\tau(k)) + \mathcal{E}\sum_{s=1}^{+\infty}\kappa_s\vec{x}(k-s) + \mathcal{H}_1(k) - \mathcal{K}_2\varepsilon(k)\Big]\\
&\quad -\vec{x}^T(k)P\vec{x}(k)\Big\}\\
&= \mathbb{E}\Big\{\Big[\mathcal{K}_1\vec{x}(k) + \mathcal{B}\vec{x}(k-\tau(k)) + \mathcal{E}\sum_{s=1}^{+\infty}\kappa_s\vec{x}(k-s) - \mathcal{K}_2\varepsilon(k)\Big]^T P\\
&\quad \times\Big[\mathcal{K}_1\vec{x}(k) + \mathcal{B}\vec{x}(k-\tau(k)) + \mathcal{E}\sum_{s=1}^{+\infty}\kappa_s\vec{x}(k-s) - \mathcal{K}_2\varepsilon(k)\Big]\\
&\quad +\sigma^T\Big(AS\vec{x}(k) + \alpha(k)BS\vec{x}(k-\tau(k)) + \beta(k)ES\sum_{s=1}^{+\infty}\kappa_s\vec{x}(k-s)\Big)P_1\\
&\quad \times\sigma\Big(AS\vec{x}(k) + \alpha(k)BS\vec{x}(k-\tau(k)) + \beta(k)ES\sum_{s=1}^{+\infty}\kappa_s\vec{x}(k-s)\Big)\\
&\quad -\vec{x}^T(k)P\vec{x}(k)\Big\}.
\end{aligned}
\tag{5.19}
$$

From Lemmas 5.1 and 5.2, it is easily known that

$$
\begin{aligned}
&\mathbb{E}\{\Delta V_1(k)\}\\
&= \mathbb{E}\Big\{\Big(\sum_{j=1}^{2^{nx}}\delta_j\,[T_j\aleph^{-1}(AS\vec{x}(k) + \alpha(k)BS\vec{x}(k-\tau(k)) + \beta(k)E\sum_{s=1}^{+\infty}\kappa_sS\vec{x}(k-s))\\
&\quad +T_j^-G\aleph^{-1}S\vec{x}(k)]^T\Big)\aleph P_1\aleph\Big(\sum_{j=1}^{2^{nx}}\delta_j\,[T_j\aleph^{-1}(AS\vec{x}(k) + \alpha(k)BS\vec{x}(k-\tau(k))\\
&\quad +\beta(k)E\sum_{s=1}^{+\infty}\kappa_sS\vec{x}(k-s)) + T_j^-G\aleph^{-1}S\vec{x}(k)]\Big) + \Big[\mathcal{K}_1\vec{x}(k) + \mathcal{B}\vec{x}(k-\tau(k))\\
&\quad +\mathcal{E}\sum_{s=1}^{+\infty}\kappa_s\vec{x}(k-s) - \mathcal{K}_2\varepsilon(k)\Big]^T P\Big[\mathcal{K}_1\vec{x}(k) + \mathcal{B}\vec{x}(k-\tau(k))\\
&\quad +\mathcal{E}\sum_{s=1}^{+\infty}\kappa_s\vec{x}(k-s) - \mathcal{K}_2\varepsilon(k)\Big] - \vec{x}^T(k)P\vec{x}(k)\Big\}
\end{aligned}
$$

$$\leq \max_{j\in[1,2^{n_x}]} \mathbb{E}\Big\{\Big[T_j\aleph^{-1}AS\vec{x}(k)+T_j^{-}G\aleph^{-1}S\vec{x}(k)+\alpha T_j\aleph^{-1}BS\vec{x}(k-\tau(k))$$
$$+\beta T_j\aleph^{-1}ES\sum_{s=1}^{+\infty}\kappa_s\vec{x}(k-s)\Big]^T\aleph P_1\aleph\Big[T_j\aleph^{-1}AS\vec{x}(k)+T_j^{-}G\aleph^{-1}S\vec{x}(k)$$
$$+\alpha T_j\aleph^{-1}BS\vec{x}(k-\tau(k))+\beta T_j\aleph^{-1}ES\sum_{s=1}^{+\infty}\kappa_s\vec{x}(k-s)\Big]+\alpha(1-\alpha)$$
$$\times\vec{x}^T(k-\tau(k))S^TB^TT_jP_1T_jBS\vec{x}(k-\tau(k))+\beta(1-\beta)\Big(\sum_{s=1}^{+\infty}\kappa_s\vec{x}(k-s)\Big)^T$$
$$\times S^TE^TT_jP_1T_jES\Big(\sum_{s=1}^{+\infty}\kappa_s\vec{x}(k-s)\Big)-\vec{x}^T(k)P\vec{x}(k)$$
$$+\Big[\mathcal{K}_1\vec{x}(k)+\mathcal{B}\vec{x}(k-\tau(k))+\mathcal{E}\sum_{s=1}^{+\infty}\kappa_s\vec{x}(k-s)-\mathcal{K}_2\varepsilon(k)\Big]^TP$$
$$\times\Big[\mathcal{K}_1\vec{x}(k)+\mathcal{B}\vec{x}(k-\tau(k))+\mathcal{E}\sum_{s=1}^{+\infty}\kappa_s\vec{x}(k-s)-\mathcal{K}_2\varepsilon(k)\Big]\Big\} \tag{5.20}$$

with $\delta_j>0$ *and* $\sum_{j=1}^{2^{n_x}}\delta_j=1$.

Similarly, we obtain

$$\begin{aligned}\mathbb{E}\{\Delta V_2(k)\}=&\mathbb{E}\{V_2(k+1)-V_2(k)\}\\=&\mathbb{E}\Big\{\bar{\kappa}\vec{x}^T(k)Q\vec{x}(k)-\sum_{s=1}^{+\infty}\kappa_s\vec{x}^T(k-s)Q\vec{x}(k-s)\Big\},\end{aligned} \tag{5.21}$$

and

$$\begin{aligned}\mathbb{E}\{\Delta V_3(k)\}=&\mathbb{E}\{V_3(k+1)-V_3(k)\}\\\leq&\mathbb{E}\Big\{\vec{x}^T(k)R\vec{x}(k)-\vec{x}^T(k-\tau(k))R\vec{x}(k-\tau(k))\\&+\sum_{i=k-\tau_M+1}^{k-\tau_m}\vec{x}^T(i)R\vec{x}(i)+(\tau_M-\tau_m)\vec{x}^T(k)R\vec{x}(k)\\&-\sum_{i=k-\tau_M+1}^{k-\tau_m}\vec{x}^T(i)R\vec{x}(i)\Big\}.\end{aligned} \tag{5.22}$$

Noting Lemma 5.3 and condition (5.6), one has

$$\begin{aligned}&-\sum_{s=1}^{+\infty}\kappa_s\vec{x}^T(k-s)Q\vec{x}(k-s)\\&\leq-\frac{1}{\bar{\kappa}}\Big(\sum_{s=1}^{+\infty}\kappa_s\vec{x}(k-s)\Big)^TQ\Big(\sum_{s=1}^{+\infty}\kappa_s\vec{x}(k-s)\Big),\end{aligned} \tag{5.23}$$

and

$$\varepsilon^T(k)\varepsilon(k) - \vec{x}^T(k)\mathcal{C}^T\Lambda\mathcal{C}\vec{x}(k) \leq 0. \tag{5.24}$$

By considering (5.20)-(5.24), it can be obtained that

$$\begin{aligned}
&\mathbb{E}\{\Delta V(k)\} \\
&= \mathbb{E}\{V(k+1) - V(k)\} \\
&= \mathbb{E}\{\Delta V_1(k) + \Delta V_2(k) + \Delta V_3(k)\} \\
&\leq \max_{j\in[1,2^{n_x}]} \mathbb{E}\Big\{\Big[T_j\aleph^{-1}AS\vec{x}(k) + T_j^- G\aleph^{-1}S\vec{x}(k) + \alpha T_j\aleph^{-1}BS\vec{x}(k-\tau(k)) \\
&\quad +\beta T_j\aleph^{-1}ES\sum_{s=1}^{+\infty}\kappa_s\vec{x}(k-s)\Big]^T \aleph P_1\aleph\Big[T_j\aleph^{-1}AS\vec{x}(k) + T_j^- G\aleph^{-1}S\vec{x}(k) \\
&\quad +\alpha T_j\aleph^{-1}BS\vec{x}(k-\tau(k)) + \beta T_j\aleph^{-1}ES\sum_{s=1}^{+\infty}\kappa_s\vec{x}(k-s)\Big] + \alpha(1-\alpha) \\
&\quad \times\vec{x}^T(k-\tau(k))S^TB^TT_jP_1T_jBS\vec{x}(k-\tau(k)) + \beta(1-\beta)\Big(\sum_{s=1}^{+\infty}\kappa_s\vec{x}(k-s)\Big)^T \\
&\quad \times S^TE^TT_jP_1T_jES\Big(\sum_{s=1}^{+\infty}\kappa_s\vec{x}(k-s)\Big) - \vec{x}^T(k)P\vec{x}(k) + \Big[\mathcal{K}_1\vec{x}(k) - \mathcal{K}_2\varepsilon(k) \\
&\quad +\mathcal{B}\vec{x}(k-\tau(k)) + \mathcal{E}\sum_{s=1}^{+\infty}\kappa_s\vec{x}(k-s)\Big]^T P\Big[\mathcal{K}_1\vec{x}(k) + \mathcal{B}\vec{x}(k-\tau(k)) - \mathcal{K}_2\varepsilon(k) \\
&\quad +\mathcal{E}\sum_{s=1}^{+\infty}\kappa_s\vec{x}(k-s)\Big] + \bar{\kappa}\vec{x}^T(k)Q\vec{x}(k) + \vec{x}^T(k)R\vec{x}(k) - \lambda\varepsilon^T(k)\varepsilon(k) \\
&\quad -\frac{1}{\bar{\kappa}}\Big(\sum_{s=1}^{+\infty}\kappa_s\vec{x}(k-s)\Big)^T Q\Big(\sum_{s=1}^{+\infty}\kappa_s\vec{x}(k-s)\Big) + \lambda\vec{x}^T(k)\mathcal{C}^T\Lambda\mathcal{C}\vec{x}(k) \\
&\quad -\vec{x}^T(k-\tau(k))R\vec{x}(k-\tau(k)) + (\tau_M - \tau_m)\vec{x}^T(k)R\vec{x}(k)\Big\} \\
&= \max_{j\in[1,2^{n_x}]} \mathbb{E}\Big\{(\xi^T(k)(\Phi_{22} + \Pi_1^TP\Pi_1 + \Pi_{2j}^TP_1\Pi_{2j})\xi(k))\Big\} \\
&= \max_{j\in[1,2^{n_x}]} \mathbb{E}\Big\{\xi^T(k)\Omega_j\xi(k)\Big\}
\end{aligned} \tag{5.25}$$

where

$$\begin{aligned}
&\Omega_j = \Phi_{22} + \Pi_1^T P \Pi_1 + \Pi_{2j}^T P_1 \Pi_{2j},\\
&\mathcal{Z} = -P + \bar{\kappa} Q + (1 + \tau_M - \tau_m) R + \lambda C^T \Lambda C,\\
&\Pi_{2j} = \begin{bmatrix} T_j A S + \aleph T_j^- G \aleph^{-1} S & \alpha T_j B S & \beta T_j E S & 0 \end{bmatrix},\\
&\Phi_{22} = \mathrm{diag}\{\mathcal{Z}, \mathcal{R}, \mathcal{Q}, -\lambda I\}, \quad \Pi_1 = \begin{bmatrix} \mathcal{K}_1 & \mathcal{B} & \mathcal{E} & -\mathcal{K}_2 \end{bmatrix},\\
&\xi(k) = \begin{bmatrix} \vec{x}^T(k) & \vec{x}^T(k-\tau(k)) & \sum_{s=1}^{+\infty} \kappa_s \vec{x}^T(k-s) & \varepsilon^T(k) \end{bmatrix}^T.
\end{aligned}$$

By using Schur complement, it follows from (5.16) that $\Omega_j < 0$. Then, based on the analysis in [175], the exponentially mean-square stability of the augmented system (5.12) can be guaranteed.

Next, for all nonzero $w(k) \in l_2[0, \infty)$ and $v(k) \in l_2[0, \infty)$, it follows from (5.16)–(5.25) that

$$\begin{aligned}
&\mathbb{E}\left\{\vec{z}^T(k)\vec{z}(k) - \gamma^2 w^T(k)w(k) - \gamma^2 v^T(k)v(k) + \Delta V(k)\right\}\\
=&\mathbb{E}\left\{\vec{x}^T(k)\mathcal{M}^T\mathcal{M}\vec{x}(k) - \gamma^2 w^T(k)w(k) - \gamma^2 v^T(k)v(k) + \Delta V(k)\right\}\\
\leq&\mathbb{E}\left\{\bar{\xi}^T(k)(\hat{\Phi}_{22} + \bar{\Pi}_1^T P \bar{\Pi}_1 + \bar{\Pi}_{2j}^T P_1 \bar{\Pi}_{2j})\bar{\xi}(k)\right\}\\
<&0
\end{aligned} \tag{5.26}$$

where $\bar{\xi}(k) = \begin{bmatrix} \xi^T(k) & w^T(k) & v^T(k) \end{bmatrix}^T$.

Summing up (5.26) from 0 to $+\infty$ with respect to k yields

$$\begin{aligned}
&\sum_{k=0}^{+\infty} \mathbb{E}\{\vec{z}^T(k)\vec{z}(k)\}\\
<&\gamma^2 \sum_{k=0}^{+\infty} w^T(k)w(k) + \gamma^2 \sum_{k=0}^{+\infty} v^T(k)v(k) + \mathbb{E}\{V(0)\} - \mathbb{E}\{V(\infty)\}.
\end{aligned} \tag{5.27}$$

Under the zero-initial condition and considering the stability of the system (5.12), it is easy to see that

$$\sum_{k=0}^{+\infty} \mathbb{E}\{\vec{z}^T(k)\vec{z}(k)\} < \gamma^2 \sum_{k=0}^{+\infty} w^T(k)w(k) + \gamma^2 \sum_{k=0}^{+\infty} v^T(k)v(k). \tag{5.28}$$

Then, the H_∞ performance requirement (5.14) is satisfied and, therefore, the proof of Theorem 5.1 is complete.

Next, according to the results obtained in Theorem 5.1, the design method of the desired estimators is given in the following theorem.

Theorem 5.2 *For the given $\gamma > 0$, if there exist matrices $P = \mathrm{diag}\{P_1, P_2\} > 0$, $P_2 = \mathrm{diag}\{P_{21}, \ldots, P_{2N}\} > 0$, $Q > 0$, $R > 0$, G, $\bar{X} = \mathrm{diag}\{\bar{X}_1, \ldots, \bar{X}_N\}$ and $X \in \mathcal{L}_{n_x \times n_y}$, and a scalar $\lambda > 0$ satisfying*

$$\|G\|_\infty \leq 1, \tag{5.29}$$

$$\breve{\Phi}_j = \begin{bmatrix} -P & 0 & \breve{\Pi}_1 \\ * & -P_1 & P_1\bar{\Pi}_{2j} \\ * & * & \hat{\Phi}_{22} \end{bmatrix} < 0, \quad j \in [1, 2^{n_x}] \tag{5.30}$$

where

$$\breve{\Pi}_1 = \begin{bmatrix} \bar{\mathcal{X}}_1 & P\mathcal{B} & P\mathcal{E} & -\bar{\mathcal{X}}_2 & 0 & \bar{\mathcal{X}}_3 \end{bmatrix}, \quad \bar{\mathcal{X}}_3 = \begin{bmatrix} 0 \\ (\bar{X} + X)D \end{bmatrix},$$

$$\bar{\mathcal{X}}_1 = \begin{bmatrix} 0 & 0 \\ (\bar{X} + X)C & P_2\bar{A} - (\bar{X} + X)\bar{C} \end{bmatrix}, \quad \bar{\mathcal{X}}_2 = \begin{bmatrix} 0 \\ X \end{bmatrix},$$

and $\bar{\Pi}_{2j}$ and $\hat{\Phi}_{22}$ are defined in (5.17), then the design problem of the desired distributed estimators is solved for the state-saturated systems (5.1). In this case, the distributed estimator gains are given by

$$\bar{K} = P_2^{-1}\bar{X}, \quad K = P_2^{-1}X. \tag{5.31}$$

Accordingly, the desired estimator parameters K_i and K_{ij} ($i \in \mathcal{V}$, $j \in \mathcal{N}_i$) can be obtained.

Proof *Pre- and post-multiplying the inequality (5.16) by* $\mathrm{diag}\{P, P_1, I\}$, *we can obtain that*

$$\begin{bmatrix} -P & 0 & P\bar{\Pi}_1 \\ * & -P_1 & P_1\bar{\Pi}_{2j} \\ * & * & \hat{\Phi}_{22} \end{bmatrix} < 0, \quad j \in [1, 2^{n_x}]. \tag{5.32}$$

Then, by letting $\bar{K} = P_2^{-1}\bar{X}$ and $K = P_2^{-1}X$, the inequality (5.30) can be obtained readily. Moreover, it follows from Lemma 5.4 that $K \in \mathcal{L}_{n_x \times n_y}$. The proof of this theorem is now complete.

Remark 5.1 *In Theorem 5.2, a sufficient condition is derived under which the expected stability and the H_∞ performance constraint on the augmented system (5.12) are guaranteed. The corresponding solvability conditions for the desired estimator gains are expressed in terms of the feasibility of certain matrix inequalities (5.29) and (5.30) that can be solved by using the available software package. It is worth mentioning that the conservativeness of the derived condition could be reduced further by using the advanced technologies to tackle time-delays such as the Wirtinger-based summation inequality approach proposed in [130].*

Remark 5.2 *In Theorems 5.1 and 5.2, the ET distributed H_∞ state estimation issue has been solved for a class of discrete-time state-saturated systems subject to ROMDs over sensor networks and an algorithm for designing the desired estimators has been given. Note that our main results obtained in Theorems 5.1 and 5.2 are comprehensive since all the information on the target plant and the sensor network (i.e. the plant parameters, the network topology, the levels of the saturation, the thresholds of the event-triggers as well as the bounds and occurrence probabilities of the mixed delays) have been reflected.*

5.2 Event-Triggered H_∞ State Estimation in Complex Networks

5.2.1 Problem Formulation

Consider a class of complex dynamical networks consisting of N coupled nodes as follows:

$$\begin{cases} x_i(k+1) = \sigma\Big(A_i x_i(k) + \sum_{j=1}^{N} w_{ij}\Gamma x_j(k) + \alpha_i(k) E_i \sum_{l=1}^{+\infty} k_l \\ \qquad \times x_i(k-l) + B_i w(k)\Big) \\ z_i(k) = F_i x_i(k), \quad i = 1, 2, \ldots, N \end{cases} \tag{5.33}$$

where, for the ith node, $x_i(k) \in \mathbb{R}^{n_x}$ and $z_i(k) \in \mathbb{R}^{n_z}$ are the state vector and the signal to be estimated, respectively. The exogenous disturbance $w(k) \in \mathbb{R}^{n_w}$ belongs to $l_2[0, +\infty)$. $W = [w_{ij}]_{N\times N}$ is the coupled configuration matrix of the network with $w_{ij} \geq 0$ $(i \neq j)$ but not all zeros, and W is assumed to be symmetric and satisfy $w_{ii} = -\sum_{j=1, j\neq i}^{N} w_{ij}$ for all $i = 1, 2, \ldots, N$. $\Gamma = \mathrm{diag}\{\iota_1, \iota_2, \ldots, \iota_{n_x}\}$ is an inner-coupling matrix. A_i, B_i, E_i and F_i are known matrices of appropriate dimensions. The initial value is set as $x_i(-l) = 0$ for $l = 1, 2, \ldots$ and $x_i(0) = \bar{x}_i$. The non-negative constants k_l $(l = 1, 2, \ldots)$ satisfy

$$\bar{k} = \sum_{l=1}^{+\infty} k_l < +\infty \quad \text{and} \quad \sum_{l=1}^{+\infty} l k_l < +\infty. \tag{5.34}$$

The saturated function $\sigma(\cdot) : \mathbb{R}^{n_x} \to \mathbb{R}^{n_x}$ is defined as follows:

$$\sigma(\mu) = \begin{bmatrix} \sigma_1(\mu_1) & \sigma_2(\mu_2) & \cdots & \sigma_{n_x}(\mu_{n_x}) \end{bmatrix}^T, \quad \forall \mu \in \mathbb{R}^{n_x} \tag{5.35}$$

with $\sigma_i(\mu_i) = \text{sign}(\mu_i)\min\{\delta_i, |\mu_i|\}$, where "sign" means the signum function, μ_i is the ith element of the vector μ and δ_i stands for the saturation level. $\alpha_i(k) \in \mathbb{R}$, which governs the phenomenon of randomly occurring distributed delays, is a Bernoulli distributed random variable satisfying

$$\text{Prob}\{\alpha_i(k) = 1\} = \bar{\alpha}_i \quad \text{and} \quad \text{Prob}\{\alpha_i(k) = 0\} = 1 - \bar{\alpha}_i \tag{5.36}$$

with $\bar{\alpha}_i \in [0, 1]$ being a given scalar. In this section, we assume that the phenomenon of randomly occurring distributed delays for each node occurs independently, that is, the random variables $\alpha_1(k), \alpha_2(k), \dots, \alpha_N(k)$ are mutually independent.

Remark 5.3 *For complex networks, the distributed delays are likely to take place due to the spatial nature of information transmissions, see e.g [106, 166, 221]. Recently, the random occurrence of such distributed delays has been considered in the context of fuzzy systems [184] and neural networks [192]. In this case, there is a practical need to introduce the randomly occurring distributed delays into the complex networks. Therefore, in this section, a series of distributed delays having different occurrence probabilities are proposed.*

In this section, the quantization effect on measurement is considered as follows:

$$y_i(k) = q\left(C_i x_i(k)\right) + D_i v(k), \quad i = 1, 2, \dots, N \tag{5.37}$$

where $y_i(k) \in \mathbb{R}^{n_y}$ is the measurement of the ith node and $v(k) \in \mathbb{R}^{n_v}$ is the measurement noise belonging to $l_2[0, +\infty)$. C_i and D_i are known constant matrices. $q(\cdot) : \mathbb{R}^{n_y} \to \mathbb{R}^{n_y}$ is the quantization function given by

$$q(\vartheta) = \begin{bmatrix} q_1(\vartheta_1) & q_2(\vartheta_2) & \dots & q_{n_y}(\vartheta_{n_y}) \end{bmatrix}^T, \quad \forall \vartheta \in \mathbb{R}^{n_y}. \tag{5.38}$$

The set of quantization levels, for each $q_j(\cdot)$ $(j = 1, 2, \dots, n_y)$, is described by

$$\mathcal{U}_j = \{\pm u_i^{(j)}, u_i^{(j)} = \rho_j^i u_0^{(j)}, i = 0, \pm 1, \pm 2, \cdots\} \cup \{0\}, \quad 0 < \rho_j < 1, \quad u_0^{(j)} > 0,$$

and the logarithmic quantizer $q_j(\cdot)$ is defined as

$$q_j(\vartheta_j) = \begin{cases} u_i^{(j)}, & \text{if } \dfrac{1}{1+\delta_j} u_i^{(j)} \le \vartheta_j \le \dfrac{1}{1-\delta_j} u_i^{(j)} \\ 0, & \text{if } \vartheta_j = 0 \\ -q_j(-\vartheta_j), & \text{if } \vartheta_j < 0 \end{cases}$$

where $\delta_j = (1 - \rho_j)/(1 + \rho_j)$. It follows from [141] that $q_j(\vartheta_j) = (1 + \theta_j)\vartheta_j$ with $\theta_j \in [-\delta_j, \delta_j]$.

For the sake of mitigating network communication burden, an ET transmission mechanism is used before the quantized measurement enters into the estimator. Denote by $0 = t_0^i < t_1^i < \ldots < t_s^i < \ldots$ the triggering instant sequence of the ith node that is determined by:

$$t_{s+1}^i = \min\left\{k | k > t_s^i,\ \sigma_i y_i^T(k) y_i(k) < (y_i(k) - y_i(t_s^i))^T (y_i(k) - y_i(t_s^i))\right\} \tag{5.39}$$

where σ_i is a known positive scalar and $y_i(t_s^i)$ is the transmitted measurement at latest event time.

Based on the ET measurement, the state estimator on the ith node is constructed as follows:

$$\begin{cases} \hat{x}_i(k+1) = A_i\hat{x}_i(k) + \sum_{j=1}^{N} w_{ij}\Gamma\hat{x}_j(k) + \bar{\alpha}_i E_i \sum_{l=1}^{+\infty} k_l \hat{x}_i(k-l) \\ \qquad\qquad + K_i(y_i(t_s^i) - C_i\hat{x}_i(k)) \\ \hat{z}_i(k) = F_i\hat{x}_i(k) \end{cases} \tag{5.40}$$

for $k \in [t_s^i, t_{s+1}^i)$, $s = 0, 1, 2, \ldots, \infty$, where $\hat{x}_i(k)$ and $\hat{z}_i(k)$ are the estimates of state $x_i(k)$ and output $z_i(k)$ for node i, respectively, and K_i is the estimator parameter to be designed. The initial value of estimator is chosen as $\hat{x}_i(l) = 0$ for $l \in \mathbb{Z}_-$.

Defining $\varepsilon_i(k) = y_i(k) - y_i(t_s^i)$ for $k \in [t_s^i, t_{s+1}^i)$ $(s = 0, 1, 2, \ldots, \infty)$, (5.40) can be rewritten as follows:

$$\begin{cases} \hat{x}_i(k+1) = A_i\hat{x}_i(k) + \sum_{j=1}^{N} w_{ij}\Gamma\hat{x}_j(k) + \bar{\alpha}_i E_i \sum_{l=1}^{+\infty} k_l \hat{x}_i(k-l) \\ \qquad\qquad + K_i\big(q(C_i x_i(k)) + D_i v(k) - \varepsilon_i(k) - C_i\hat{x}_i(k)\big) \\ \hat{z}_i(k) = F_i\hat{x}_i(k). \end{cases} \tag{5.41}$$

For convenience of later analysis, we denote

$$\begin{aligned}
&A = \mathrm{diag}\{A_1, A_2, \ldots, A_N\}, \quad x(k) = \begin{bmatrix} x_1^T(k) & x_2^T(k) & \cdots & x_N^T(k)\end{bmatrix}^T, \\
&E = \mathrm{diag}\{E_1, E_2, \ldots, E_N\}, \quad z(k) = \begin{bmatrix} z_1^T(k) & z_2^T(k) & \cdots & z_N^T(k)\end{bmatrix}^T, \\
&C = \mathrm{diag}\{C_1, C_2, \ldots, C_N\}, \quad \hat{x}(k) = \begin{bmatrix} \hat{x}_1^T(k) & \hat{x}_2^T(k) & \cdots & \hat{x}_N^T(k)\end{bmatrix}^T, \\
&K = \mathrm{diag}\{K_1, K_2, \ldots, K_N\}, \quad \varepsilon(k) = \begin{bmatrix} \varepsilon_1^T(k) & \varepsilon_2^T(k) & \cdots & \varepsilon_N^T(k)\end{bmatrix}^T, \\
&D = \begin{bmatrix} D_1^T & D_2^T & \ldots & D_N^T\end{bmatrix}^T, \quad \hat{z}(k) = \begin{bmatrix} \hat{z}_1^T(k) & \hat{z}_2^T(k) & \cdots & \hat{z}_N^T(k)\end{bmatrix}^T, \\
&B = \begin{bmatrix} B_1^T & B_2^T & \cdots & B_N^T\end{bmatrix}^T, \quad F = \mathrm{diag}\{F_1, F_2, \ldots, F_N\}, \\
&\alpha = \mathrm{diag}\{\bar{\alpha}_1 I, \bar{\alpha}_2 I, \ldots, \bar{\alpha}_N I\}, \quad \Im = \mathrm{diag}\{\delta_1, \ldots, \delta_{Nn_x}\}, \\
&I_i = \mathrm{diag}\{\underbrace{0, \cdots, 0}_{i-1}, I, \underbrace{0, \cdots, 0}_{N-i}\}.
\end{aligned}$$

Letting $\vec{x}(k) = \begin{bmatrix} x^T(k) & \hat{x}^T(k) \end{bmatrix}^T$ and $\vec{z}(k) = z(k) - \hat{z}(k)$, an augmented system is obtained as follows:

$$\begin{cases} \vec{x}(k+1) = \mathcal{K}_1\vec{x}(k) + \mathcal{E}\sum_{l=1}^{+\infty} k_l\vec{x}(k-l) + \mathcal{K}_2 v(k) \\ \qquad\qquad - \mathcal{K}_3\varepsilon(k) + \mathcal{K}_3 q(\mathcal{C}\vec{x}(k)) + \mathcal{L}(k) \\ \vec{z}(k) = \mathcal{F}\vec{x}(k) \end{cases} \tag{5.42}$$

where

$$\mathcal{E} = \text{diag}\{0, \alpha E\}, \quad \mathcal{F} = \begin{bmatrix} F & -F \end{bmatrix}, \quad \mathcal{C} = \begin{bmatrix} C & 0 \end{bmatrix}, \quad \mathcal{L}(k) = \begin{bmatrix} L(k) \\ 0 \end{bmatrix},$$

$$L(k) = \sigma\Big((A + W\otimes\Gamma)x(k) + Bw(k) + \sum_{i=1}^{N}(\alpha_i(k) - \bar{\alpha}_i)I_i E\sum_{l=1}^{+\infty} k_l x(k-l) + \alpha E\sum_{l=1}^{+\infty} k_l x(k-l)\Big),$$

$$\mathcal{K}_1 = \begin{bmatrix} 0 & 0 \\ 0 & A + W\otimes\Gamma - KC \end{bmatrix}, \quad \mathcal{K}_2 = \begin{bmatrix} 0 \\ KD \end{bmatrix}, \quad \mathcal{K}_3 = \begin{bmatrix} 0 \\ K \end{bmatrix}.$$

By defining $\Theta = \text{diag}\{\theta_1, \theta_2, \ldots, \theta_{Nn_y}\}$, $\Upsilon = \text{diag}\{\delta_1, \delta_2, \ldots, \delta_{Nn_y}\}$ and $\mathcal{T} = \Theta\Upsilon^{-1}$, the quantization effect can be expressed as $q(\mathcal{C}\vec{x}(k)) = (I + \mathcal{T}\Upsilon)\mathcal{C}\vec{x}(k)$ with $\mathcal{T}^T\mathcal{T} = \mathcal{T}\mathcal{T}^T \leq I$. Then, we rewrite (5.42) as follows:

$$\begin{cases} \vec{x}(k+1) = (\mathcal{K}_1 + \mathcal{K}_3\mathcal{C} + \mathcal{K}_3\mathcal{T}\Upsilon\mathcal{C})\vec{x}(k) + \mathcal{E}\sum_{l=1}^{+\infty} k_l\vec{x}(k-l) \\ \qquad\qquad - \mathcal{K}_3\varepsilon(k) + \mathcal{L}(k) + \mathcal{K}_2 v(k) \\ \vec{z}(k) = \mathcal{F}\vec{x}(k). \end{cases} \tag{5.43}$$

In this section, we aim at designing a set of state estimator parameters K_i $(i = 1, 2, \ldots, N)$ for the complex dynamical network (5.33) to achieve the following two requirements.

a) The augmented dynamics (5.43) with $w(k) = 0$ and $v(k) = 0$ is exponentially stable in the mean square.

b) For all nonzero $w(k)$ and $v(k)$, under the zero-initial condition, the output estimation error satisfies the following H_∞ performance constraint:

$$\mathbb{E}\left\{\sum_{k=0}^{+\infty}\|\vec{z}(k)\|^2\right\} < \gamma^2\sum_{k=0}^{+\infty}\|v(k)\|^2 + \gamma^2\sum_{k=0}^{+\infty}\|w(k)\|^2 \tag{5.44}$$

with $\gamma > 0$ being a given disturbance attenuation level.

5.2.2 Main Results

In this subsection, a sufficient condition is first derived to make sure that the augmented dynamics (5.43) is exponentially mean-square stable with the given H_∞ disturbance attenuation level γ, and a set of ET state estimators is then designed for the complex dynamical network (5.33). In preparation for stating our main results, the following lemma is introduced that would be used in sequel.

Lemma 5.5 *[65] Let $\mathcal{Y}_{Nn_x}$ be the set of $Nn_x \times Nn_x$ diagonal matrices whose diagonal elements are either 1 or 0. 1) There are 2^{Nn_x} elements in $\mathcal{Y}_{Nn_x}$, where its ith element is denoted as Y_i, $i \in [1, 2^{Nn_x}]$. 2) By defining $Y_i^- = I - Y_i$ and letting $G \in \mathbb{R}^{Nn_x \times Nn_x}$ satisfy $\|G\|_\infty \leq 1$, for any vector $\nu \in \mathbb{R}^{Nn_x}$, we have*

$$\Im^{-1}\sigma(Ax(k)+\nu) \in co\left\{Y_i\Im^{-1}(Ax(k)+\nu) + Y_i^- G\Im^{-1}x(k), i \in [1, 2^{Nn_x}]\right\},$$

where co$\{\ldots\}$ is the convex hull of a set.

In the following theorem, a sufficient condition is provided that ensures the exponentially mean-square stability with H_∞ performance of the augmented dynamics (5.43).

Theorem 5.3 *Let the estimator gain K and the disturbance level $\gamma > 0$ be given. The augmented dynamics (5.43) (with $w(k) = 0$ and $v(k) = 0$) is exponentially stable in the mean square and the output estimation error satisfies the H_∞ performance constraint (5.44) if there exist matrices $P = \text{diag}\{P_1, P_2\} > 0$, $Q > 0$ and G satisfying*

$$\|G\|_\infty \leq 1, \tag{5.45}$$

$$\Phi_j = \begin{bmatrix} -\Lambda^{-1} & 0 & 0 & \bar{\Xi}_1 \\ * & -P & 0 & P\bar{\Xi}_2 \\ * & * & -P_1 & P_1\bar{\Xi}_{3j} \\ * & * & * & \bar{\Phi}_{44} \end{bmatrix} < 0, \quad j \in [1, 2^{Nn_x}] \tag{5.46}$$

where

$$\begin{aligned}
\bar{\Xi}_1 &= \begin{bmatrix}(I+\mathcal{T}\Upsilon)\mathcal{C} & 0 & 0 & 0 & D\end{bmatrix}, \\
\Lambda &= \text{diag}\{\sigma_1 I, \sigma_2 I, \ldots, \sigma_N I\}, \quad S = \begin{bmatrix} I & 0\end{bmatrix}, \\
\bar{\Xi}_2 &= \begin{bmatrix}\mathcal{K}_1 + \mathcal{K}_3\mathcal{C} + \mathcal{K}_3\mathcal{T}\Upsilon\mathcal{C} & \mathcal{E} & -\mathcal{K}_3 & 0 & \mathcal{K}_2\end{bmatrix}, \\
\mathcal{Q} &= -\frac{1}{\bar{k}}Q + \sum_{i=1}^{N}\bar{\alpha}_i(1-\bar{\alpha}_i)S^T E^T I_i Y_j P_1 Y_j I_i E S, \\
\bar{\mathcal{Z}}_1 &= -P + \bar{k}Q + \mathcal{F}^T\mathcal{F}, \quad \bar{\Phi}_{44} = \text{diag}\{\bar{\mathcal{Z}}_1, \mathcal{Q}, -I, -\gamma^2 I, -\gamma^2 I\}, \\
\bar{\Xi}_{3j} &= \left[Y_j(A + W\otimes\Gamma)S + \Im Y_j^- G\Im^{-1}S \quad Y_j\alpha E S \quad 0 \quad Y_j B \quad 0\right].
\end{aligned} \tag{5.47}$$

Proof *Consider the following Lyapunov functional candidate:*

$$V(k) = \vec{x}^T(k)P\vec{x}(k) + \sum_{l=1}^{+\infty} k_l \sum_{i=k-l}^{k-1} \vec{x}^T(i)Q\vec{x}(i) \tag{5.48}$$

with $P = \text{diag}\{P_1, P_2\}$.

Let $\mathcal{L}_1(k) = \begin{bmatrix} L_1^T(k) & 0 \end{bmatrix}^T$ *with* $L_1(k) = \sigma\big((A + W \otimes \Gamma)x(k) + \alpha E \sum_{l=1}^{+\infty} k_l x(k-l) + \sum_{i=1}^{N}(\alpha_i(k) - \alpha_i)I_i E \sum_{l=1}^{+\infty} k_l x(k-l)\big)$. *Calculating the difference of* $V(k)$ *along the dynamics (5.43) (with* $w(k) = 0$ *and* $v(k) = 0$*) and taking the mathematical expectation, we have*

$$\begin{aligned}
&\mathbb{E}\{\Delta V(k)\}\\
&= \mathbb{E}\{V(k+1) - V(k)\}\\
&= \mathbb{E}\Big\{\Big[(\mathcal{K}_1 + \mathcal{K}_3\mathcal{C} + \mathcal{K}_3\mathcal{T}\Upsilon\mathcal{C})\vec{x}(k) + \mathcal{E}\sum_{l=1}^{+\infty} k_l\vec{x}(k-l) - \mathcal{K}_3\varepsilon(k) + \mathcal{L}_1(k)\Big]^T P\\
&\quad \times\Big[(\mathcal{K}_1 + \mathcal{K}_3\mathcal{C} + \mathcal{K}_3\mathcal{T}\Upsilon\mathcal{C})\vec{x}(k) + \mathcal{E}\sum_{l=1}^{+\infty} k_l\vec{x}(k-l) - \mathcal{K}_3\varepsilon(k) + \mathcal{L}_1(k)\Big]\\
&\quad -\vec{x}^T(k)P\vec{x}(k) + \bar{k}\vec{x}^T(k)Q\vec{x}(k) - \sum_{l=1}^{+\infty} k_l\vec{x}^T(k-l)Q\vec{x}(k-l)\Big\}\\
&= \mathbb{E}\Big\{\Big[(\mathcal{K}_1 + \mathcal{K}_3\mathcal{C} + \mathcal{K}_3\mathcal{T}\Upsilon\mathcal{C})\vec{x}(k) + \mathcal{E}\sum_{l=1}^{+\infty} k_l\vec{x}(k-l) - \mathcal{K}_3\varepsilon(k)\Big]^T P\\
&\quad \times\Big[(\mathcal{K}_1 + \mathcal{K}_3\mathcal{C} + \mathcal{K}_3\mathcal{T}\Upsilon\mathcal{C})\vec{x}(k) + \mathcal{E}\sum_{l=1}^{+\infty} k_l\vec{x}(k-l) - \mathcal{K}_3\varepsilon(k))\Big]\\
&\quad -\vec{x}^T(k)P\vec{x}(k) + \sigma^T\Big((A + W \otimes \Gamma)x(k) + \alpha E\sum_{l=1}^{+\infty} k_l x(k-l)\\
&\quad + \sum_{i=1}^{N}(\alpha_i(k) - \bar{\alpha}_i)I_i E\sum_{l=1}^{+\infty} k_l x(k-l)\Big)P_1\sigma\Big((A + W \otimes \Gamma)x(k)\\
&\quad +\alpha E\sum_{l=1}^{+\infty} k_l x(k-l) + \sum_{i=1}^{N}(\alpha_i(k) - \bar{\alpha}_i)I_i E\sum_{l=1}^{+\infty} k_l x(k-l)\Big)\\
&\quad +\bar{k}\vec{x}^T(k)Q\vec{x}(k) - \sum_{l=1}^{+\infty} k_l\vec{x}^T(k-l)Q\vec{x}(k-l)\Big\}.
\end{aligned} \tag{5.49}$$

In terms of Lemma 5.5 and Lemma 5.2, one has

$$\begin{aligned}
&\mathbb{E}\{\Delta V(k)\}\\
&= \mathbb{E}\Big\{\Big(\sum_{j=1}^{2^{Nn_x}} \delta_j\big[Y_j\Im^{-1}\big((A + W \otimes \Gamma)S\vec{x}(k) + \sum_{i=1}^{N}(\alpha_i(k) - \bar{\alpha}_i)I_i E\sum_{l=1}^{+\infty} k_l
\end{aligned}$$

$$
\begin{aligned}
&\times S\vec{x}(k-l)+\alpha E\sum_{l=1}^{+\infty}k_sS\vec{x}(k-s))+Y_j^-G\Im^{-1}S\vec{x}(k)]^T\Big)\Im P_1\Im\\
&\times\Big(\sum_{j=1}^{2^{Nn_x}}\delta_j\big[Y_j\Im^{-1}\big((W\otimes\Gamma+A)S\vec{x}(k)+\alpha E\sum_{l=1}^{+\infty}k_lS\vec{x}(k-l)\\
&+\sum_{i=1}^{N}(\alpha_i(k)-\bar{\alpha}_i)I_iE\sum_{l=1}^{+\infty}k_lS\vec{x}(k-l)\big)+Y_j^-G\Im^{-1}S\vec{x}(k)\big]\Big)\\
&-\vec{x}^T(k)P\vec{x}(k)+\bar{k}\vec{x}^T(k)Q\vec{x}(k)-\sum_{l=1}^{+\infty}k_l\vec{x}^T(k-l)Q\vec{x}(k-l)\\
&+\Big[(\mathcal{K}_1+\mathcal{K}_3\mathcal{C}+\mathcal{K}_3\mathcal{T}\Upsilon\mathcal{C})\vec{x}(k)+\mathcal{E}\sum_{l=1}^{+\infty}k_l\vec{x}(k-l)-\mathcal{K}_3\varepsilon(k)\Big]^T\\
&\times P\Big[(\mathcal{K}_1+\mathcal{K}_3\mathcal{C}+\mathcal{K}_3\mathcal{T}\Upsilon\mathcal{C})\vec{x}(k)+\mathcal{E}\sum_{l=1}^{+\infty}k_l\vec{x}(k-l)-\mathcal{K}_3\varepsilon(k)\Big]\Big\}\\
\leq&\max_{j\in[1,2^{Nn_x}]}\mathbb{E}\Big\{\Big[Y_j\Im^{-1}(A+W\otimes\Gamma)S\vec{x}(k)+Y_j\Im^{-1}\alpha E\sum_{l=1}^{+\infty}k_lS\vec{x}(k-l)\\
&+Y_j^-G\Im^{-1}S\vec{x}(k)\Big]^T\Im P_1\Im\Big[Y_j\Im^{-1}(A+W\otimes\Gamma)S\vec{x}(k)\\
&+Y_j\Im^{-1}\alpha E\sum_{l=1}^{+\infty}k_lS\vec{x}(k-l)+Y_j^-G\Im^{-1}S\vec{x}(k)\Big]-\vec{x}^T(k)P\vec{x}(k)\\
&+\Big[(\mathcal{K}_1+\mathcal{K}_3\mathcal{C}+\mathcal{K}_3\mathcal{T}\Upsilon\mathcal{C})\vec{x}(k)+\mathcal{E}\sum_{l=1}^{+\infty}k_l\vec{x}(k-l)-\mathcal{K}_3\varepsilon(k)\Big]^TP\\
&\times\Big[(\mathcal{K}_1+\mathcal{K}_3\mathcal{C}+\mathcal{K}_3\mathcal{T}\Upsilon\mathcal{C})\vec{x}(k)+\mathcal{E}\sum_{l=1}^{+\infty}k_l\vec{x}(k-l)-\mathcal{K}_3\varepsilon(k)\Big]\\
&+\sum_{i=1}^{N}\bar{\alpha}_i(1-\bar{\alpha}_i)\Big(Y_jI_iE\sum_{l=1}^{+\infty}k_lS\vec{x}(k-l)\Big)^TP_1\Big(Y_jI_iE\sum_{l=1}^{+\infty}k_lS\vec{x}(k-l)\Big)\\
&+\bar{k}\vec{x}^T(k)Q\vec{x}(k)-\sum_{l=1}^{+\infty}k_l\vec{x}^T(k-s)Q\vec{x}(k-s)\Big\}
\end{aligned}\tag{5.50}
$$

with $\delta_j>0$ and $\sum_{j=1}^{2^{Nn_x}}\delta_j=1$.

It follows from Lemma 5.3 and inequality (5.39) that

$$
-\sum_{l=1}^{+\infty}k_l\vec{x}^T(k-l)Q\vec{x}(k-l)
$$

$$\leq -\frac{1}{\bar{k}}\Big(\sum_{l=1}^{+\infty} k_l\vec{x}(k-l)\Big)^T Q\Big(\sum_{l=1}^{+\infty} k_l\vec{x}(k-l)\Big), \tag{5.51}$$

and

$$\varepsilon^T(k)\varepsilon(k) - \vec{x}^T(k)\mathcal{C}^T(I+\mathcal{T}\Upsilon)^T\Lambda(I+\mathcal{T}\Upsilon)\mathcal{C}\vec{x}(k) \leq 0. \tag{5.52}$$

By considering (5.51)–(5.52), we further have

$$\begin{aligned}
&\mathbb{E}\{\Delta V(k)\}\\
\leq\ & \max_{j\in[1,2^{Nn_x}]}\mathbb{E}\Big\{\Big[Y_j\Im^{-1}(A+W\otimes\Gamma)S\vec{x}(k)+Y_j\Im^{-1}\alpha E\sum_{l=1}^{+\infty}k_lS\vec{x}(k-l)\\
&+Y_j^-G\Im^{-1}S\vec{x}(k)\Big]^T\Im P_1\Im\Big[Y_j\Im^{-1}(A+W\otimes\Gamma)S\vec{x}(k)+Y_j\Im^{-1}\alpha\\
&\times E\sum_{l=1}^{+\infty}k_lS\vec{x}(k-l)+Y_j^-G\Im^{-1}S\vec{x}(k)\Big]+\bar{k}\vec{x}^T(k)Q\vec{x}(k)-\vec{x}^T(k)P\vec{x}(k)\\
&+\sum_{i=1}^{N}\bar{\alpha}_i(1-\bar{\alpha}_i)\Big(Y_jI_iE\sum_{l=1}^{+\infty}k_lS\vec{x}(k-l)\Big)^TP_1\Big(Y_jI_iE\sum_{l=1}^{+\infty}k_lS\vec{x}(k-l)\Big)\\
&+\Big[(\mathcal{K}_1+\mathcal{K}_3\mathcal{C}+\mathcal{K}_3\mathcal{T}\Upsilon\mathcal{C})\vec{x}(k)+\mathcal{E}\sum_{l=1}^{+\infty}k_l\vec{x}(k-l)-\mathcal{K}_3\varepsilon(k)\Big]^TP\\
&\times\Big[(\mathcal{K}_1+\mathcal{K}_3\mathcal{C}+\mathcal{K}_3\mathcal{T}\Upsilon\mathcal{C})\vec{x}(k)+\mathcal{E}\sum_{l=1}^{+\infty}k_l\vec{x}(k-l)-\mathcal{K}_3\varepsilon(k)\Big]\\
&-\frac{1}{\bar{k}}\Big(\sum_{l=1}^{+\infty}k_l\vec{x}(k-l)\Big)^TQ\Big(\sum_{l=1}^{+\infty}k_l\vec{x}(k-l)\Big)-\varepsilon^T(k)\varepsilon(k)\\
&+\vec{x}^T(k)\mathcal{C}^T(I+\mathcal{T}\Upsilon)^T\Lambda(I+\mathcal{T}\Upsilon)\mathcal{C}\vec{x}(k)\Big\}\\
=\ & \max_{j\in[1,2^{Nn_x}]}\mathbb{E}\Big\{\xi^T(k)(\Phi_{44}+\Xi_2^TP\Xi_2+\Xi_{3j}^TP_1\Xi_{3j}+\Xi_1^T\Lambda\Xi_1)\xi(k)\Big\}\\
=\ & \max_{j\in[1,2^{Nn_x}]}\mathbb{E}\Big\{\xi^T(k)\Omega_j\xi(k)\Big\}
\end{aligned} \tag{5.53}$$

where

$$\begin{aligned}
\Xi_2 &= \big[\mathcal{K}_1+\mathcal{K}_3\mathcal{C}+\mathcal{K}_3\mathcal{T}\Upsilon\mathcal{C}\quad \mathcal{E}\quad -\mathcal{K}_3\big],\\
\Omega_j &= \Phi_{44}+\Xi_2^TP\Xi_2+\Xi_{3j}^TP_1\Xi_{3j}+\Xi_1^T\Lambda\Xi_1,\\
\xi(k) &= \Big[\vec{x}^T(k)\quad \textstyle\sum_{l=1}^{+\infty}k_l\vec{x}^T(k-l)\quad \varepsilon^T(k)\Big]^T,\\
\Xi_{3j} &= \Big[Y_j(A+W\otimes\Gamma)S+\Im Y_j^-G\Im^{-1}S\quad Y_j\alpha ES\quad 0\Big],
\end{aligned}$$

$$\Xi_1 = \begin{bmatrix}(I+\mathcal{T}\Upsilon)\mathcal{C} & 0 & 0\end{bmatrix},\ \Phi_{44} = \mathrm{diag}\{-P+\bar{k}Q, \mathcal{Q}, -I\}.$$

Using Schur complement, we have $\Omega_j < 0$ ($j \in [1, 2^{Nn_x}]$) if (5.45) and (5.46) hold. Then, along the similar line of the proof of Theorem 1 in [175], the exponentially mean-square stability of the dynamics (5.43) can be confirmed.

To analyze the H_∞ performance, we introduce the following index function:

$$\mathcal{J} = \mathbb{E}\left\{\sum_{k=0}^{n}(\vec{z}^T(k)\vec{z}(k) - \gamma^2 w^T(k)w(k) - \gamma^2 v^T(k)v(k))\right\} \tag{5.54}$$

where n is a non-negative integer. It is obtained from (5.43) and (5.50)–(5.52) that

$$\begin{aligned}\mathbb{E}\{\Delta V(k)\} =& \mathbb{E}\{V(k+1) - V(k)\}\\ \leq& \max_{j\in[1,2^{Nn_x}]} \mathbb{E}\left\{\bar{\xi}^T(k)(\hat{\Phi}_{44} + \bar{\Xi}_2^T P\bar{\Xi}_2 + \bar{\Xi}_{3j}^T P_1\bar{\Xi}_{3j} + \bar{\Xi}_1^T \Lambda\bar{\Xi}_1)\bar{\xi}(k)\right\}\end{aligned} \tag{5.55}$$

where $\bar{\xi}(k) = \begin{bmatrix}\xi^T(k) & w^T(k) & v^T(k)\end{bmatrix}^T$ and $\hat{\Phi}_{44} = \mathrm{diag}\{-P+\bar{k}Q, \mathcal{Q}, -I, 0, 0\}$.

Under the zero-initial condition, it can be seen from (5.46) that

$$\begin{aligned}\mathcal{J} =& \sum_{k=0}^{n}\Big(\mathbb{E}\{\vec{z}^T(k)\vec{z}(k)\} - \gamma^2 w^T(k)w(k) - \gamma^2 v^T(k)v(k) + \mathbb{E}\{\Delta V(k)\}\Big)\\ &- \mathbb{E}\{V(k+1)\}\\ \leq& \sum_{k=0}^{n}\Big(\mathbb{E}\{\vec{z}^T(k)\vec{z}(k)\} - \gamma^2 w^T(k)w(k) - \gamma^2 v^T(k)v(k) + \mathbb{E}\{\Delta V(k)\}\Big)\\ \leq& \sum_{k=0}^{n}\max_{j\in[1,2^{Nn_x}]} \mathbb{E}\left\{\bar{\xi}^T(k)(\bar{\Phi}_{44} + \bar{\Xi}_1^T\Lambda\bar{\Xi}_1 + \bar{\Xi}_2^T P\bar{\Xi}_2 + \bar{\Xi}_{3j}^T P_1\bar{\Xi}_{3j})\bar{\xi}(k)\right\}\\ <& 0\end{aligned} \tag{5.56}$$

for all nonzero $w(k) \in l_2[0,\infty)$ and $v(k) \in l_2[0,\infty)$. Letting $n \to +\infty$, it follows immediately from (5.56) that the performance constraint (5.44) is satisfied and, therefore, the proof of Theorem 5.3 is complete.

Next, in light of the obtained result of Theorem 5.3, the design method of the desired estimators is given in the following theorem.

Theorem 5.4 *For the given disturbance level $\gamma > 0$, the augmented dynamics (5.43) (with $w(k) = 0$ and $v(k) = 0$) is exponentially stable in the mean square and the output estimation error satisfies the H_∞ performance constraint (5.44) if there exist matrices $P = \mathrm{diag}\{P_1, P_2\} > 0$, $P_2 = \mathrm{diag}\{P_{21}, \ldots, P_{2N}\} > 0$, $Q > 0$, G, $X = \mathrm{diag}\{X_1, \ldots, X_N\}$ and positive scalars λ_1, λ_2 satisfying*

$$\|G\|_\infty \leq 1, \tag{5.57}$$

$$\check{\Phi}_j = \begin{bmatrix} -\lambda_1 I & 0 & \Pi_1 \\ * & -\lambda_2 I & \check{\Pi}_2 \\ * & * & \tilde{\Phi}_j \end{bmatrix} < 0, \quad j \in [1, 2^{Nn_x}] \tag{5.58}$$

where

$$\begin{aligned}
\check{\Pi}_2 &= \begin{bmatrix} 0 & \bar{X}_3^T & 0 & 0 & 0 & 0 & 0 & 0 \end{bmatrix}, \\
\check{\Phi}_{44} &= \mathrm{diag}\{\check{\mathcal{Z}}_1, \mathcal{Q}, -I, -\gamma^2 I, -\gamma^2 I\}, \\
\tilde{\Phi}_j &= \begin{bmatrix} -\Lambda^{-1} & 0 & 0 & \check{\Xi}_1 \\ * & -P & 0 & \check{\Xi}_2 \\ * & * & -P_1 & P_1\bar{\Xi}_{3j} \\ * & * & * & \check{\Phi}_{44} \end{bmatrix}, \\
\check{\Xi}_2 &= \begin{bmatrix} \bar{X}_1 + \bar{X}_3\mathcal{C} & P\mathcal{E} & -\bar{X}_3 & 0 & \bar{X}_2 \end{bmatrix}, \\
\check{\mathcal{Z}}_1 &= -P + \bar{k}Q + \mathcal{F}^T\mathcal{F} + (\lambda_1 + \lambda_2)\mathcal{C}^T\Upsilon^T\Upsilon\mathcal{C}, \\
\check{\Xi}_1 &= \begin{bmatrix} \mathcal{C} & 0 & 0 & 0 & D \end{bmatrix}, \quad \Pi_1 = \begin{bmatrix} I & 0 & 0 & 0 & 0 & 0 & 0 & 0 \end{bmatrix}, \\
\bar{X}_1 &= \begin{bmatrix} 0 & 0 \\ 0 & P_2(A + W \otimes \Gamma) - XC \end{bmatrix}, \quad \bar{X}_2 = \begin{bmatrix} 0 \\ XD \end{bmatrix}, \quad \bar{X}_3 = \begin{bmatrix} 0 \\ X \end{bmatrix},
\end{aligned}$$

and $\bar{\Xi}_{3j}$ is defined in (5.47). Furthermore, if inequalities (5.57) and (5.58) are feasible, the desired estimator parameters are given by

$$K_i = P_{2i}^{-1} X_i, \quad i = 1, 2, \ldots, N. \tag{5.59}$$

Proof *Note that $\Phi_j = \check{\Phi}_j + \Delta\check{\Phi}$, where*

$$\begin{aligned}
\Pi_3 &= \begin{bmatrix} 0 & 0 & 0 & \Upsilon\mathcal{C} & 0 & 0 & 0 & 0 \end{bmatrix}, \\
\Pi_2 &= \begin{bmatrix} 0 & \mathcal{K}_3^T P & 0 & 0 & 0 & 0 & 0 & 0 \end{bmatrix}, \\
\tilde{\Xi}_2 &= \begin{bmatrix} \mathcal{K}_1 + \mathcal{K}_3\mathcal{C} & \mathcal{E} & -\mathcal{K}_3 & 0 & \mathcal{K}_2 \end{bmatrix}, \\
\check{\Phi}_j &= \begin{bmatrix} -\Lambda^{-1} & 0 & 0 & \check{\Xi}_1 \\ * & -P & 0 & P\tilde{\Xi}_2 \\ * & * & -P_1 & P_1\bar{\Xi}_{3j} \\ * & * & * & \bar{\Phi}_{44} \end{bmatrix}, \\
\Delta\check{\Phi} &= (\Pi_1^T \mathcal{T} \Pi_3)^T + \Pi_1^T \mathcal{T} \Pi_3 + (\Pi_2^T \mathcal{T} \Pi_3)^T + \Pi_2^T \mathcal{T} \Pi_3.
\end{aligned}$$

Considering (5.46), we have

$$\Phi_j = \check{\Phi}_j + (\Pi_1^T \mathcal{T} \Pi_3)^T + \Pi_1^T \mathcal{T} \Pi_3 + (\Pi_2^T \mathcal{T} \Pi_3)^T + \Pi_2^T \mathcal{T} \Pi_3 < 0. \tag{5.60}$$

From Lemma 3.2, it is easy to see that (5.60) is true if there exist positive scalars λ_1 and λ_2 such that

$$\check{\Phi}_j + (\lambda_1 + \lambda_2)\Pi_3^T \Pi_3 + \lambda_1^{-1}\Pi_1^T \Pi_1 + \lambda_2^{-1}\Pi_2^T \Pi_2 < 0. \tag{5.61}$$

In addition, noting $P_{2i}K_i = X_i$, we can conclude that inequality (5.61) is satisfied if (5.58) holds, which completes the proof.

Remark 5.4 *For the complex dynamical network (5.33) with the measurement output described by (5.37), multiple phenomena have been taken into account that include state saturations, quantization effects, randomly occurring distributed delays and external disturbances, which complicate the design of the state estimators considerably. In Theorem 5.3, a sufficient condition involving all of the information about the aforementioned phenomena is established to guarantee both the desired stability of and the H_∞ constraint on the augmented dynamics (5.43). Then, a set of ET state estimators in the form of (5.59) is designed in Theorem 5.4 by solving some matrix inequalities via the use of available software packages. It should be pointed out that the inequality (5.57) in Theorem 5.4 is given in terms of infinite norm which is nonlinear. Hence, in the practical implementation, the matrix G is usually given before the design of the state estimators.*

5.3 Illustrative Examples

Two simulation examples are presented in this section to illustrate the merits of the proposed ET H_∞ state estimation algorithms of this chapter.

5.3.1 Example 1

The sensor network is represented by a graph $\mathcal{G} = (\mathcal{V}, \mathcal{Q}, \mathcal{W})$ with the set of nodes $\mathcal{V} = \{1, 2, 3\}$, the set of edges $\mathcal{Q} = \{(2, 1), (3, 2)\}$ and the following adjacency matrix:

$$\mathcal{W} = \begin{bmatrix} 0 & 0 & 0 \\ 1 & 0 & 0 \\ 0 & 1 & 0 \end{bmatrix}.$$

The parameters of the plant (5.1) and the measurement (5.5) are given as follows:

$$A = \begin{bmatrix} -0.3 & 0.3 \\ -0.2 & 0.4 \end{bmatrix}, \quad B = \begin{bmatrix} -0.2 & 0.1 \\ -0.2 & 0.4 \end{bmatrix}, \quad E = \begin{bmatrix} -0.15 & 0.1 \\ -0.1 & 0.1 \end{bmatrix},$$
$$L = \begin{bmatrix} 0.2 \\ -0.1 \end{bmatrix}, \quad M = \begin{bmatrix} -0.4 & 0.5 \end{bmatrix}, \quad C_1 = \begin{bmatrix} 0.4 & 0.3 \end{bmatrix}, \quad D_1 = 0.8,$$
$$C_2 = \begin{bmatrix} 0.35 & 0.5 \end{bmatrix}, \quad C_3 = \begin{bmatrix} 0.2 & 0.25 \end{bmatrix}, \quad D_2 = 1, \quad D_3 = 1.2.$$

Let the TVD be $\tau(k) = 1 + \cos(k\pi)$ and the constant sequence be taken as $\kappa(s) = 2^{-(3+s)}$. It is easy to obtain that $\tau_M = 2$, $\tau_m = 0$ and $\breve{\kappa} = 0.125$. The probabilities α and β are taken as 0.7 and 0.8, respectively.

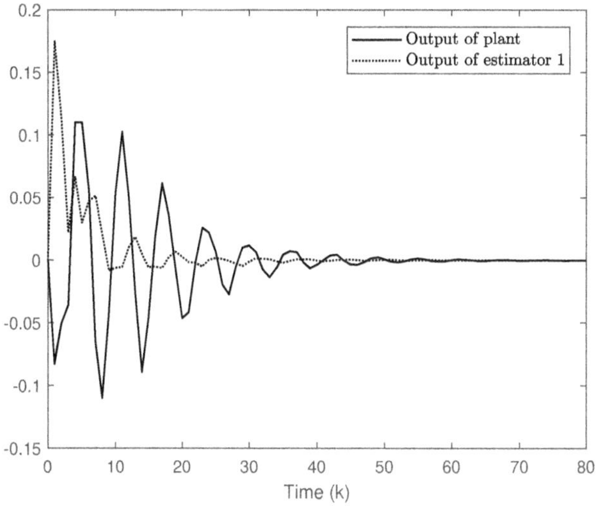

FIGURE 5.1
Output $z(k)$ and its estimate from estimator 1.

The H_∞ performance attenuation level is set as $\gamma = 0.6$ and the saturation level is $\aleph = \text{diag}\{0.15, 0.1\}$. The thresholds σ_i $(i = 1, 2, 3)$ are taken as $\sigma_1 = 0.4$, $\sigma_2 = 0.5$, and $\sigma_3 = 0.4$, respectively. Furthermore, we set $G = 0.1I$ to satisfy the inequality (5.29). Based on the above parameters, the inequality (5.30) is solved by using Matlab software. Then, according to (5.31), the parameters of the desired state estimators are obtained as follows:

$$K_1 = \begin{bmatrix} 0.0118 \\ 0.0355 \end{bmatrix}, \quad K_{21} = \begin{bmatrix} -0.0046 \\ -0.0206 \end{bmatrix}, \quad K_2 = \begin{bmatrix} 0.0211 \\ 0.0568 \end{bmatrix},$$
$$K_{32} = \begin{bmatrix} 0.0083 \\ 10.0023 \end{bmatrix}, \quad K_3 = \begin{bmatrix} 0.0022 \\ 0.021 \end{bmatrix}.$$

In the simulation, we set $w(k) = 3\text{sin}(k)e^{-0.1k}$ and $v(k) = 5\text{cos}(k)e^{-0.1k}$. The initial value of the state $\bar{x}$ is taken randomly by following the uniform distribution over $[-1, 1]$. Simulation results are shown in Figs. 5.1–5.5. Figs. 5.1–5.3 show the output $z(k)$ and its estimate from estimators 1, 2, and 3, respectively. In Fig. 5.4, the estimation errors $\vec{z}_i(k)$ $(i = 1, 2, 3)$ are depicted and the triggering instants of measurement outputs for sensor nodes 1, 2, and 3 are shown in Fig. 5.5. The simulation results have verified the usefulness of our distributed estimation algorithm developed in Section 5.1.

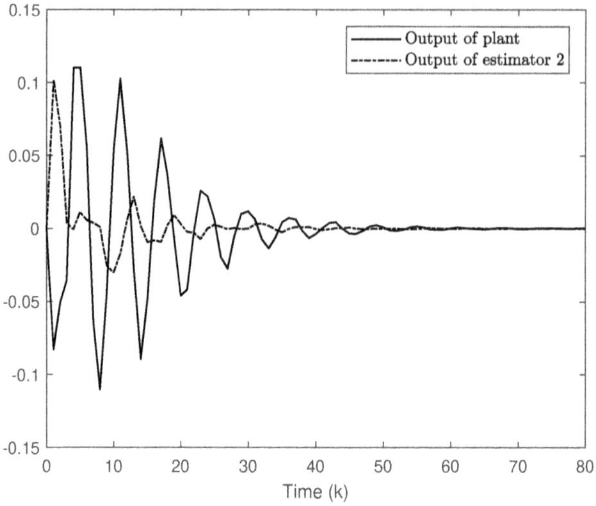

FIGURE 5.2
Output $z(k)$ and its estimate from estimator 2.

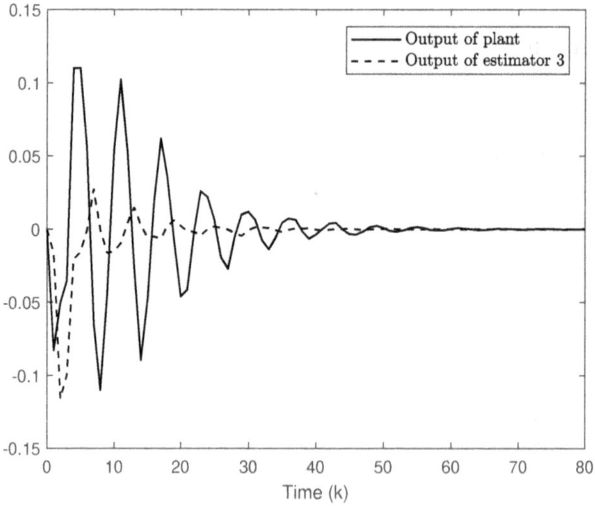

FIGURE 5.3
Output $z(k)$ and its estimate from estimator 3.

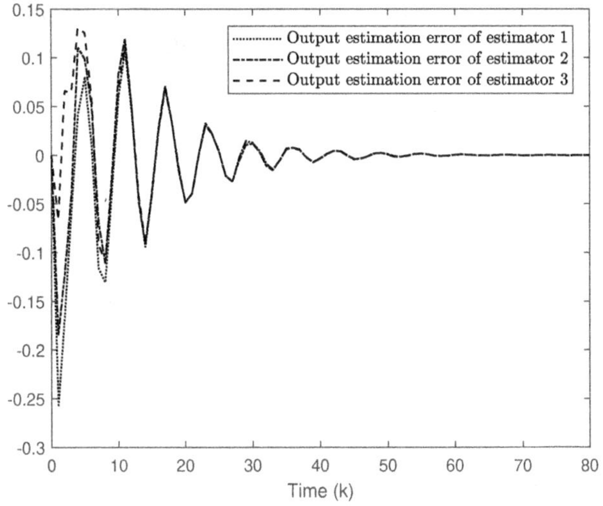

FIGURE 5.4
Estimation errors $\vec{z}_i(k)$ $(i = 1, 2, 3)$.

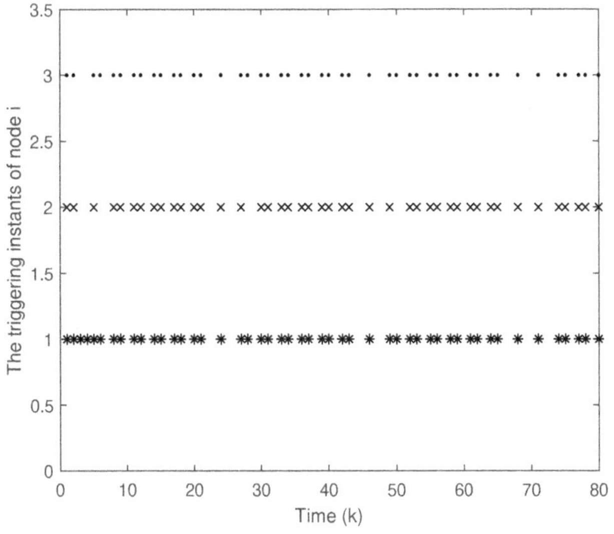

FIGURE 5.5
The triggering instants.

5.3.2 Example 2

This example considers the design of ET H_∞ state estimators for state-saturated complex networks with quantization effects as well as randomly occurring distributed delays.

Consider the complex network (5.33) consisting of three nodes with the following parameters:

$$B_1 = \begin{bmatrix} 0.2 & 0.1 \end{bmatrix}, \quad \Gamma = \begin{bmatrix} 1 & 0 \\ 0 & 1 \end{bmatrix}, \quad W = \begin{bmatrix} -0.2 & 0.1 & 0.1 \\ 0.1 & -0.2 & 0.1 \\ 0.1 & 0.1 & -0.2 \end{bmatrix},$$
$$B_2 = \begin{bmatrix} 0.15 & 0.2 \end{bmatrix}, \quad B_3 = \begin{bmatrix} 0.35 & 0.15 \end{bmatrix}, \quad F_1 = \begin{bmatrix} 0.5 & 0.6 \end{bmatrix},$$
$$F_2 = \begin{bmatrix} 0.48 & 0.5 \end{bmatrix}, \quad F_3 = \begin{bmatrix} 0.45 & 0.3 \end{bmatrix}, \quad k(l) = 2^{-(3+l)},$$
$$A_i = \begin{bmatrix} 0.4 & 0 \\ 0 & 0.5 \end{bmatrix}, \quad E_i = \begin{bmatrix} 0.5 & 0 \\ -0.4 & 0.35 \end{bmatrix}, \quad i = 1, 2, 3.$$

It is easy to obtain that $\bar{k} = 0.125$. The probabilities $\bar{\alpha}_i$ $(i = 1, 2, 3)$ are taken as 0.7, 0.8, and 0.8, respectively. The parameters of the measurement output (5.37) are given as follows:

$$C_1 = \begin{bmatrix} 0.21 & 0.14 \end{bmatrix}, \quad C_2 = \begin{bmatrix} 0.14 & -0.21 \end{bmatrix}, \quad D_1 = 0.5,$$
$$C_3 = \begin{bmatrix} 0.21 & 0.245 \end{bmatrix}, \quad D_2 = 0.4, \quad D_3 = 0.4,$$

and the quantizer parameters are set as $\delta_1 = \delta_2 = \delta_3 = 0.25$ and $u^{(1)} = u^{(2)} = u^{(3)} = 3$.

Assume that the thresholds σ_i $(i = 1, 2, 3)$ are taken as $\sigma_1 = 0.3$, $\sigma_2 = 0.3$, and $\sigma_3 = 0.2$, respectively. The H_∞ performance attenuation level is set as $\gamma = 0.8$ and $\Im = \text{diag}\{0.2, 0.15, 0.2, 0.15, 0.2, 0.15\}$. Furthermore, we set $G = 0.1I$ to satisfy the inequality (5.57). Based on the above parameters, we solve the inequality (5.58) by using the Matlab software and obtain a set of feasible solutions as follows:

$$P_{21} = \begin{bmatrix} 1.093 & 0.5573 \\ 0.5573 & 2.5401 \end{bmatrix}, \quad X_1 = \begin{bmatrix} 0.1540 \\ 0.1817 \end{bmatrix}, \quad P_{22} = \begin{bmatrix} 1.5253 & 0.4738 \\ 0.4738 & 1.7226 \end{bmatrix},$$
$$X_2 = \begin{bmatrix} 0.0345 \\ 0.0531 \end{bmatrix}, \quad P_{23} = \begin{bmatrix} 1.0257 & 0.1777 \\ 0.1777 & 0.7502 \end{bmatrix}, \quad X_3 = \begin{bmatrix} 0.0288 \\ 0.0331 \end{bmatrix}.$$

Then, the parameters of the desired state estimators can be obtained as follows:

$$K_1 = \begin{bmatrix} 0.0349 \\ 0.0161 \end{bmatrix}, \quad K_2 = \begin{bmatrix} 0.0142 \\ 0.0269 \end{bmatrix}, \quad K_3 = \begin{bmatrix} 0.0213 \\ 0.0391 \end{bmatrix},$$

In the simulation, the exogenous disturbances are, respectively, chosen as $w(k) = 3\sin(k)e^{-0.1k}$ and $v(k) = 5\cos(k)e^{-0.1k}$. Simulation results are shown in Figs. 5.6–5.10. Figs. 5.6–5.8 plot the output $z_i(k)$ and its estimates $\hat{z}_i(k)$ $(i = 1, 2, 3)$. The estimation errors $\vec{z}_i(k)$ $(i = 1, 2, 3)$ are depicted in Fig. 5.9. In addition, Fig. 5.10 shows the triggering instants of measurement outputs of nodes 1, 2, and 3. The simulation results show that the ET state estimation scheme presented in Section 5.2 is indeed effective.

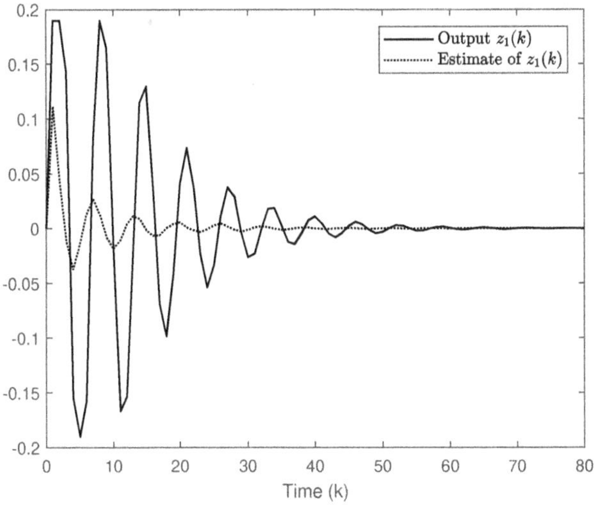

FIGURE 5.6
Output $z_1(k)$ and its estimate.

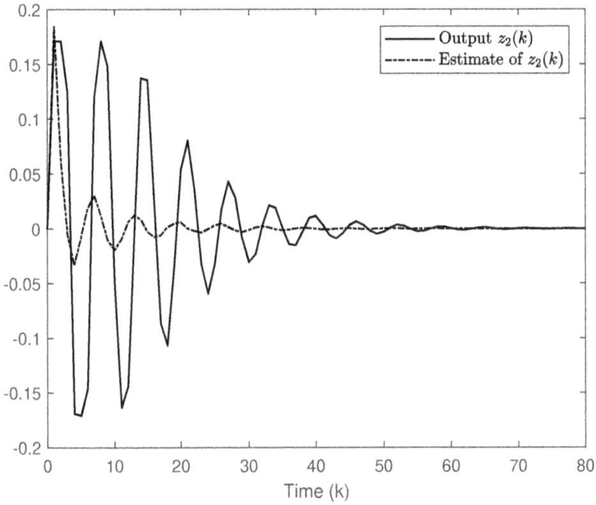

FIGURE 5.7
Output $z_2(k)$ and its estimate.

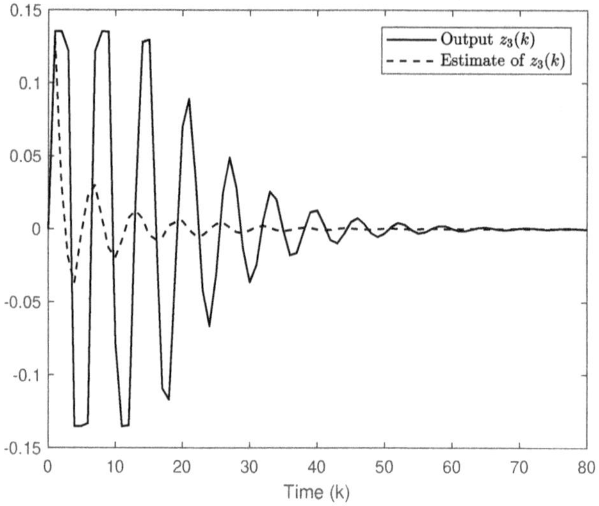

FIGURE 5.8
Output $z_3(k)$ and its estimate.

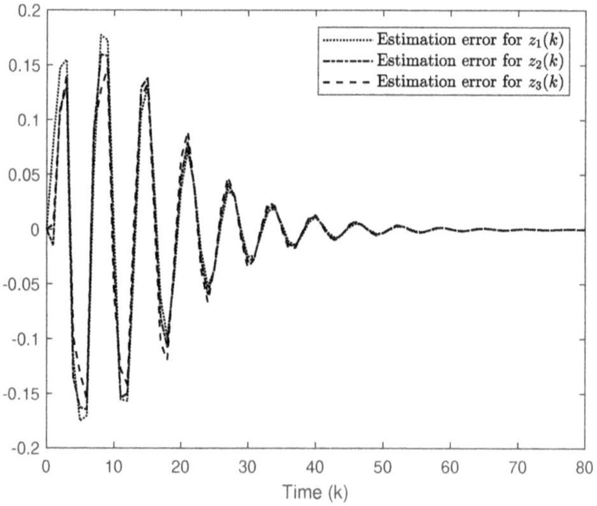

FIGURE 5.9
Estimation errors $\vec{z}_i(k)$ $(i = 1, 2, 3)$.

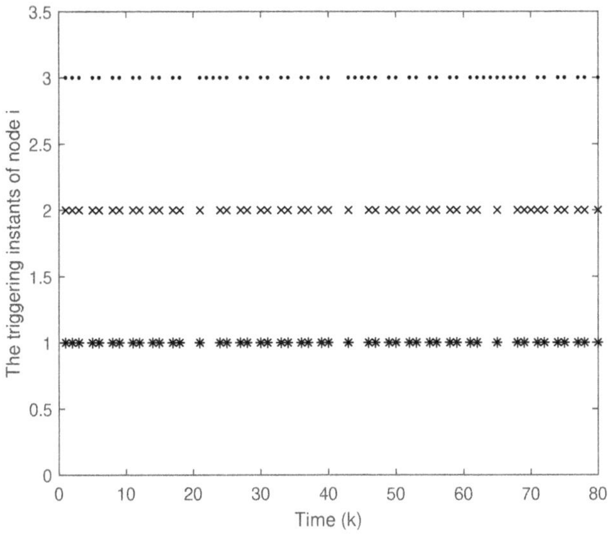

FIGURE 5.10
The triggering instants.

5.4 Summary

In this chapter, we have studied the ET H_∞ state estimation problem for state-saturated systems. Firstly, the distributed ET H_∞ state estimation problem has been considered for a class of state-saturated systems subject to ROMDs over sensor networks. The mixed delays comprising both discrete and distributed delays have been allowed to occur in a random way governed by two Bernoulli distributed random variables. Sufficient conditions for the exponential mean-square stability of the estimation error dynamics have been established and, at the same time, the prescribed H_∞ performance level has been guaranteed. Then, some parallel results have also been derived for a class of state-saturated complex networks in the simultaneous presence of quantization effects and randomly occurring distributed delays by using similar analysis techniques. Finally, we have used some illustrative examples to show the validity of the ET estimator design algorithm proposed in this chapter.

6

Event-Triggered State Estimation for Discrete-Time Neural Networks

In this chapter, the ET state estimation problem is investigated for discrete-time neural networks. For the purpose of energy saving, the ET mechanism is adopted and the measurement outputs are only transmitted to the estimator when a certain triggered condition is met. Firstly, we develop a new ET estimation technique for the delayed neural networks with SPs and IMs. In order to cater for more realistic transmission process of the neural signals, we make the first attempt to introduce a set of stochastic variables to characterize the random fluctuations of system parameters. The incomplete information under consideration includes randomly occurring sensor saturations and quantizations. A Lyapunov functional is constructed to obtain sufficient conditions under which the estimation error dynamics is exponentially ultimately bounded in the mean square. It is worth noting that the ultimate boundedness of the error dynamics is explicitly estimated. The other research focus of this chapter is to design the ET H_∞ state estimators for a class of discrete-time stochastic GRNs (that can be reviewed as a special case of neural networks) such that, in the presence of MJPs and TVDs, the estimation error dynamics is stochastically stable with a prescribed H_∞ performance level. Finally, some numerical examples are presented to illustrate the effectiveness of the propose ET state estimation scheme.

6.1 Event-Triggered State Estimation with Stochastic Parameters

This section is concerned with the ET state estimation problem for a class of discrete-time multi-delayed neural networks with SPs and IMs.

DOI: 10.1201/9781003307648-6

6.1.1 Problem Formulation

Consider the following class of discrete-time multi-delayed neural networks with n neurons

$$\begin{aligned} x_i(k+1) =& a_i(k)x_i(k) + \sum_{j=1}^{n} w_{ij}^0 g_j(x_j(k)) \\ &+ \sum_{j=1}^{n} w_{ij}^1 g_j(x_j(k-\tau_{ij})) + b_i\omega_i(k), \end{aligned} \tag{6.1}$$

for $i = 1, 2, \cdots, n$, where $x_i(k) \in \mathbb{R}$ is the state variable of neuron i; w_{ij}^0 and w_{ij}^1 are the interconnection strength and the delayed interconnection strength between neurons i and j, respectively; τ_{ij} is a constant representing the delay from neuron i to j; $g_j(\cdot)$ denotes the neuron activation function; $\omega_i(k)$ is a zero mean Gaussian white-noise process; and b_i is a deterministic constant while $a_i(k)$ is a *random* variable satisfying $0 < a_i(k) < 1$.

Remark 6.1 *The neural networks given by (6.1) is in nature a Hopfield neural network. Differently, we make the first attempt to introduce a set of stochastic variables $a_i(k)$ ($i = 1, 2, \cdots, n$) to describe the fluctuations of the values of the capacitance and resistance. Moreover, the model of neural networks given by (6.1) admits different time delays for different interconnections. Such kind of multiple delays is more general than those in the existing literature.*

Remark 6.2 *In general, the model of neural network contains an external input which, in many neural network applications, is held constant over a time interval of interest [115]. By shifting the equilibrium point to the origin, the external input can be eliminated and, the estimate of the real state of neural networks can be obtained by re-shifting to the equilibrium point. In order to avoid unnecessary mathematical complexity, in this section, we consider the neural networks without external inputs.*

The neuron activation function g_i satisfies $g_i(0) = 0$ and the following Lipschitz condition:

$$|g_i(x) - g_i(y)| \leq m_i|x - y|. \tag{6.2}$$

The random variables $\omega_i(k)$ and $a_i(k)$ have the following statistical properties

$$\mathbb{E}\{a_i(k)\} = \bar{a}_i,\ \mathbb{E}\{a_i(k)a_j(k)\} = \tilde{a}_{ij},\ \mathbb{E}\{\omega_i(k)\omega_j(k)\} = q_{ij}, \tag{6.3}$$

where $\bar{a}_i$, $\tilde{a}_{ij}$ and q_{ij} are known constants. Moreover, $a_i(k)$ is assumed to be uncorrelated with the initial state $x_i(0)$ and the Gaussian white-noise noise $w_i(k)$.

Set

$$
\begin{aligned}
x(k) &= \begin{bmatrix} x_1(k) & x_2(k) & \cdots & x_n(k) \end{bmatrix}^T, \\
w(k) &= \begin{bmatrix} \omega_1(k) & \omega_2(k) & \cdots & \omega_n(k) \end{bmatrix}^T, \\
A(k) &= \operatorname{diag}\{a_1(k), a_2(k), \cdots, a_n(k)\}, \\
B &= \operatorname{diag}\{b_1, b_2, \cdots, b_n\}, \quad W_0 = [w_{ij}^0]_{n\times n}, \\
E_i &= \operatorname{diag}\{\underbrace{0, \cdots, 0}_{i-1}, 1, \underbrace{0, \cdots, 0}_{n-i}\}, \quad W_1 = [w_{ij}^1]_{n\times n}, \\
g(x(k)) &= \begin{bmatrix} g_1(x_1(k)) & g_2(x_2(k)) & \cdots & g_n(x_n(k)) \end{bmatrix}^T.
\end{aligned} \tag{6.4}
$$

Then, the neural networks given by (6.1) can be written as the following compact form

$$
\begin{aligned}
x(k+1) =& A(k)x(k) + W_0 g(x(k)) \\
&+ \sum_{i=1}^{n}\sum_{j=1}^{n} E_i W_1 E_j g(x(k-\tau_{ij})) + Bw(k).
\end{aligned} \tag{6.5}
$$

In practice, the information about the neuron states is often incomplete from the network measurements and, meanwhile, the network measurements might be subject to the issues induced by limited communication. In this section, both the quantization effects and sensor saturations are taken into account and the network measurement model is given as follows [135]:

$$
\begin{aligned}
y(k) =& \delta(\alpha(k), 1)Cx(k) + \delta(\alpha(k), 2)s(Cx(k)) \\
&+ \delta(\alpha(k), 3)q(Cx(k)) + v(k)
\end{aligned} \tag{6.6}
$$

where $y(k) \in \mathbb{R}^m$ is the measurement output; $\delta(\cdot,\cdot)$ is the Kronecker delta function whose value is 1 if the two variables are equal, and 0 otherwise; $s(\cdot)$ is the saturation nonlinear function; $q(\cdot)$ is the quantization function; C is a deterministic matrix with appropriate dimensions; $v(k) \in \mathbb{R}^m$ is the measurement noise, which is a zero mean Gaussian white-noise process with $\mathbb{E}\{v(k)v^T(k)\} = Q_v$; and $\alpha(k)$ is a stochastic variable satisfying the following probability distribution:

$$
\operatorname{Prob}\{\alpha(k) = i\} = \beta_i, \quad i = 1, 2, 3. \tag{6.7}
$$

Here, $\alpha(k)$ is uncorrelated with other noise signals and $\beta_i \in [0, 1]$ $(i = 1, 2, 3)$ are constants satisfying $\beta_1 + \beta_2 + \beta_3 = 1$.

The saturation function $s(\cdot)$ is defined by

$$
s(\vartheta) = \begin{bmatrix} s(\vartheta_1) & s(\vartheta_2) & \ldots & s(\vartheta_m) \end{bmatrix}, \quad \forall \vartheta \in \mathbb{R}^m \tag{6.8}
$$

where $s(\vartheta_i) = \text{sign}(\vartheta_i)\min\{\vartheta_{\max}, |\vartheta_i|\}$ for each $i = 1, 2, \ldots, m$, with $\vartheta_{\max}$ denoting the saturation level. The saturation function defined above is actually a nonlinear function and, for a given scalar τ, we assume that

$$[s(\tau) - \bar{k}\tau][s(\tau) - \tau] \leq 0 \tag{6.9}$$

where $\bar{k}$ is a positive scalar satisfying $0 < \bar{k} < 1$.

Remark 6.3 *According to the definition of saturation function, it is easily seen that the saturation function $s(\cdot)$ satisfies the sector-bounded condition (6.9). The parameter $\bar{k}$ is used to characterize the lower bound of the sector-bounded nonlinearity and, in theory, parameter $\bar{k}$ should be taken as zero. However, this may be conservative. In practical applications, we usually choose the parameter $\bar{k}$ as a known scalar, which can be determined by estimating the practical value of the measurement $y(k)$.*

For the quantization function $q(\cdot)$, we adopt the logarithmic-type quantizer defined by

$$q(\vartheta) = \begin{bmatrix} q_1(\vartheta_1) & q_2(\vartheta_2) & \cdots & q_m(\vartheta_m) \end{bmatrix}^T, \quad \forall \vartheta \in \mathbb{R}^m.$$

For each $q_j(\cdot)$ $(1 \leq j \leq m)$, the set of quantization levels is described by

$$\mathcal{U}_j = \left\{ \pm u_i^{(j)}, u_i^{(j)} = \rho_j^i u_0^{(j)}, i = 0, \ \pm 1, \ \pm 2, \ \cdots \right\} \cup \{0\},$$
$$0 < \rho_j < 1, \quad u_0^{(j)} > 0,$$

and $q_j(\cdot)$ is given by

$$q_j(\vartheta_j) = \begin{cases} u_i^{(j)}, & \frac{1}{1+\kappa_j} u_i^{(j)} < \vartheta_j \leq \frac{1}{1-\kappa_j} u_i^{(j)}, \\ 0, & \vartheta_j = 0, \\ -q_j(-\vartheta_j), & \vartheta_j \leq 0 \end{cases}$$

with $\kappa_j = (1 - \rho_j)/(1 + \rho_j)$.

In this section, we would like to estimate the neuron states by using the available network measurements given by (6.6). In order to save energy, we consider the ET estimation scheme. Denote by $\{0 = r_0 < r_1 < r_2 < \cdots\}$ the sequence of event triggering instants determined by

$$r_{l+1} = \min\{k \in \mathbb{N} | k > r_l, \xi^T(k)\xi(k) - \delta > 0\}$$

where $\xi(k) = y(k) - y(r_l)$ and δ is the triggering threshold.

The estimator structure adopted is given as follows:

$$\begin{aligned} \hat{x}(k+1) = & \bar{A}\hat{x}(k) + W_0 g(\hat{x}(k)) + G(y(r_l) - C\hat{x}(k)) \\ & + \sum_{i=1}^{n}\sum_{j=1}^{n} E_i W_1 E_j g(\hat{x}(k - \tau_{ij})), \end{aligned} \tag{6.10}$$

for $k \in [r_l, r_{l+1})$, where $\bar{A} = \text{diag}\{\bar{a}_1, \bar{a}_2, \cdots, \bar{a}_n\}$ and G is the estimator gain to be designed.

Remark 6.4 *The Hopfield neural network is actually a modelling of neural circuits where each neuron is a simple analogue processor, while the rich connectivity provided in real neural circuits by the synapses formed between neurons are provided by the parallel communication lines in the value-passing analogue processor networks [155]. This kind of neural networks happens to fall into the category of the artificial neural networks mentioned in this section, and the energy-saving problem seems to be important when the neurons' states are estimated. Therefore, the ET estimation scheme could be a good candidate for the energy-saving purpose.*

By letting the estimation error be $e(k) = x(k) - \hat{x}(k)$, it follows from (6.5) and (6.10) that

$$\begin{aligned} e(k+1) =& \tilde{A}(k)x(k) + (1-\beta_1)GCx(k) + (\bar{A} - GC)e(k) + W_0\tilde{g}(k) \\ &+ G\xi(k) + \bar{B}\varpi(k) + \sum_{i=1}^{n}\sum_{j=1}^{n} E_iW_1E_j\tilde{g}(k-\tau_{ij}) - \beta_2 Gs(Cx(k)) \\ &- \beta_3 Gq(Cx(k)) - \tilde{\delta}_1^{\alpha}(k)GCx(k) - \tilde{\delta}_2^{\alpha}(k)Gs(Cx(k)) \\ &- \tilde{\delta}_3^{\alpha}(k)Gq(Cx(k)), \quad k \in [r_l, r_{l+1}), \end{aligned} \tag{6.11}$$

where

$$\begin{aligned} &\tilde{A}(k) = A(k) - \bar{A}, \quad \tilde{g}(k) = g(x(k)) - g(\hat{x}(k)), \quad \bar{B} = \begin{bmatrix} B & -G \end{bmatrix}, \\ &\varpi(k) = \begin{bmatrix} w^T(k) & v^T(k) \end{bmatrix}^T, \quad \tilde{\delta}_i^{\alpha}(k) = \delta(\alpha(k), i) - \beta_i, \quad i = 1, 2, 3. \end{aligned} \tag{6.12}$$

Furthermore, set $\eta(k) = \begin{bmatrix} x^T(k) & e^T(k) \end{bmatrix}^T$. Then, the dynamics of the neural network (6.5) and the error system (6.11) can be expressed by the following augmented system

$$\begin{aligned} \eta(k+1) =& \mathcal{A}\eta(k) + \bar{W}_0\mathcal{G}(k) + \sum_{i=1}^{n}\sum_{j=1}^{n} \bar{W}_{ij}^1\mathcal{G}(k-\tau_{ij}) + \mathcal{B}\varpi(k) + H_1G\xi(k) \\ &+ \tilde{\mathcal{A}}(k)H_2\eta(k) - \beta_1 H_1GCH_2\eta(k) - \beta_2 H_1Gs(CH_2\eta(k)) \\ &- \beta_3 H_1Gq(CH_2\eta(k)) - \tilde{\delta}_1^{\alpha}(k)H_1GCH_2\eta(k) \\ &- \tilde{\delta}_2^{\alpha}(k)H_1Gs(CH_2\eta(k)) - \tilde{\delta}_3^{\alpha}(k)H_1Gq(CH_2\eta(k)) \end{aligned} \tag{6.13}$$

where

$$\begin{aligned} &\mathcal{A} = \begin{bmatrix} \bar{A} & 0 \\ GC & \bar{A} - GC \end{bmatrix}, \ H_1 = \begin{bmatrix} 0 \\ I \end{bmatrix}, \ \bar{W}_0 = \mathrm{diag}_2\{W_0\}, \\ &\mathcal{B} = \begin{bmatrix} B & 0 \\ B & -G \end{bmatrix}, \ \tilde{\mathcal{A}}(k) = \begin{bmatrix} \tilde{A}(k) \\ \tilde{A}(k) \end{bmatrix}, \ \mathcal{G}(k) = \begin{bmatrix} g(x(k)) \\ \tilde{g}(k) \end{bmatrix}, \\ &\bar{W}_{ij}^1 = \mathrm{diag}_2\{E_iW_1E_j\}, \ H_2 = \begin{bmatrix} I & 0 \end{bmatrix}. \end{aligned} \tag{6.14}$$

Definition 6.1 *[176] The solution of a dynamical system is exponentially ultimately bounded in the mean square if there exist constants* $0 < \mu < 1$, $\nu > 0$, $\bar{\kappa} > 0$ *such that*

$$\mathbb{E}\{\|\eta(k)\|^2\} \leq \mu^k \nu + \kappa(k), \quad and \quad \lim_{k \to +\infty} \kappa(k) = \bar{\kappa}. \tag{6.15}$$

The aim of this section is to design an ET estimator with the form (6.10) for the multi-delayed neural networks (6.1) with IMs described by (6.6). More specifically, we are interested in looking for the estimator parameter G such that the dynamics of the augmented system (6.13) is exponentially mean-square ultimately bounded with a satisfactory bound.

6.1.2 Main Results

In this subsection, the boundedness issue is first analyzed for the augmented system (6.13). Then, according to the conducted analysis results, we shall investigate the design problem of the state estimator and give the desired estimator gain in terms of the solution to a certain matrix inequality.

For the logarithmic-type quantization function $q(\cdot)$, it is shown in [38] that $q_j(\vartheta_j) = (1+\Delta_j)\vartheta_j$ such that $|\Delta_j| \leq \kappa_j$. Setting $\Delta = \text{diag}\{\Delta_1, \ldots, \Delta_m\}$, $\Lambda_q = \text{diag}\{\kappa_1, \ldots, \kappa_m\}$ and $F = \Delta\Lambda_q^{-1}$, the quantization effect can be described as

$$q(CH_2\eta(k)) = U\eta(k) \tag{6.16}$$

where $U = (I + F\Lambda_q)CH_2$ and $FF^T = F^TF \leq I$.

For the purpose of the notation simplicity, we denote $s(CH_2\eta(k))$ by $\bar{s}(k)$ and set

$$\begin{aligned}
&\bar{\mathcal{G}}_i(k) = \begin{bmatrix} \mathcal{G}^T(k-\tau_{i1}) & \mathcal{G}^T(k-\tau_{i2}) & \cdots & \mathcal{G}^T(k-\tau_{in}) \end{bmatrix}^T, \\
&\bar{\mathcal{G}}(k) = \begin{bmatrix} \bar{\mathcal{G}}_1^T(k) & \bar{\mathcal{G}}_2^T(k) & \cdots & \bar{\mathcal{G}}_n^T(k) \end{bmatrix}^T, \\
&\bar{W}_i^1 = \begin{bmatrix} \bar{W}_{i1}^1 & \bar{W}_{i2}^1 & \cdots & \bar{W}_{in}^1 \end{bmatrix}, \, \bar{W}_1 = \begin{bmatrix} \bar{W}_1^1 & \bar{W}_2^1 & \cdots & \bar{W}_n^1 \end{bmatrix}.
\end{aligned}$$

Then, the augmented system given by (6.13) can be rewritten as the following concise form:

$$\begin{aligned}
\eta(k+1) =& \mathcal{A}\eta(k) + \bar{W}_0\mathcal{G}(k) + \bar{W}_1\bar{\mathcal{G}}(k) + \mathcal{B}\bar{w}(k) + H_1G\xi(k) \\
&+ \tilde{\mathcal{A}}(k)H_2\eta(k) - \beta_1H_1GCH_2\eta(k) - \beta_2H_1G\bar{s}(k) \\
&- \beta_3H_1GU\eta(k) - \tilde{\delta}_1^\alpha(k)H_1GCH_2\eta(k) \\
&- \tilde{\delta}_2^\alpha(k)H_1G\bar{s}(k) - \tilde{\delta}_3^\alpha(k)H_1GU\eta(k).
\end{aligned} \tag{6.17}$$

The following lemmas will be used in the proof of our main results.

Lemma 6.1 *Under the condition (6.2), we have*

$$\mathcal{G}^T(k)\mathcal{G}(k) - \eta^T(k)\bar{M}\eta(k) \leq 0 \tag{6.18}$$

where $\bar{M} = \text{diag}_2\{M\}$ *and* $M = \text{diag}\{m_1^2, m_2^2, \cdots, m_n^2\}$.

Proof *From the definitions of $e(k)$ and $\eta(k)$ together with (6.4), (6.12), and (6.14), it can be obtained that*

$$\begin{aligned}&\mathcal{G}^T(k)\mathcal{G}(k) - \eta^T(k)\bar{M}\eta(k)\\ =&g^T(x(k))g(x(k)) - x^T(k)Mx(k) + \tilde{g}^T(k)\tilde{g}(k) - e^T(k)Me(k)\\ =&\sum_{i=1}^{n}\Big(\left(g_i(x_i(k)) - g_i(\hat{x}_i(k))\right)^2 - m_i^2\left(x_i(k) - \hat{x}_i(k)\right)^2\Big)\\ &+\sum_{i=1}^{n}\Big(g_i^2(x_i(k)) - m_i^2 x_i^2(k)\Big).\end{aligned}$$

It is easily seen from (6.2) that $g_i^2(x_i(k)) - m_i^2x_i^2(k) < 0$ and $(g_i(x_i(k)) - g_i(\hat{x}_i(k)))^2 - m_i^2\left(x_i(k) - \hat{x}_i(k)\right)^2 < 0$, from which the inequality (6.18) follows directly. Therefore, the proof of this lemma is complete.

Setting $\hat{M} = \text{diag}_{n^2}\{\bar{M}\}$, we further have

$$\bar{\mathcal{G}}^T(k)\bar{\mathcal{G}}(k) - \eta_d^T(k)\hat{M}\eta_d(k) \leq 0 \tag{6.19}$$

where

$$\begin{aligned}\eta_d(k) &= \begin{bmatrix}\bar{\eta}_1^T(k) & \bar{\eta}_2^T(k) & \cdots & \bar{\eta}_n^T(k)\end{bmatrix}^T,\\ \bar{\eta}_i(k) &= \begin{bmatrix}\eta^T(k-\tau_{i1}) & \eta^T(k-\tau_{i2}) & \cdots & \eta^T(k-\tau_{in})\end{bmatrix}^T.\end{aligned}$$

Lemma 6.2 *The saturation function $\bar{s}(k)$ satisfies*

$$\bar{s}^T(k)\bar{s}(k) - \eta^T(k)H_2^TC^T(K^T+I)\bar{s}(k) + \eta^T(k)H_2^TC^TK^TCH_2\eta(k) \leq 0 \tag{6.20}$$

where $K = \text{diag}_m\{\bar{k}\}$.

Proof *According to (6.9), we have*

$$\begin{aligned}\left(s(\vartheta) - K\vartheta\right)^T\left(s(\vartheta) - \vartheta\right) =&\sum_{i=1}^{m}\Big(\left(s(\vartheta_i) - \bar{k}\vartheta_i\right)^T\left(s(\vartheta_i) - \vartheta_i\right)\Big)\\ \leq&0.\end{aligned}$$

Letting ϑ be $\vartheta = CH_2\eta(k)$ and noting $\bar{s}(k) = s(CH_2\eta(k))$, it immediately follows from the above inequality that

$$\left(\bar{s}(k) - KCH_2\eta(k)\right)^T\left(\bar{s}(k) - CH_2\eta(k)\right) \leq 0$$

which is the exactly same as inequality (6.20) and hence the proof of this lemma is accomplished.

Lemma 6.3 *[53] For a stochastic variable $\alpha(k)$ satisfying the probability distribution given by (6.7), we have*

$$\mathbb{E}\{\tilde{\delta}_i^\alpha(k)\tilde{\delta}_j^\alpha(k)\} = \begin{cases} \beta_i(1-\beta_i), & i=j, \\ -\beta_i\beta_j, & i \neq j \end{cases} \tag{6.21}$$

for $i, j = 1, 2, 3$.

In the following theorem, a sufficient condition is provided under which the augmented system (6.13) is exponentially ultimately bounded in the mean square.

Theorem 6.1 *For the given estimator parameter G, the augmented system (6.13) is exponentially ultimately bounded in the mean square if there exist positive definite matrices $P = [P_{ij}]_{2\times 2}$, Q_{ij} $(i, j = 1, 2, \cdots, n)$ and positive scalars λ_1, λ_2, λ_3, λ_4 satisfying the following inequality:*

$$\Phi = \begin{bmatrix} \Theta_{11} & \Theta_{12} & \Theta_{13} & \Theta_{14} & \Theta_{15} & 0 \\ * & \Theta_{22} & \Theta_{23} & \Theta_{24} & \Theta_{25} & 0 \\ * & * & \Theta_{33} & \Theta_{34} & \Theta_{35} & 0 \\ * & * & * & \Theta_{44} & 0 & 0 \\ * & * & * & * & \Theta_{55} & 0 \\ * & * & * & * & * & \Theta_{66} \end{bmatrix} < 0 \tag{6.22}$$

where

$$\begin{aligned} \Theta_{11} &= (1+\beta_2+\beta_3)\mathcal{A}^T P \mathcal{A} + H_2^T \tilde{P} H_2 + 5\beta_3 U^T G^T H_1^T P H_1 G U \\ &\quad + \beta_1 H_2^T C^T G^T H_1^T P H_1 G C H_2 - \beta_1 \mathcal{A}^T P H_1 G C H_2 - P + \lambda_1 \bar{M} \\ &\quad - \beta_1 H_2^T C^T G^T H_1^T P \mathcal{A} + \sum_{i=1}^{n}\sum_{j=1}^{n} Q_{ij} - \lambda_3 H_2^T C^T K^T C H_2, \\ \Theta_{12} &= \mathcal{A}^T P \bar{W}_0 - \beta_1 H_2^T C^T G^T H_1^T P \bar{W}_0, \quad Q_i = \mathrm{diag}\{Q_{i1}, Q_{i2}, \cdots, Q_{in}\}, \\ \Theta_{13} &= \mathcal{A}^T P \bar{W}_1 - \beta_1 H_2^T C^T G^T H_1^T P \bar{W}_1, \quad E = [\tilde{a}_{ij} - \bar{a}_i \bar{a}_j]_{n\times n}, \\ \Theta_{14} &= \mathcal{A}^T P H_1 G - \beta_1 H_2^T C^T G^T H_1^T P H_1 G, \quad \Theta_{15} = \frac{1}{2}\lambda_3 H_2^T C^T (K^T + I), \\ \Theta_{22} &= (1+\beta_3)\bar{W}_0^T P \bar{W}_0 - \lambda_1 I, \quad \Theta_{23} = \bar{W}_0^T P \bar{W}_1, \quad \Theta_{24} = \bar{W}_0^T P H_1 G, \\ \Theta_{25} &= -\beta_2 \bar{W}_0^T P H_1 G, \quad \Theta_{33} = (1+\beta_3)\bar{W}_1^T P \bar{W}_1 - \lambda_2 I, \\ \Theta_{34} &= \bar{W}_1^T P H_1 G, \quad \Theta_{35} = -\beta_2 \bar{W}_1^T P H_1 G, \quad Q = \mathrm{diag}\{Q_1, Q_2, \cdots, Q_n\}, \\ \Theta_{44} &= (1+\beta_2+\beta_3) G^T H_1^T P H_1 G - \lambda_4 I, \quad \Theta_{55} = 3\beta_2 G^T H_1^T P H_1 G - \lambda_3 I, \\ \Theta_{66} &= -Q + \lambda_2 \hat{M}, \quad \tilde{P} = E \circ (P_{11} + P_{12} + P_{21} + P_{22}). \end{aligned}$$

Furthermore, if the inequality (6.22) is solvable, the ultimate bound of the dynamics of the augmented system (6.13) is given by

$$\bar{\kappa} = \frac{\alpha_0}{(\alpha_0 - 1)\lambda_{\min}(P)}(\lambda_4 \delta + \vartheta) \tag{6.23}$$

where $\vartheta = \lambda_{\max}(\mathcal{B}^T P \mathcal{B})(\sum_{i=1}^{n} q_{ii} + trace(Q_v))$ *and* $\alpha_0 > 1$ *satisfies*

$$\phi(\alpha_0) + \frac{1}{\alpha_0 - 1}\sum_{i=1}^{n}\sum_{j=1}^{n}\varphi_{ij}(\alpha_0)(\alpha_0^{\tau_{ij}+1} + \alpha_0^{\tau_{ij}} - 2\alpha_0) = 0 \tag{6.24}$$

with

$$\begin{aligned}\phi(\alpha_0) &= (\alpha_0 - 1)\lambda_{\max}(P) - \alpha_0\lambda_{\min}(-\Phi),\\ \varphi_{ij}(\alpha_0) &= (\alpha_0 - 1)\lambda_{\max}(Q_{ij}).\end{aligned} \tag{6.25}$$

Proof *In this proof, the exponentially mean-square ultimate boundedness of the augmented system (6.13) is shown with the help of Lyapynov functional-like theory.*

Define the following functional

$$V(k) = \eta^T(k)P\eta(k) + \sum_{i=1}^{n}\sum_{j=1}^{n}\sum_{l=k-\tau_{ij}}^{k-1}\eta^T(l)Q_{ij}\eta(l) \tag{6.26}$$

and calculate the difference of $V(k)$ *along the system (6.17) as follows:*

$$\begin{aligned}
&\mathbb{E}\{V(k+1) - V(k)\}\\
={}& \mathbb{E}\Big\{\eta^T(k)\mathcal{A}^T P\mathcal{A}\eta(k) + \mathcal{G}^T(k)\bar{W}_0^T P\bar{W}_0\mathcal{G}(k) + \bar{\mathcal{G}}^T(k)\bar{W}_1^T P\bar{W}_1\bar{\mathcal{G}}(k)\\
&+\xi^T(k)G^T H_1^T PH_1 G\xi(k) + \eta^T(k)H_2^T\tilde{\mathcal{A}}^T(k)P\tilde{\mathcal{A}}(k)H_2\eta(k)\\
&+\beta_1^2\eta^T(k)H_2^T C^T G^T H_1^T PH_1 GCH_2\eta(k) + \bar{\varpi}^T(k)\mathcal{B}^T P\mathcal{B}\bar{\varpi}(k)\\
&+\beta_2^2\bar{s}^T(k)G^T H_1^T PH_1 G\bar{s}(k) + \beta_3^2\eta^T(k)U^T G^T H_1^T PH_1 GU\eta(k)\\
&+\tilde{\delta}_1^{\alpha 2}(k)\eta^T(k)H_2^T C^T G^T H_1^T PH_1 GCH_2\eta(k) + 2\eta^T(k)\mathcal{A}^T P\bar{W}_0\mathcal{G}(k)\\
&+\tilde{\delta}_2^{\alpha 2}(k)\bar{s}^T(k)G^T H_1^T PH_1 G\bar{s}(k) + 2\eta^T(k)\mathcal{A}^T P\bar{W}_1\bar{\mathcal{G}}(k)\\
&+\tilde{\delta}_3^{\alpha 2}(k)\eta^T(k)U^T G^T H_1^T PH_1 GU\eta(k) + 2\eta^T(k)\mathcal{A}^T PH_1 G\xi(k)\\
&-2\beta_1\eta^T(k)\mathcal{A}^T PH_1 GCH_2\eta(k) - 2\beta_2\eta^T(k)\mathcal{A}^T PH_1 G\bar{s}(k)\\
&-2\beta_3\eta^T(k)\mathcal{A}^T PH_1 GU\eta(k) + 2\mathcal{G}^T(k)\bar{W}_0^T P\bar{W}_1\bar{\mathcal{G}}(k)\\
&+2\mathcal{G}^T(k)\bar{W}_0^T PH_1 G\xi(k) - 2\beta_1\mathcal{G}^T(k)\bar{W}_0^T PH_1 GCH_2\eta(k)\\
&-2\beta_2\mathcal{G}^T(k)\bar{W}_0^T PH_1 G\bar{s}(k) - 2\beta_3\mathcal{G}^T(k)\bar{W}_0^T PH_1 GU\eta(k)\\
&+2\bar{\mathcal{G}}^T(k)\bar{W}_1^T PH_1 G\xi(k) - 2\beta_1\bar{\mathcal{G}}^T(k)\bar{W}_1^T PH_1 GCH_2\eta(k)\\
&-2\beta_2\bar{\mathcal{G}}^T(k)\bar{W}_1^T PH_1 G\bar{s}(k) - 2\beta_3\bar{\mathcal{G}}^T(k)\bar{W}_1^T PH_1 GU\eta(k)\\
&-2\beta_1\xi^T(k)G^T H_1^T PH_1 GCH_2\eta(k) - 2\beta_2\xi^T(k)G^T H_1^T PH_1 G\bar{s}(k)\\
&-2\beta_3\xi^T(k)G^T H_1^T PH_1 GU\eta(k) + 2\beta_1\beta_2\eta^T(k)H_2^T C^T G^T H_1^T PH_1 G\bar{s}(k)\\
&+2\beta_1\beta_3\eta^T(k)H_2^T C^T G^T H_1^T PH_1 GU\eta(k) + 2\beta_2\beta_3\bar{s}^T(k)G^T H_1^T\\
&\times PH_1 GU\eta(k) + 2\tilde{\delta}_2^{\alpha}(k)\tilde{\delta}_3^{\alpha}(k)\bar{s}^T(k)G^T H_1^T PH_1 GU\eta(k)
\end{aligned}$$

$$+2\tilde{\delta}_1^{\alpha}(k)\tilde{\delta}_3^{\alpha}(k)\eta^T(k)H_2^TC^TG^TH_1^TPH_1GU\eta(k)+2\tilde{\delta}_1^{\alpha}(k)\tilde{\delta}_2^{\alpha}(k)$$
$$\times\eta^T(k)H_2^TC^TG^TH_1^TPH_1G\bar{s}(k)+\sum_{i=1}^{n}\sum_{j=1}^{n}\eta^T(k)Q_{ij}\eta(k)$$
$$-\eta_d^T(k)Q\eta_d(k)-\eta^T(k)P\eta(k)\Big\}. \tag{6.27}$$

By noting (6.3) and (6.14), the term containing $\tilde{\mathcal{A}}^T(k)P\tilde{\mathcal{A}}(k)$ can be computed as follows:

$$\begin{aligned}
&\mathbb{E}\{\eta^T(k)H_2^T\tilde{\mathcal{A}}^T(k)P\tilde{\mathcal{A}}(k)H_2\eta(k)\}\\
=&\mathbb{E}\{\eta^T(k)H_2^T\begin{bmatrix}\tilde{A}(k)\\ \tilde{A}(k)\end{bmatrix}^T\begin{bmatrix}P_{11} & P_{12}\\ P_{21} & P_{22}\end{bmatrix}\begin{bmatrix}\tilde{A}(k)\\ \tilde{A}(k)\end{bmatrix}H_2\eta(k)\}\\
=&\mathbb{E}\{\eta^T(k)H_2^T(\tilde{A}^T(k)P_{11}\tilde{A}(k)+2\tilde{A}^T(k)P_{12}\tilde{A}(k)\\
&+\tilde{A}^T(k)P_{22}\tilde{A}(k))H_2\eta(k)\}\\
=&\mathbb{E}\{\eta^T(k)H_2^T(E\circ(P_{11}+P_{12}+P_{21}+P_{22}))H_2\eta(k)\}\\
=&\mathbb{E}\{\eta^T(k)H_2^T\tilde{P}H_2\eta(k)\}.
\end{aligned} \tag{6.28}$$

For the term $\bar{w}^T(k)\mathcal{B}^TP\mathcal{B}\bar{w}(k)$, we have

$$\begin{aligned}
&\mathbb{E}\{\bar{w}^T(k)\mathcal{B}^TP\mathcal{B}\bar{w}(k)\}\\
\leq&\lambda_{\max}(\mathcal{B}^TP\mathcal{B})\mathbb{E}\{w^T(k)w(k)+v^T(k)v(k)\}\\
=&\vartheta.
\end{aligned} \tag{6.29}$$

By using the elementary inequality $-2ab\leq a^2+b^2$, it can be obtained that

$$\begin{aligned}
&\mathbb{E}\{-2\beta_3\eta^T(k)\mathcal{A}^TPH_1GU\eta(k)\}\\
\leq&\ \mathbb{E}\{\beta_3\eta^T(k)\mathcal{A}^TP\mathcal{A}\eta(k)+\beta_3\eta^T(k)U^TG^TH_1^TPH_1GU\eta(k)\},
\end{aligned} \tag{6.30}$$

$$\begin{aligned}
&\mathbb{E}\{-2\beta_2\eta^T(k)\mathcal{A}^TPH_1G\bar{s}(k)\}\\
\leq&\ \mathbb{E}\{\beta_2\eta^T(k)\mathcal{A}^TP\mathcal{A}\eta(k)+\beta_2\bar{s}^T(k)G^TH_1^TPH_1G\bar{s}(k)\},
\end{aligned} \tag{6.31}$$

$$\begin{aligned}
&\mathbb{E}\{-2\beta_2\xi^T(k)G^TH_1^TPH_1G\bar{s}(k)\}\\
\leq&\ \mathbb{E}\{\beta_2\xi^T(k)G^TH_1^TPH_1G\xi(k)+\beta_2\bar{s}^T(k)G^TH_1^TPH_1G\bar{s}(k)\},
\end{aligned} \tag{6.32}$$

$$\begin{aligned}
&\mathbb{E}\{-2\beta_3\mathcal{G}^T(k)\bar{W}_0^TPH_1GU\eta(k)\}\\
\leq&\ \mathbb{E}\{\beta_3\mathcal{G}^T(k)\bar{W}_0^TP\bar{W}_0\mathcal{G}(k)+\beta_3\eta^T(k)U^TG^TH_1^TPH_1GU\eta(k)\},
\end{aligned} \tag{6.33}$$

$$\begin{aligned}
&\mathbb{E}\{-2\beta_3\bar{\mathcal{G}}^T(k)\bar{W}_1^TPH_1GU\eta(k)\}\\
\leq&\ \mathbb{E}\{\beta_3\bar{\mathcal{G}}^T(k)\bar{W}_1^TP\bar{W}_1\bar{\mathcal{G}}(k)+\beta_3\eta^T(k)U^TG^TH_1^TPH_1GU\eta(k)\},
\end{aligned} \tag{6.34}$$

$$\begin{aligned}
&\mathbb{E}\{-2\beta_3\xi^T(k)G^TH_1^TPH_1GU\eta(k)\}\\
\leq&\ \mathbb{E}\{\beta_3\xi^T(k)G^TH_1^TPH_1G\xi(k)\\
&+\beta_3\eta^T(k)U^TG^TH_1^TPH_1GU\eta(k)\}.
\end{aligned} \tag{6.35}$$

Substituting (6.28)-(6.35) into (6.27) and using Lemma 6.3 yield

$$\begin{aligned}
&\mathbb{E}\{V(k+1)-V(k)\}\\
=&\,\mathbb{E}\Big\{(1+\beta_2+\beta_3)\eta^T(k)\mathcal{A}^TP\mathcal{A}\eta(k)+(1+\beta_3)\mathcal{G}^T(k)\bar{W}_0^TP\bar{W}_0\mathcal{G}(k)\\
&+(1+\beta_3)\bar{\mathcal{G}}^T(k)\bar{W}_1^TP\bar{W}_1\bar{\mathcal{G}}(k)+(1+\beta_2+\beta_3)\xi^T(k)G^TH_1^TPH_1G\xi(k)\\
&+\eta^T(k)H_2^T\tilde{P}H_2\eta(k)+\beta_1\eta^T(k)H_2^TC^TG^TH_1^TPH_1GCH_2\eta(k)\\
&+3\beta_2\bar{s}^T(k)G^TH_1^TPH_1G\bar{s}(k)+5\beta_3\eta^T(k)U^TG^TH_1^TPH_1GU\eta(k)\\
&+2\eta^T(k)\mathcal{A}^TP\bar{W}_0\mathcal{G}(k)+2\eta^T(k)\mathcal{A}^TP\bar{W}_1\bar{\mathcal{G}}(k)+2\eta^T(k)\mathcal{A}^TPH_1G\xi(k)\\
&-2\beta_1\eta^T(k)\mathcal{A}^TPH_1GCH_2\eta(k)+2\mathcal{G}^T(k)\bar{W}_0^TP\bar{W}_1\bar{\mathcal{G}}(k)\\
&+2\mathcal{G}^T(k)\bar{W}_0^TPH_1G\xi(k)-2\beta_1\mathcal{G}^T(k)\bar{W}_0^TPH_1GCH_2\eta(k)\\
&-2\beta_2\mathcal{G}^T(k)\bar{W}_0^TPH_1G\bar{s}(k)+2\bar{\mathcal{G}}^T(k)\bar{W}_1^TPH_1G\xi(k)\\
&-2\beta_1\bar{\mathcal{G}}^T(k)\bar{W}_1^TPH_1GCH_2\eta(k)-2\beta_2\bar{\mathcal{G}}^T(k)\bar{W}_1^TPH_1G\bar{s}(k)\\
&-2\beta_1\xi^T(k)G^TH_1^TPH_1GCH_2\eta(k)+\sum_{i=1}^{n}\sum_{j=1}^{n}\eta^T(k)Q_{ij}\eta(k)\\
&-\eta_d^T(k)Q\eta_d(k)-\eta^T(k)P\eta(k)+\vartheta\Big\}\\
\leq&\,\mathbb{E}\{\zeta^T(k)\bar{\Phi}\zeta(k)+\vartheta\}
\end{aligned}$$

where

$$\begin{aligned}
\zeta(k)&=\begin{bmatrix}\eta^T(k) & \mathcal{G}^T(k) & \bar{\mathcal{G}}^T(k) & \xi^T(k) & \bar{s}^T(k) & \eta_d^T(k)\end{bmatrix}^T\\
\bar{\Phi}&=\begin{bmatrix}\bar{\Theta}_{11} & \Theta_{12} & \Theta_{13} & \Theta_{14} & 0 & 0\\ * & \bar{\Theta}_{22} & \Theta_{23} & \Theta_{24} & \Theta_{25} & 0\\ * & * & \bar{\Theta}_{33} & \Theta_{34} & \Theta_{35} & 0\\ * & * & * & \bar{\Theta}_{44} & 0 & 0\\ * & * & * & * & \bar{\Theta}_{55} & 0\\ * & * & * & * & * & -Q\end{bmatrix},\\
\bar{\Theta}_{11}&=(1+\beta_2+\beta_3)\mathcal{A}^TP\mathcal{A}+H_2^T\tilde{P}H_2+5\beta_3U^TG^TH_1^TPH_1GU\\
&\quad+\beta_1H_2^TC^TG^TH_1^TPH_1GCH_2-\beta_1\mathcal{A}^TPH_1GCH_2\\
&\quad-\beta_1H_2^TC^TG^TH_1^TP\mathcal{A}-P+\sum_{i=1}^{n}\sum_{j=1}^{n}Q_{ij},\\
\bar{\Theta}_{22}&=(1+\beta_3)\bar{W}_0^TP\bar{W}_0,\quad \bar{\Theta}_{44}=(1+\beta_2+\beta_3)G^TH_1^TPH_1G,\\
\bar{\Theta}_{33}&=(1+\beta_3)\bar{W}_1^TP\bar{W}_1,\quad \bar{\Theta}_{55}=3\beta_2G^TH_1^TPH_1G.
\end{aligned}$$

Under the triggering condition, it follows from (6.18), (6.19), and (6.20) that

$$\begin{aligned}
&\mathbb{E}\{V(k+1)-V(k)\}\\
\leq&\mathbb{E}\{\zeta^T(k)\bar{\Phi}\zeta(k)-\lambda_1(\mathcal{G}^T(k)\mathcal{G}(k)-\eta^T(k)\bar{M}\eta(k))
\end{aligned}$$

$$
\begin{aligned}
& - \lambda_2(\bar{\mathcal{G}}^T(k)\bar{\mathcal{G}}(k) - \eta_d^T(k)\hat{M}\eta_d(k)) \\
& - \lambda_3(\bar{s}^T(k)\bar{s}(k) - \eta^T(k)H_2^T C^T(K^T + I)\bar{s}(k) \\
& + \eta^T(k)H_2^T C^T K^T C H_2\eta(k)) - \lambda_4(\xi^T(k)\xi(k) - \delta) + \vartheta\} \\
=& \mathbb{E}\{\zeta^T(k)\Phi\zeta(k)\}\} + \lambda_4\delta + \vartheta \\
\leq & - \lambda_{\min}(-\Phi)\mathbb{E}\{\eta^T(k)\eta(k)\} + \lambda_4\delta + \vartheta.
\end{aligned} \tag{6.36}
$$

Next, let's estimate the upper bound of $\mathbb{E}\{\|\eta(T)\|^2\}$. *According to the definition of functional* $V(k)$, *it can be obtained that*

$$
V(k) \leq \lambda_{\max}(P)\|\eta(k)\|^2 + \sum_{i=1}^{n}\sum_{j=1}^{n}\lambda_{\max}(Q_{ij})\sum_{l=k-\tau_{ij}}^{k-1}\|\eta(l)\|^2. \tag{6.37}
$$

Introducing a scalar $\alpha > 1$, *we compute*

$$
\begin{aligned}
& \mathbb{E}\{\alpha^{k+1}V(k+1) - \alpha^k V(k)\} \\
=& \mathbb{E}\{\alpha^{k+1}(V(k+1) - V(k)) + \alpha^k(\alpha - 1)V(k)\} \\
\leq & \alpha^k\phi(\alpha)\mathbb{E}\{\|\eta(k)\|^2\} + \alpha^{k+1}(\lambda_4\delta + \vartheta) \\
& + \alpha^k\sum_{i=1}^{n}\sum_{j=1}^{n}\varphi_{ij}(\alpha)\sum_{l=k-\tau_{ij}}^{k-1}\mathbb{E}\{\|\eta(l)\|^2\},
\end{aligned} \tag{6.38}
$$

where $\phi(\alpha)$ *and* $\varphi_{ij}(\alpha)$ *are defined in (6.25).*

Denote $d = \max_{1\leq i,j\leq n}\{\tau_{ij}\}$. *For any integer* $T \geq d+1$, *summing up both sides of (6.38) from 0 to* $T-1$ *with respect to* k, *we have*

$$
\begin{aligned}
& \mathbb{E}\{\alpha^T V(T)\} - \mathbb{E}\{V(0)\} \\
\leq & \phi(\alpha)\sum_{k=0}^{T-1}\alpha^k\mathbb{E}\{\|\eta(k)\|^2\} + \frac{\alpha(1-\alpha^T)}{1-\alpha}(\lambda_4\delta + \vartheta) \\
& + \sum_{i=1}^{n}\sum_{j=1}^{n}\varphi_{ij}(\alpha)\sum_{k=0}^{T-1}\sum_{l=k-\tau_{ij}}^{k-1}\alpha^k\mathbb{E}\{\|\eta(l)\|^2\}.
\end{aligned} \tag{6.39}
$$

For the last term in (6.39), we further have

$$
\begin{aligned}
& \sum_{k=0}^{T-1}\sum_{l=k-\tau_{ij}}^{k-1}\alpha^k\mathbb{E}\{\|\eta(l)\|^2\} \\
\leq & \Big(\sum_{l=-\tau_{ij}}^{-1}\sum_{k=0}^{l+\tau_{ij}} + \sum_{l=0}^{T-\tau_{ij}-1}\sum_{k=l+1}^{l+\tau_{ij}} + \sum_{l=T-\tau_{ij}}^{T-1}\sum_{k=l+1}^{T-1}\Big)\alpha^k\mathbb{E}\{\|\eta(l)\|^2\} \\
\leq & \frac{\alpha^{\tau_{ij}} - 1}{\alpha - 1}\sum_{l=-\tau_{ij}}^{-1}\mathbb{E}\{\|\eta(l)\|^2\} + \frac{\alpha(\alpha^{\tau_{ij}} - 1)}{\alpha - 1}\sum_{l=0}^{T-1}\alpha^l\mathbb{E}\{\|\eta(l)\|^2\}
\end{aligned}
$$

$$+\frac{\alpha(\alpha^{\tau_{ij}-1}-1)}{\alpha-1}\sum_{l=0}^{T-1}\alpha^{l}\mathbb{E}\{\|\eta(l)\|^2\}. \tag{6.40}$$

From (6.39) and (6.40), it is easily known that

$$\begin{aligned}
&\mathbb{E}\{\alpha^{T}V(T)\}-\mathbb{E}\{V(0)\}\\
\le\ &\phi(\alpha)\sum_{k=0}^{T-1}\alpha^{k}\mathbb{E}\{\|\eta(k)\|^2\}+\frac{\alpha(1-\alpha^{T})}{1-\alpha}(\lambda_4\delta+\vartheta)\\
&+\sum_{i=1}^{n}\sum_{j=1}^{n}\varphi_{ij}(\alpha)\Big(\frac{\alpha^{\tau_{ij}}-1}{\alpha-1}\sum_{l=-\tau_{ij}}^{-1}\mathbb{E}\{\|\eta(l)\|^2\}\\
&+\frac{\alpha(\alpha^{\tau_{ij}}-1)}{\alpha-1}\sum_{l=0}^{T-1}\alpha^{l}\mathbb{E}\{\|\eta(l)\|^2\}+\frac{\alpha(\alpha^{\tau_{ij}-1}-1)}{\alpha-1}\sum_{l=0}^{T-1}\alpha^{l}\mathbb{E}\{\|\eta(l)\|^2\}\Big)\\
\le\ &\zeta(\alpha)\sum_{k=0}^{T-1}\alpha^{k}\mathbb{E}\{\|\eta(k)\|^2\}+\frac{\alpha(1-\alpha^{T})}{1-\alpha}(\lambda_4\delta+\vartheta)\\
&+\frac{1}{\alpha-1}\sum_{i=1}^{n}\sum_{j=1}^{n}\varphi_{ij}(\alpha)\tau_{ij}(\alpha^{\tau_{ij}}-1)\max_{-\tau_{ij}\le l\le 0}\mathbb{E}\{\|\eta(l)\|^2\}
\end{aligned} \tag{6.41}$$

where

$$\zeta(\alpha)=\phi(\alpha)+\frac{1}{\alpha-1}\sum_{i=1}^{n}\sum_{j=1}^{n}\varphi_{ij}(\alpha)(\alpha^{\tau_{ij}+1}+\alpha^{\tau_{ij}}-2\alpha). \tag{6.42}$$

Note that $\zeta(1)=-\lambda_{\min}(-\Phi)<0$ *and* $\lim_{\alpha\to+\infty}\zeta(\alpha)=+\infty$. *Therefore, there exists a scalar* $\alpha_0>1$ *such that* $\zeta(\alpha_0)=0$. *Then, it follows from (6.41) that*

$$\begin{aligned}
&\mathbb{E}\{\alpha_0^{T}V(T)\}-\mathbb{E}\{V(0)\}\\
\le&\frac{\alpha_0(1-\alpha_0^{T})}{1-\alpha_0}(\lambda_4\delta+\vartheta)+\frac{1}{\alpha_0-1}\sum_{i=1}^{n}\sum_{j=1}^{n}\tau_{ij}\\
&\times\varphi_{ij}(\alpha_0)(\alpha_0^{\tau_{ij}}-1)\max_{-\tau_{ij}\le l\le 0}\mathbb{E}\{\|\eta(l)\|^2\}.
\end{aligned} \tag{6.43}$$

Considering

$$\begin{aligned}
\mathbb{E}\{V(0)\}\le\ &\lambda_{\max}(P)\mathbb{E}\{\|\eta(0)\|^2\}+\sum_{i=1}^{n}\sum_{j=1}^{n}\lambda_{\max}(Q_{ij})\sum_{l=-\tau_{ij}}^{-1}\|\eta(l)\|^2\\
\le&\sum_{i=1}^{n}\sum_{j=1}^{n}\frac{1}{n^2}\lambda_{\max}(P)\max_{-\tau_{ij}\le l\le 0}\mathbb{E}\{\|\eta(l)\|^2\}
\end{aligned}$$

$$
\begin{aligned}
&+\sum_{i=1}^{n}\sum_{j=1}^{n}\tau_{ij}\lambda_{\max}(Q_{ij})\max_{-\tau_{ij}\le l\le 0}\mathbb{E}\{\|\eta(l)\|^2\}\\
\le&\sum_{i=1}^{n}\sum_{j=1}^{n}\max\big(\frac{1}{n^2}\lambda_{\max}(P),\tau_{ij}\lambda_{\max}(Q_{ij})\big)\\
&\times\max_{-\tau_{ij}\le l\le 0}\mathbb{E}\{\|\eta(l)\|^2\} \qquad (6.44)
\end{aligned}
$$

and

$$\mathbb{E}\{\alpha_0^T V(T)\}\ge\lambda_{\min}(P)\alpha_0^T\mathbb{E}\{\|\eta(T)\|^2\}, \qquad (6.45)$$

we have

$$\mathbb{E}\{\|\eta(T)\|^2\}\le\frac{\alpha_0^T-1}{\alpha_0^{T-1}(\alpha_0-1)\lambda_{\min}(P)}(\lambda_4\delta+\vartheta)+\frac{\varpi(\alpha_0)}{\alpha_0^T\lambda_{\min}(P)} \qquad (6.46)$$

where

$$
\begin{aligned}
\varpi(\alpha_0)=&\sum_{i=1}^{n}\sum_{j=1}^{n}\Big(\frac{\tau_{ij}\varphi_{ij}(\alpha_0)(\alpha_0^{\tau_{ij}}-1)}{\alpha_0-1}\\
&+\max\big(\frac{1}{n^2}\lambda_{\max}(P),\tau_{ij}\lambda_{\max}(Q_{ij})\big)\Big)\max_{-\tau_{ij}\le l\le 0}\mathbb{E}\{\|\eta(l)\|^2\}.
\end{aligned} \qquad (6.47)
$$

By taking $\mu=\frac{1}{\alpha_0}$, $\kappa(T)=\frac{\alpha_0^T-1}{\alpha_0^{T-1}(\alpha_0-1)\lambda_{\min}(P)}(\lambda_4\delta+\vartheta)$, *and* $\nu=\frac{\varpi(\alpha_0)}{\lambda_{\min}(P)}$, *it follows easily from Definition 6.1 that the augmented system (6.13) is exponentially mean-square ultimately bounded and the ultimate bounded is given by*

$$\bar{\kappa}=\lim_{T\to+\infty}\kappa(T)=\frac{\alpha_0}{(\alpha_0-1)\lambda_{\min}(P)}(\lambda_4\delta+\vartheta). \qquad (6.48)$$

The proof of Theorem 6.1 is complete.

Remark 6.5 *In Theorem 6.1, a sufficient condition is provided under which the dynamics of the augmented system (6.13) is exponentially mean-square ultimately bounded and the ultimate bound of the error dynamics is given. It can be seen from (6.23) that the ultimate bound is dependent on both the event-triggering threshold and the variances of external/measurement noises. Such a bound can be reduced by decreasing the event-triggering threshold or depressing the noises' variances. Note that the ultimate boundedness is also related to other matrix parameters.*

Remark 6.6 *The measurement model (6.6) has been firstly proposed in [135]. In the proof of the main results of [135], the last three terms of (6.17) were assumed to be unrelated, which is actually not the case. Fortunately, by using Lemma 6.3 and following the lines in the proof of Theorem 6.1 in this section, the main results in [135] can be easily rectified.*

By means of Theorem 6.1, the design problem of the desired estimator shall be investigated. For the convenience of design, the positive definite matrix P is taken as $P = \text{diag}\{P_{11}, P_{22}\}$, where P_{11} and P_{22} are positive definite matrices.

In the following theorem, the design method of the desired estimator gain is given in terms of the solution to an LMI.

Theorem 6.2 *Consider the stochastic neural network given by (6.1) with the IMs described by (6.6). If there exist positive definite matrices* $P = \text{diag}\{P_{11}, P_{22}\}$, Q_{ij} $(i,j = 1,2,\cdots,n)$, *matrix* Y *and positive scalars* λ_1, λ_2, λ_3, λ_4, ϵ *satisfying the following inequality:*

$$\Upsilon = \begin{bmatrix} \Xi_{11} & 0 & \Xi_{13} & 0 & \Xi_{15} & \Xi_{16} & \Xi_{17} & 0 & 0 & 0 \\ * & \Xi_{22} & \Xi_{23} & 0 & \Xi_{25} & 0 & 0 & \Xi_{28} & 0 & 0 \\ * & * & \Xi_{33} & 0 & 0 & 0 & 0 & 0 & \Xi_{39} & 0 \\ * & * & * & \Xi_{44} & 0 & 0 & 0 & 0 & 0 & 0 \\ * & * & * & * & \Xi_{55} & 0 & 0 & 0 & 0 & 0 \\ * & * & * & * & * & \Xi_{66} & 0 & 0 & 0 & 0 \\ * & * & * & * & * & * & \Xi_{77} & 0 & 0 & \Xi_{70} \\ * & * & * & * & * & * & * & \Xi_{88} & 0 & 0 \\ * & * & * & * & * & * & * & * & \Xi_{99} & 0 \\ * & * & * & * & * & * & * & * & * & \Xi_{00} \end{bmatrix} < 0 \quad (6.49)$$

where

$$\begin{aligned}
\Xi_{11} &= H_2^T \tilde{P} H_2 - P + \lambda_1 \bar{M} + \sum_{i=1}^{n}\sum_{j=1}^{n} Q_{ij} - \lambda_3 H_2^T C^T K C H_2 \\
&\quad + \epsilon H_2^T C^T \Lambda_q^2 C H_2, \\
\Xi_{13} &= \frac{1}{2}\lambda_3 H_2^T C^T (K^T + I), \quad \Xi_{15} = \mathcal{A}_Y^T - \beta_1 H_2^T C^T G_Y^T, \\
\Xi_{16} &= \left[\sqrt{\beta_2+\beta_3}\mathcal{A}_Y^T \quad \sqrt{\beta_1(1-\beta_1)} H_2^T C^T G_Y^T\right], \\
\Xi_{17} &= \sqrt{5\beta_3} H_2^T C^T G_Y^T, \quad \Xi_{23} = \left[-\beta_2 G_Y^T \bar{W}_0 \quad -\beta_2 G_Y^T \bar{W}_1 \quad 0\right]^T, \\
\Xi_{25} &= \left[P\bar{W}_0 \quad P\bar{W}_1 \quad G_Y\right]^T, \quad \Xi_{28} = \left[0 \quad 0 \quad \sqrt{\beta_2+\beta_3} G_Y\right]^T, \\
\Xi_{22} &= \text{diag}\{\beta_3 \bar{W}_0^T P \bar{W}_0 - \lambda_1 I, \beta_3 \bar{W}_1^T P \bar{W}_1 - \lambda_2 I, -\lambda_4 I\}, \\
\Xi_{33} &= -\lambda_3 I, \quad \Xi_{39} = \sqrt{3\beta_2} G_Y^T, \quad \Xi_{44} = -Q + \lambda_2 \hat{M}, \\
\Xi_{55} &= \Xi_{77} = \Xi_{88} = \Xi_{99} = -P, \quad \Xi_{66} = \text{diag}_2\{-P\}, \\
\Xi_{70} &= \sqrt{5\beta_3} G_Y, \quad \Xi_{00} = -\epsilon I, \quad G_Y = \left[0 \quad Y^T\right]^T, \\
\mathcal{A}_Y &= \begin{bmatrix} P_{11}\bar{A} & 0 \\ YC & P_{22}\bar{A} - YC \end{bmatrix},
\end{aligned}$$

then, the design problem of the desired state estimator (6.10) is solvable. Furthermore, if the inequality (6.49) is feasible, the desired state estimator gain is given by

$$G = P_{22}^{-1}Y. \tag{6.50}$$

Proof *In terms of Theorem 6.1, we just need to show that inequality (6.49) in Theorem 6.2 holds implies inequality (6.22) in Therem 6.1 holds.*

By using the Schur Complement Lemma, it is easily known that $\Phi < 0$ *is equivalent to*

$$\tilde{\Psi} = \begin{bmatrix} \bar{\Xi}_{11} & 0 & \Xi_{13} & 0 & \bar{\Xi}_{15} & \bar{\Xi}_{16} & \bar{\Xi}_{17} & 0 & 0 \\ * & \Xi_{22} & \bar{\Xi}_{23} & 0 & \bar{\Xi}_{25} & 0 & 0 & \bar{\Xi}_{28} & 0 \\ * & * & \Xi_{33} & 0 & 0 & 0 & 0 & 0 & \bar{\Xi}_{39} \\ * & * & * & \Xi_{44} & 0 & 0 & 0 & 0 & 0 \\ * & * & * & * & \Xi_{55} & 0 & 0 & 0 & 0 \\ * & * & * & * & * & \Xi_{66} & 0 & 0 & 0 \\ * & * & * & * & * & * & \Xi_{77} & 0 & 0 \\ * & * & * & * & * & * & * & \Xi_{88} & 0 \\ * & * & * & * & * & * & * & * & \Xi_{99} \end{bmatrix} < 0$$

where

$$\begin{aligned}
\bar{\Xi}_{11} &= H_2^T \tilde{P} H_2 - P + \lambda_1 \bar{M} + \sum_{i=1}^{n}\sum_{j=1}^{n} Q_{ij} - \lambda_3 H_2^T C^T K C H_2, \\
\bar{\Xi}_{15} &= \mathcal{A}^T P - \beta_1 H_2^T C^T G^T H_1^T P, \ \bar{\Xi}_{17} = \sqrt{5\beta_3} U^T G^T H_1^T P, \\
\bar{\Xi}_{16} &= \left[\sqrt{\beta_2+\beta_3}\mathcal{A}^T P \quad \sqrt{\beta_1(1-\beta_1)} H_2^T C^T G^T H_1^T P\right], \\
\bar{\Xi}_{23} &= \left[-\beta_2 G^T H_1^T P \bar{W}_0 \quad -\beta_2 G^T H_1^T P \bar{W}_1 \quad 0\right]^T, \\
\bar{\Xi}_{25} &= \left[P\bar{W}_0 \quad P\bar{W}_1 \quad P H_1 G\right]^T, \quad \bar{\Xi}_{39} = \sqrt{3\beta_2} G^T H_1^T P, \\
\bar{\Xi}_{28} &= \left[0 \quad 0 \quad \sqrt{\beta_2+\beta_3} P H_1 G\right]^T.
\end{aligned}$$

Let's now deal with the uncertainty induced by quantization effect. Set

$$\begin{aligned}
T &= \left[0 \quad 0 \quad 0 \quad 0 \quad 0 \quad 0 \quad \sqrt{5\beta_3} G^T H_1^T P \quad 0 \quad 0\right]^T, \\
N &= \left[\Lambda_q C H_2 \quad 0 \quad 0 \quad 0 \quad 0 \quad 0 \quad 0 \quad 0 \quad 0\right].
\end{aligned}$$

The matrix $\tilde{\Psi}$ *can then be written as*

$$\tilde{\Psi} = \bar{\Psi} + TFN + (TFN)^T$$

where

$$\bar{\Psi} = \begin{bmatrix} \bar{\Xi}_{11} & 0 & \Xi_{13} & 0 & \bar{\Xi}_{15} & \bar{\Xi}_{16} & \tilde{\Xi}_{17} & 0 & 0 \\ * & \Xi_{22} & \bar{\Xi}_{23} & 0 & \bar{\Xi}_{25} & 0 & 0 & \bar{\Xi}_{28} & 0 \\ * & * & \Xi_{33} & 0 & 0 & 0 & 0 & 0 & \bar{\Xi}_{39} \\ * & * & * & \Xi_{44} & 0 & 0 & 0 & 0 & 0 \\ * & * & * & * & \Xi_{55} & 0 & 0 & 0 & 0 \\ * & * & * & * & * & \Xi_{66} & 0 & 0 & 0 \\ * & * & * & * & * & * & \Xi_{77} & 0 & 0 \\ * & * & * & * & * & * & * & \Xi_{88} & 0 \\ * & * & * & * & * & * & * & * & \Xi_{99} \end{bmatrix}$$

with $\tilde{\Xi}_{17} = \sqrt{5\beta_3} H_2^T C^T G^T H_1^T P$.

By using Lemma 3.2 and the Schur Complement Lemma again, it is shown that $\tilde{\Psi} < 0$ *if and only if*

$$\tilde{\Upsilon} = \begin{bmatrix} \Xi_{11} & 0 & \Xi_{13} & 0 & \bar{\Xi}_{15} & \bar{\Xi}_{16} & \tilde{\Xi}_{17} & 0 & 0 & 0 \\ * & \Xi_{22} & \bar{\Xi}_{23} & 0 & \bar{\Xi}_{25} & 0 & 0 & \bar{\Xi}_{28} & 0 & 0 \\ * & * & \Xi_{33} & 0 & 0 & 0 & 0 & 0 & \bar{\Xi}_{39} & 0 \\ * & * & * & \Xi_{44} & 0 & 0 & 0 & 0 & 0 & 0 \\ * & * & * & * & \Xi_{55} & 0 & 0 & 0 & 0 & 0 \\ * & * & * & * & * & \Xi_{66} & 0 & 0 & 0 & 0 \\ * & * & * & * & * & * & \Xi_{77} & 0 & 0 & \bar{\Xi}_{70} \\ * & * & * & * & * & * & * & \Xi_{88} & 0 & 0 \\ * & * & * & * & * & * & * & * & \Xi_{99} & 0 \\ * & * & * & * & * & * & * & * & * & \Xi_{00} \end{bmatrix} < 0 \quad (6.51)$$

where $\bar{\Xi}_{70} = \sqrt{5\beta_3} P H_1 G$.

By considering the relation $Y = P_{22}G$*, it can be seen that matrix* Υ *defined in Theorem 6.2 is the exactly same as matrix* $\tilde{\Upsilon}$ *given in (6.51) which means that inequality (6.49) holds implies inequality (6.22) holds. Therefore, the rest of the proof follows from that of Theorem 6.1 immediately.*

Remark 6.7 *In this section, the ET state estimation problem is investigated for a class of multi-delayed neural networks with SPs and IMs. The main novelty can be summarized as follows: 1) the SPs are introduced for the first time to account for the fluctuations of the capacitance and resistance in the neural circuits, 2) different time delays for different interconnections are taken into account which extends ones in the existing literature, and 3) the event-triggering mechanism is employed to estimate the neuron's states of multi-delayed neural networks with SPs. Moreover, we adopt a unified measurement model, which is capable of reflecting randomly occurring sensor saturations and randomly occurring quantizations. Finally, the state estimator is designed such that, for all admissible IMs as well as the SPs, the estimation error dynamics is exponentially mean-square ultimately bounded.*

6.2 Event-Triggered H_∞ State Estimation in Genetic Regulatory Networks

In this section, we discuss the ET H_∞ state estimation for a class of discrete-time GRNs with MJPs and TVDs.

6.2.1 Problem Formulation

Consider the following discrete-time GRN with TVDs:

$$\begin{aligned} m(k+1) &= Am(k) + Bf(p(k-\tau(k))) + E_m v(k) + L, \\ p(k+1) &= Cp(k) + Dm(k-\theta(k)) + E_p v(k) \end{aligned} \tag{6.52}$$

where $m(k) = \begin{bmatrix} m_1(k) & \dots & m_n(k) \end{bmatrix}^T \in \mathbb{R}^n$ and $p(k) = \begin{bmatrix} p_1(k) & \dots & p_n(k) \end{bmatrix}^T \in \mathbb{R}^n$ are, respectively, the concentrations of mRNA and protein at time k; the constant matrices $A = \mathrm{diag}\{a_1 \ \cdots \ a_n\}$ and $C = \mathrm{diag}\{c_1 \ \cdots \ c_n\}$ stand for the rate of degradation with entries $|a_i| < 1$ and $|c_i| < 1$; $B = (b_{ij})_{n\times n}$ is the coupling matrix of the genetic network; $D = \mathrm{diag}\{d_1 \ \cdots \ d_n\}$ is the translation rate; $f(p(k-\tau(k))) = \begin{bmatrix} f_1(p_1(k-\tau(k))) & \cdots & f_n(p_n(k-\tau(k))) \end{bmatrix}^T \in \mathbb{R}^n$ is a nonlinear function with $f_i(\cdot)$ denoting the feedback regulation of the protein on the transcription; $\tau(k)$ and $\theta(k)$ are the feedback regulation and the translation delays which satisfy $0 \le \tau_m \le \tau(k) \le \tau_M$ and $0 \le \theta_m \le \theta(k) \le \theta_M$, respectively. $L = \begin{bmatrix} l_1 & \cdots & l_n \end{bmatrix}^T$ is a constant vector with l_i standing for the basal transcriptional rate of the repressor of gene i; $v(k) \in \mathbb{R}^s$ represents the exogenous disturbance belonging to $l_2[0,\infty)$; $E_m = \begin{bmatrix} e_{m1} & \dots & e_{mn} \end{bmatrix}^T$ and $E_p = \begin{bmatrix} e_{p1} & \dots & e_{pn} \end{bmatrix}^T$ are matrices characterizing the intensities of the exogenous disturbance to the mRNA and the protein, respectively.

Assumption 6.1 *The nonlinear function $f_i(\cdot)$ $(i = 1, 2, \dots, n)$ satisfies*

$$0 \le \frac{f_i(s_1) - f_i(s_2)}{s_1 - s_2} \le u_i \tag{6.53}$$

for all $s_1, s_2 \in \mathbb{R}$ with $s_1 \neq s_2$, where u_i is a known positive constant.

Suppose that the equilibrium point of system (6.52) are $p^* = \begin{bmatrix} p_1^* & \dots & p_n^* \end{bmatrix}^T$ and $m^* = \begin{bmatrix} m_1^* & \dots & m_n^* \end{bmatrix}^T$. Letting $\bar{m}(k) = m(k) - m^*$ and $\bar{p}(k) = p(k) - p^*$, system (6.52) can be transformed into

$$\begin{aligned} \bar{m}(k+1) &= A\bar{m}(k) + B\tilde{f}(\bar{p}(k-\tau(k))) + E_m v(k), \\ \bar{p}(k+1) &= C\bar{p}(k) + D\bar{m}(k-\theta(k)) + E_p v(k) \end{aligned} \tag{6.54}$$

where $\tilde{f}(\bar{p}(k-\tau(k))) = f(\bar{p}(k-\tau(k)) + p^*) - f(p^*)$.

From Assumption 6.1, it is easily known that the nonlinear function $\tilde{f}(\cdot)$ satisfies

$$\tilde{f}^T(\bar{p})(\tilde{f}(\bar{p}) - U\bar{p}) \leq 0 \tag{6.55}$$

for all $\bar{p} \in \mathbb{R}^n$, where $U = \text{diag}\{u_1, \cdots, u_n\}$.

Remark 6.8 *In the modelling process of genetic networks, it is well-known that time delays are often unavoidable due to the finite speed in the slow process of transcription, translation, and translocation. It has been pointed out in [177] that the delays play an important role in dynamics of genetic networks, and the mathematical models without addressing the delay effects may provide wrong predictions of the mRNA and protein concentrations. Therefore, it is necessary to consider the time delays in the model of GRNs.*

In real GRNs, the system parameters may be subject to unexpected random changes and may exhibit unwanted uncertainties which may occur in a probabilistic way. In addition, due to the existence of intrinsic molecular fluctuations in real genetic networks, it is necessary to consider the stochastic perturbations in the gene regulation process. In order to study the effect of these phenomena on the estimation performance, we modify the model of GRNs given by (6.54) as follows.

Let $r(k)$ be a homogeneous Markovian chain taking values in a finite state space $\mathcal{S} = \{1, 2, \ldots, p\}$ with the transition probability matrix $\Lambda = [\pi_{ij}]_{p\times p}$ given by

$$P\{r(k+1) = j | r(k) = i\} = \pi_{ij}, \quad i, j \in \mathcal{S} \tag{6.56}$$

where $\pi_{ij} \geq 0$ and $\sum_{j=1}^{p} \pi_{ij} = 1$ and consider the following class of comprehensive GRNs:

$$\begin{aligned} \bar{m}(k+1) =& A(r(k))\bar{m}(k) + B(r(k))\tilde{f}(\bar{p}(k-\tau(k))) \\ &+ E_m(r(k))v(k) + g_m(\bar{m}(k), \bar{p}(k))w_m(k), \\ \bar{p}(k+1) =& C(r(k))\bar{p}(k) + D(r(k))\bar{m}(k-\theta(k)) \\ &+ E_p(r(k))v(k) + g_p(\bar{m}(k), \bar{p}(k))w_p(k) \end{aligned} \tag{6.57}$$

where $w_m(k) \in \mathbb{R}$ and $w_p(k) \in \mathbb{R}$ are independent stochastic perturbations satisfying

$$\begin{aligned} &\mathbb{E}\{w_m(k)\} = 0, \quad \mathbb{E}\{w_m^2(k)\} = 1, \quad \mathbb{E}\{w_m(k)w_m(l)\} = 0, \ \ k \neq l, \\ &\mathbb{E}\{w_p(k)\} = 0, \quad \mathbb{E}\{w_p^2(k)\} = 1, \quad \mathbb{E}\{w_p(k)w_p(l)\} = 0, \ \ k \neq l, \end{aligned} \tag{6.58}$$

and the intensity functions $g_m(\cdot,\cdot)$ and $g_p(\cdot,\cdot)$ satisfy

$$\begin{aligned} g_m^T(\bar{m}(k), \bar{p}(k))g_m(\bar{m}(k), \bar{p}(k)) \leq \bar{m}^T(k)G_{m1}\bar{m}(k) + \bar{p}^T(k)G_{m2}\bar{p}(k), \\ g_p^T(\bar{m}(k), \bar{p}(k))g_p(\bar{m}(k), \bar{p}(k)) \leq \bar{m}^T(k)G_{p1}\bar{m}(k) + \bar{p}^T(k)G_{p2}\bar{p}(k), \end{aligned} \tag{6.59}$$

respectively, where $G_{mi} \geq 0$ and $G_{pi} \geq 0$ $(i = 1, 2)$ are known matrices. Here, it is assumed that the Markovian chain $r(k)$ is independent of the stochastic perturbations $w_m(k)$ and $w_p(k)$.

As suggested in [138], the measurement model is described as follows:

$$y(k) = L_m(r(k))\bar{m}(k) + L_p(r(k))\bar{p}(k) \tag{6.60}$$

where $y(k) \in \mathbb{R}^q$ (usually, $q < n$) is the measurement output at time instant k and $L_m(r(k))$, $L_p(r(k))$ are two known constant matrices with appropriate dimensions.

By setting $x(k) = \begin{bmatrix}\bar{m}^T(k) & \bar{p}^T(k)\end{bmatrix}^T$, the following augmented system can be obtained from (6.57) and (6.60):

$$\begin{aligned} x(k+1) =& \tilde{A}(r(k))x(k) + \tilde{B}(r(k))\tilde{f}(Nx(k-\tau(k))) + \tilde{D}(r(k)) \\ &\times x(k-\theta(k)) + \tilde{E}(r(k))v(k) + \tilde{g}(x(k))w(k), \\ y(k) =& \tilde{L}(r(k))x(k), \end{aligned} \tag{6.61}$$

and the constraint conditions (6.55) and (6.59) become

$$\tilde{f}^T(Nx(k-\tau(k)))\tilde{f}(Nx(k-\tau(k))) \leq \tilde{f}^T(Nx(k-\tau(k))\tilde{U}x(k-\tau(k)), \tag{6.62}$$
$$\text{trace}(\tilde{g}^T(x(k))\tilde{g}(x(k))) \leq x^T(k)\tilde{G}x(k) \tag{6.63}$$

where

$$\begin{aligned} &\tilde{L}(r(k)) = \begin{bmatrix}L_m(r(k)) & L_p(r(k))\end{bmatrix}, \ \tilde{U} = \begin{bmatrix}0 & U\end{bmatrix}, \ N = \begin{bmatrix}0 & I\end{bmatrix}, \\ &\tilde{D}(r(k)) = \begin{bmatrix}0 & 0 \\ D(r(k)) & 0\end{bmatrix}, \ \tilde{A}(r(k)) = \begin{bmatrix}A(r(k)) & 0 \\ 0 & C(r(k))\end{bmatrix}, \\ &\tilde{B}(r(k)) = \begin{bmatrix}B(r(k)) \\ 0\end{bmatrix}, \ \tilde{E}(r(k)) = \begin{bmatrix}E_m(r(k)) \\ E_p(r(k))\end{bmatrix}, \ w(k) = \begin{bmatrix}w_m(k) \\ w_p(k)\end{bmatrix}, \\ &\tilde{g}(x(k)) = \begin{bmatrix}g_m(\bar{m}(k), \bar{p}(k)) & 0 \\ 0 & g_p(\bar{m}(k), \bar{p}(k))\end{bmatrix}, \ \tilde{G} = \begin{bmatrix}G_{m1} + G_{P1} & 0 \\ 0 & G_{m2} + G_{p2}\end{bmatrix}. \end{aligned}$$

In this section, the ET strategy is adopted to save computer resources. Denote by $\{r_0, r_1, \ldots\}$ the sequence of the time instants at which the measurement output is transmitted to the estimator. Each time instant r_l is defined according to the following event-triggering condition:

$$\varepsilon^T(k)\varepsilon(k) > \sigma y^T(k)y(k), \quad \sigma > 0 \tag{6.64}$$

where $\varepsilon(k) = y(k) - y(r_l)$ is the event error for $k \in [r_l, r_{l+1})$, $l = 0, 1, 2, \ldots, \infty$. Note that the measurement outputs satisfying (6.64) will be sent to the estimator.

We construct an event-based estimator as follows:

$$\hat{x}(k+1) = \hat{A}(r(k))\hat{x}(k) + \hat{G}(r(k))y(r_l) \tag{6.65}$$

for $k \in [r_l, r_{l+1})$, $l = 0, 1, 2, \ldots, \infty$, where $\hat{x}(k) \in \mathbb{R}^{2n}$ is the estimate for $x(t)$; $\hat{A}(r(k)) \in \mathbb{R}^{2n\times 2n}$ and $\hat{G}(r(k)) \in \mathbb{R}^{2n\times p}$ are the estimator parameters to be determined. The initial value of the estimator is set as $\hat{x}(k) = 0$ for $k \leq 0$.

Note that, for each possible value $r(k) = i$ ($i \in \mathcal{S}$), we denote

$$\begin{aligned}&\tilde{A}(r(k)=i)=\tilde{A}_i,\ \tilde{B}(r(k)=i)=\tilde{B}_i,\ \tilde{D}(r(k)=i)=\tilde{D}_i,\ \tilde{E}(r(k)=i)=\tilde{E}_i,\\ &\hat{A}(r(k)=i)=\hat{A}_i,\ \hat{G}(r(k)=i)=\hat{G}_i,\ \tilde{L}(r(k)=i)=\tilde{L}_i.\end{aligned} \tag{6.66}$$

Denoting $e(k) = x(k) - \hat{x}(k)$, for each $i \in \mathcal{S}$, we have the following estimation error system for the GRN (6.57):

$$\begin{aligned}x(k+1)=&\tilde{A}_i x(k)+\tilde{B}_i\tilde{f}(Nx(k-\tau(k)))+\tilde{D}_i x(k-\theta(k))\\&+E_i v(k)+\tilde{g}(x(k))w(k),\\ e(k+1)=&(\tilde{A}_i-\hat{A}_i-\hat{G}_i\tilde{L}_i)x(k)+\hat{A}_i e(k)+\hat{G}_i\varepsilon(k)+\tilde{B}_i\tilde{f}(Nx(k-\tau(k)))\\&+\tilde{D}_i x(k-\theta(k))+\tilde{E}_i v(k)+\tilde{g}(x(k))w(k).\end{aligned} \tag{6.67}$$

Definition 6.2 *[161] The system (6.67) with $v(k) = 0$ is said to be stochastically stable if, for every initial state $\Re_0 = \{x(k), e(k), k = -d, -d+1, \ldots, 0\}$ with $d = \max\{\tau_M, \theta_M\}$, the following condition holds:*

$$\mathbb{E}\{\sum_{k=0}^{\infty}(\|x(k)\|^2+\|e(k)\|^2)|\Re_0, r(0)\} < \infty. \tag{6.68}$$

The aim of this section is to design a state estimator of form (6.65) such that the following two requirements are satisfied simultaneously.
(1) The estimation error system (6.67) with $v(k) = 0$ is stochastically stable;
(2) Under the zero-initial condition, the estimation error satisfies the following H_∞ performance constraint:

$$\mathbb{E}\{\sum_{k=0}^{\infty}\|e(k)\|^2\} < \gamma^2\sum_{k=0}^{\infty}\|v(k)\|^2 \tag{6.69}$$

for all nonzero $v(k)$, where $\gamma > 0$ is a given disturbance attenuation level.

6.2.2 Main Results

In this subsection, we first deal with the stability as well as H_∞ performance analysis issue for the estimation error dynamics (6.67). Then, based on the analysis results, the ET H_∞ estimator is designed for the GRN (6.57).

In the following theorem, a sufficient condition is derived to guarantee the stochastic stability of estimation error dynamics (6.67) with $v(k) = 0$.

Theorem 6.3 *Given estimator parameters $\hat{A}_i$ and $\hat{G}_i$ ($i \in \mathcal{S}$), the system (6.67) is stochastically stable with $v(k) = 0$ if there exist positive definite matrices P_{1i}, P_{2i}, Q_1, Q_2, T_1, T_2 and a positive scalar ρ such that*

$$\tilde{P}_{1i} + \tilde{P}_{2i} \leq \rho I, \tag{6.70}$$

$$\Pi_i = \begin{bmatrix} \Upsilon_{11i} & \Upsilon_{12i}^T & \Upsilon_{13i}^T \\ * & -\tilde{P}_{1i}^{-1} & 0 \\ * & * & -\tilde{P}_{2i}^{-1} \end{bmatrix} < 0 \tag{6.71}$$

where

$$\Upsilon_{11i} = \begin{bmatrix} \Phi_{1i} & 0 & 0 & 0 & 0 & 0 & 0 & 0 \\ * & -Q_1 & 0 & 0 & 0 & 0 & 0 & \frac{1}{2}\tilde{U}^T \\ * & * & -Q_2 & 0 & 0 & 0 & 0 & 0 \\ * & * & * & \Phi_{4i} & 0 & 0 & 0 & 0 \\ * & * & * & * & -T_1 & 0 & 0 & 0 \\ * & * & * & * & * & -T_2 & 0 & 0 \\ * & * & * & * & * & * & -I & 0 \\ * & * & * & * & * & * & * & -I \end{bmatrix},$$

$$\begin{aligned}
&\tilde{P}_{1i} = \sum_{j=1}^{p} \pi_{ij} P_{1j}, \quad \tilde{P}_{2i} = \sum_{j=1}^{p} \pi_{ij} P_{2j}, \\
&\Phi_{4i} = -P_{2i} + (1 + \tau_M - \tau_m)T_1 + (1 + \theta_M - \theta_m)T_2, \\
&\Phi_{1i} = -P_{1i} + (1 + \tau_M - \tau_m)Q_1 + (1 + \theta_M - \theta_m)Q_2 + \rho\tilde{G} + \sigma\tilde{L}_i^T\tilde{L}_i, \\
&\Upsilon_{12i} = \begin{bmatrix} \tilde{A}_i & 0 & \tilde{D}_i & 0 & 0 & 0 & 0 & \tilde{B}_i \end{bmatrix}, \\
&\Upsilon_{13i} = \begin{bmatrix} \tilde{A}_i - \hat{A}_i - \hat{G}_i\tilde{L}_i & 0 & \tilde{D}_i & \hat{A}_i & 0 & 0 & \hat{G}_i & \tilde{B}_i \end{bmatrix}.
\end{aligned}$$

Proof *Denote $\Re_k = \{x(k-d), e(k-d), \ldots, x(k), e(k)\}$ and construct the following Lyapunov functional candidate for the estimation error system (6.67):*

$$\begin{aligned} V(k, \Re_k, r(k)) =& V_1(k, \Re_k, r(k)) + V_2(k, \Re_k, r(k)) + V_3(k, \Re_k, r(k)) \\ &+ V_4(k, \Re_k, r(k)) + V_5(k, \Re_k, r(k)) \end{aligned} \tag{6.72}$$

where

$$\begin{aligned}
V_1(k, \Re_k, r(k)) =& x^T(k)P_{1r(k)}x(k) + e^T(k)P_{2r(k)}e(k), \\
V_2(k, \Re_k, r(k)) =& \sum_{\mu=k-\tau(k)}^{k-1} x^T(\mu)Q_1x(\mu) + \sum_{\mu=k-\theta(k)}^{k-1} x^T(\mu)Q_2x(\mu), \\
V_3(k, \Re_k, r(k)) =& \sum_{\mu=k-\tau(k)}^{k-1} e^T(\mu)T_1e(\mu) + \sum_{\mu=k-\theta(k)}^{k-1} e^T(\mu)T_2e(\mu),
\end{aligned}$$

$$V_4(k,\Re_k,r(k)) = \sum_{\nu=-\tau_M+1}^{-\tau_m}\sum_{\mu=k+\nu}^{k-1} x^T(\mu)Q_1x(\mu) + \sum_{\nu=-\theta_M+1}^{-\theta_m}\sum_{\mu=k+\nu}^{k-1} x^T(\mu)Q_2x(\mu),$$

$$V_5(k,\Re_k,r(k)) = \sum_{\nu=-\tau_M+1}^{-\tau_m}\sum_{\mu=k+\nu}^{k-1} e^T(\mu)T_1e(\mu) + \sum_{\nu=-\theta_M+1}^{-\theta_m}\sum_{\mu=k+\nu}^{k-1} e^T(\mu)T_2e(\mu).$$

For $r(k) = i$ and $r(k+1) = j$, by noting (6.56) and calculating the difference of $V(k,\Re_k,r(k))$ along system (6.67) with $v(k) = 0$, it can be easily obtained that

$$\begin{aligned}
&\mathbb{E}\{\triangle V_1(k)|\Re_k,r(k)\}\\
&\mathbb{E}\{[V_1(k+1,\Re_{k+1},r(k+1)) - V_1(k,\Re_k,r(k))]|\Re_k,r(k)\}\\
=&\zeta^T(k)(\Upsilon_{12i}^T\tilde{P}_{1i}\Upsilon_{12i} + \Upsilon_{13i}^T\tilde{P}_{2i}\Upsilon_{13i})\zeta(k) - x^T(k)P_{1i}x(k) - e^T(k)P_{2i}e(k)\\
&+ \mathrm{trace}[\tilde{g}^T(x(k))(\tilde{P}_{1i}+\tilde{P}_{2i})\tilde{g}(x(k))],
\end{aligned} \tag{6.73}$$

$$\begin{aligned}
&\mathbb{E}\{\triangle V_2(k)|\Re_k,r(k)\}\\
=&\mathbb{E}\{[V_2(k+1,\Re_{k+1},r(k+1)) - V_2(k,\Re_k,r(k))]|\Re_k,r(k)\}\\
\le&x^T(k)Q_1x(k) - x^T(k-\tau(k))Q_1x(k-\tau(k)) + \sum_{\mu=k-\tau_M+1}^{k-\tau_m} x^T(\mu)Q_1x(\mu)\\
&+ x^T(k)Q_2x(k) - x^T(k-\theta(k))Q_2e(k-\theta(k)) + \sum_{\mu=k-\theta_M+1}^{k-\theta_m} x^T(\mu)Q_2x(\mu),
\end{aligned} \tag{6.74}$$

$$\begin{aligned}
&\mathbb{E}\{\triangle V_3(k)|\Re_k,r(k)\}\\
=&\mathbb{E}\{[V_3(k+1,\Re_{k+1},r(k+1)) - V_3(k,\Re_k,r(k))]|\Re_k,r(k)\}\\
\le&e^T(k)T_1e(k) - e^T(k-\tau(k))T_1e(k-\tau(k)) + \sum_{\mu=k-\tau_M+1}^{k-\tau_m} e^T(\mu)T_1e(\mu)\\
&+ e^T(k)T_2e(k) - e^T(k-\theta(k))T_2e(k-\theta(k)) + \sum_{\mu=k-\theta_M+1}^{k-\theta_m} e^T(\mu)T_2e(\mu),
\end{aligned} \tag{6.75}$$

$$\begin{aligned}
&\mathbb{E}\{\triangle V_4(k)|\Re_k,r(k)\}\\
=&\mathbb{E}\{[V_4(k+1,\Re_{k+1},r(k+1)) - V_4(k,\Re_k,r(k))]|\Re_k,r(k)\}\\
=&(\tau_M-\tau_m)x^T(k)Q_1x(k) + (\theta_M-\theta_m)x^T(k)Q_2x(k) - \sum_{\mu=k-\tau_M+1}^{k-\tau_m} x^T(\mu)Q_1x(\mu)\\
&- \sum_{\mu=k-\theta_M+1}^{k-\theta_m} x^T(\mu)Q_2x(\mu),
\end{aligned} \tag{6.76}$$

$$\begin{aligned}
&\mathbb{E}\{\triangle V_5(k)|\Re_k, r(k)\} \\
=&\mathbb{E}\{[V_5(k+1, \Re_{k+1}, r(k+1)) - V_5(k, \Re_k, r(k))]|\Re_k, r(k)\} \\
=&(\tau_M - \tau_m)e^T(k)T_1e(k) + (\theta_M - \theta_m)e^T(k)T_2e(k) - \sum_{\mu=k-\tau_M+1}^{k-\tau_m} e^T(\mu)T_1e(\mu) \\
&- \sum_{\mu=k-\theta_M+1}^{k-\theta_m} e^T(\mu)T_2e(\mu)
\end{aligned} \tag{6.77}$$

where

$$\zeta(k) = \begin{bmatrix} x^T(k) & x^T(k-\tau(k)) & x^T(k-\theta(k)) & e^T(k) & e^T(k-\tau(k)) \\ & & e^T(k-\theta(k)) & \varepsilon^T(k) & \tilde{f}^T(Nx(k-\tau(k))) \end{bmatrix}^T. \tag{6.78}$$

It follows from (6.63) and (6.70) that

$$\text{trace}[\tilde{g}^T(x(k))(\tilde{P}_{1i} + \tilde{P}_{2i})\tilde{g}(x(k))] \leq \rho x^T(k)\tilde{G}x(k). \tag{6.79}$$

Then, combining inequalities (6.62), (6.64), (6.71), and (6.73)–(6.79), we have

$$\begin{aligned}
\mathbb{E}\{\Delta V(k)|\Re_k, r(k)\} &= \mathbb{E}\{[V(k+1, \Re_{k+1}, r(k+1)) - V(k, \Re_k, r(k))]|\Re_k, r(k)\} \\
&\leq \zeta^T(k)\Pi_i\zeta(k) \\
&\leq \lambda_{\max}(\Pi_i)\|\zeta(k)\|^2.
\end{aligned} \tag{6.80}$$

Then, by following the analysis in [161], the stochastic stability of the system (6.67) with $v(k) = 0$ can be guaranteed and hence the proof of Theorem 6.3 is complete.

Next, the H_∞ performance analysis for the estimation error dynamics (6.67) is conducted and, in the following theorem, a sufficient condition is given under which the estimation error system is stochastically stable and the H_∞ performance constraint is met.

Theorem 6.4 *Given estimator parameters $\hat{A}_i$, $\hat{G}_i$ ($i \in \mathcal{S}$) and the H_∞ performance index γ, the system (6.67) is stochastically stable with $v(k) = 0$ and the estimation error $e(k)$ satisfies the H_∞ performance constraint (6.69) under the zero initial condition for all nonzero $v(k)$ if there exist positive definite matrices P_{1i}, P_{2i}, Q_1, Q_2, T_1, T_2 and a positive scalar ρ such that*

$$\tilde{P}_{1i} + \tilde{P}_{2i} \leq \rho I, \tag{6.81}$$

$$\tilde{\Pi}_i = \begin{bmatrix} \tilde{\Upsilon}_{11i} & \tilde{\Upsilon}_{12i}^T & \tilde{\Upsilon}_{13i}^T \\ * & -\tilde{P}_{1i}^{-1} & 0 \\ * & * & -\tilde{P}_{2i}^{-1} \end{bmatrix} < 0 \tag{6.82}$$

where

$$\tilde{\Upsilon}_{11i}=\begin{bmatrix} \Phi_{1i} & 0 & 0 & 0 & 0 & 0 & 0 & 0 & 0 \\ * & -Q_1 & 0 & 0 & 0 & 0 & 0 & \frac{1}{2}\tilde{U}^T & 0 \\ * & * & -Q_2 & 0 & 0 & 0 & 0 & 0 & 0 \\ * & * & * & \tilde{\Phi}_{4i} & 0 & 0 & 0 & 0 & 0 \\ * & * & * & * & -T_1 & 0 & 0 & 0 & 0 \\ * & * & * & * & * & -T_2 & 0 & 0 & 0 \\ * & * & * & * & * & * & -I & 0 & 0 \\ * & * & * & * & * & * & * & -I & 0 \\ * & * & * & * & * & * & * & * & -\gamma^2 I \end{bmatrix},$$

$$\tilde{\Phi}_{4i}=-P_{2i}+(1+\tau_M-\tau_m)T_1+(1+\theta_M-\theta_m)T_2+I,$$
$$\tilde{\Upsilon}_{12i}=\begin{bmatrix}\tilde{A}_i & 0 & \tilde{D}_i & 0 & 0 & 0 & 0 & \tilde{B}_i & \tilde{E}_i\end{bmatrix},$$
$$\tilde{\Upsilon}_{13i}=\begin{bmatrix}\tilde{A}_i-\hat{A}_i-\hat{G}_i\tilde{L}_i & 0 & \tilde{D}_i & \hat{A}_i & 0 & 0 & \hat{G}_i & \tilde{B}_i & \tilde{E}_i\end{bmatrix}.$$

Proof *It is easy to verify that (6.82) implies (6.71). Therefore, according to Theorem 6.3, the stochastic stability of the system (6.67) with $v(k)=0$ is guaranteed.*

For the H_∞ performance analysis, we introduce an index function as follows:

$$J=\mathbb{E}\{\sum_{k=0}^{M}[e^T(k)e(k)-\gamma^2 v^T(k)v(k)]|\Re_k,r(k)\} \tag{6.83}$$

where M is a nonnegative integer. Obviously, our aim is to show $J<0$ under the zero-initial condition. Firstly, by choosing the same Lyapunov functional and using the similar techniques, $\mathbb{E}\{\triangle V(k)|\Re_k,r(k)\}$ can be computed as follows:

$$\begin{aligned}&\mathbb{E}\{\triangle V(k)|\Re_k,r(k)\}\\ =&\mathbb{E}\{[V(k+1,\Re_{k+1},r(k+1))-V(k,\Re_k,r(k))]|\Re_k,r(k)\}\\ =&\mathbb{E}\{[\triangle V_1(k)+\triangle V_2(k)+\triangle V_3(k)+\triangle V_4(k)+\triangle V_5(k)]|\Re_k,r(k)\}\\ \le&\tilde{\zeta}^T(k)\hat{\Pi}_i\tilde{\zeta}(k)\end{aligned} \tag{6.84}$$

where $\tilde{\zeta}(k)=\begin{bmatrix}\zeta^T(k) & v^T(k)\end{bmatrix}^T$ and

$$\hat{\Pi}_i=\begin{bmatrix}\hat{\Upsilon}_{11i} & \tilde{\Upsilon}_{12i}^T & \tilde{\Upsilon}_{13i}^T \\ * & -\tilde{P}_{1i}^{-1} & 0 \\ * & * & -\tilde{P}_{2i}^{-1}\end{bmatrix}, \tag{6.85}$$

$$\hat{\Upsilon}_{11i} = \begin{bmatrix} \Phi_{1i} & 0 & 0 & 0 & 0 & 0 & 0 & 0 & 0 \\ * & -Q_1 & 0 & 0 & 0 & 0 & 0 & \frac{1}{2}\tilde{U}^T & 0 \\ * & * & -Q_2 & 0 & 0 & 0 & 0 & 0 & 0 \\ * & * & * & \Phi_{4i} & 0 & 0 & 0 & 0 & 0 \\ * & * & * & * & -T_1 & 0 & 0 & 0 & 0 \\ * & * & * & * & * & -T_2 & 0 & 0 & 0 \\ * & * & * & * & * & * & -I & 0 & 0 \\ * & * & * & * & * & * & * & -I & 0 \\ * & * & * & * & * & * & * & * & 0 \end{bmatrix}.$$

Then, under the zero-initial condition, it follows readily from (6.84) that

$$\begin{aligned} J =& \mathbb{E}\{\sum_{k=0}^{M}[e^T(k)e(k) - \gamma^2 v^T(k)v(k) + \triangle V(k)]|\Re_k, r(k)\} \\ & - \mathbb{E}\{\sum_{k=0}^{M}\triangle V(k)|\Re_k, r(k)\} \\ \leq& \mathbb{E}\{\sum_{k=0}^{M}[e^T(k)e(k) - \gamma^2 v^T(k)v(k) + \triangle V(k)]|\Re_k, r(k)\} \\ =& \sum_{k=0}^{M}\tilde{\zeta}^T(k)\tilde{\Pi}_i\tilde{\zeta}(k). \end{aligned} \tag{6.86}$$

On the other hand, it can be concluded from (6.82) that $J < 0$. Finally, by letting $M \to \infty$, it is easy to see that

$$\mathbb{E}\{\sum_{k=0}^{\infty} e^T(k)e(k)\} < \sum_{k=0}^{\infty}\gamma^2 v^T(k)v(k), \tag{6.87}$$

which accomplishes the proof of Theorem 6.4.

Remark 6.9 *In Theorem 6.4, a sufficient condition is derived under which the estimation error system is stochastically stable and the H_∞ performance constraint is met. Such a sufficient condition may be a bit conservative. The conservatism could be reduced further by involving other advance technologies to deal with time-delays such as delay-fractioning approach.*

Finally, according to the above analysis, the design method of the desired H_∞ estimator parameters is provided in the following theorem.

Theorem 6.5 *Given the H_∞ performance index γ, the event-based H_∞ estimation problem is solvable for the GRN (6.57) if there exist matrices $P_{1i} > 0$, $P_{2i} > 0$, $Q_1 > 0$, $Q_2 > 0$, $T_1 > 0$, $T_2 > 0$, X_i, Y_i ($i \in \mathcal{S}$) and a positive scalar ρ such that*

$$\tilde{P}_{1i} + \tilde{P}_{2i} \leq \rho I, \tag{6.88}$$

$$\check{\Pi}_i = \begin{bmatrix} \check{\Upsilon}_{11i} & \check{\Upsilon}_{12i}^T & \check{\Upsilon}_{13i}^T \\ * & -\tilde{P}_{1i} & 0 \\ * & * & -\tilde{P}_{2i} \end{bmatrix} < 0 \tag{6.89}$$

where

$$\check{\Upsilon}_{12i} = \begin{bmatrix} \tilde{P}_{1i}\tilde{A}_i & 0 & \tilde{P}_{1i}\tilde{D}_i & 0 & 0 & 0 & 0 & \tilde{P}_{1i}\tilde{B}_i & \tilde{P}_{1i}\tilde{E}_i \end{bmatrix},$$
$$\check{\Upsilon}_{13i} = \begin{bmatrix} \tilde{P}_{2i}\tilde{A}_i - X_i - Y_i\tilde{L}_i & 0 & \tilde{P}_{2i}\tilde{D}_i & X_i & 0 & 0 & Y_i & \tilde{P}_{2i}\tilde{B}_i & \tilde{P}_{2i}\tilde{E}_i \end{bmatrix}.$$

Furthermore, if the LMIs (6.88)-(6.89) are feasible, the desired estimator gain matrices are given by

$$X_i = \tilde{P}_{2i}\hat{A}_i, \quad Y_i = \tilde{P}_{2i}\hat{G}_i. \tag{6.90}$$

Proof *By noting the relations* $X_i = \tilde{P}_{2i}\hat{A}_i$ *and* $Y_i = \tilde{P}_{2i}\hat{G}_i$ *and performing a congruence transformation of* $\mathrm{diag}\{I, \tilde{P}_{1i}, \tilde{P}_{2i}\}$, *it can be seen that inequality (6.89) is equivalent to (6.82) in Theorem 6.4 and hence the proof of Theorem 6.5 follows immediately from Theorem 6.4.*

To this end, the ET H_∞ state estimation problem has been solved and the design method of the desired H_∞ estimator has been given in terms of the solution to the LMIs (6.88) and (6.89).

6.3 Illustrative Examples

In this section, some simulation examples are given to demonstrate the ET state estimation approaches developed in this chapter.

6.3.1 Example 1

We present a numerical simulation example to demonstrate the effectiveness of the ET state estimation scheme proposed in Section 6.1 for the multi-delayed neural networks with the SPs, randomly occurring sensor saturations as well as randomly occurring quantizations.

Consider a class of discrete-time multi-delayed neural networks described by (6.1) with three neurons. The deterministic system parameters are chosen as $b_1 = b_3 = 0.2$ and $b_2 = 0.1$ and the expectations of the SPs are set as $\bar{a}_1 = 0.2$, $\bar{a}_2 = 0.3$, and $\bar{a}_3 = 0.4$. The delays among the interconnections are taken as $\tau_{11} = \tau_{13} = \tau_{22} = \tau_{23} = \tau_{31} = \tau_{32} = \tau_{33} = 1$ and $\tau_{12} = \tau_{21} = 2$. The coupled configuration matrix and the delayed coupled configuration matrix are, respectively, given by

$$W_0 = \begin{bmatrix} -0.3 & 0.1 & 0.2 \\ 0.1 & -0.2 & 0.1 \\ 0.2 & 0.1 & -0.3 \end{bmatrix}, \quad W_1 = \begin{bmatrix} -0.2 & 0.1 & 0.1 \\ 0.1 & -0.2 & 0.1 \\ 0.1 & 0.1 & -0.2 \end{bmatrix}.$$

The neuron activation functions are assumed to be of the following form:

$$\begin{aligned} g_1(x_1(k)) &= \tanh(0.2x_1(k)), \\ g_2(x_2(k)) &= \tanh(-0.15x_2(k)), \\ g_3(x_3(k)) &= \tanh(0.23x_3(k)) \end{aligned}$$

which satisfy (6.2) with

$$m_1 = 0.2, \quad m_2 = 0.15, \quad m_3 = 0.23.$$

In the measurement model described by (6.6), the parameter matrix C is taken as $C = \begin{bmatrix} 0.5 & -0.3 & 0.4 \end{bmatrix}$, the parameters of saturation function are selected as $k = 0.4$ and $\vartheta_{\max} = 0.05$, the quantizer parameters are set as $\kappa = 0.25$ and $u_0^{(1)} = 3$, and the probabilities are given by $\beta_1 = 0.5$, $\beta_2 = 0.2$, and $\beta_3 = 0.3$.

In this example, the bound of the triggering condition is chosen as $\delta = 1$. With the above parameters, by using the Matlab with LMI Toolbox, we solve the matrix inequality (6.49) and obtain a feasible solution as follows (only the main parameters are listed):

$$P_{22} = \begin{bmatrix} 4.4917 & 0.0902 & 0.8268 \\ 0.0902 & 4.8111 & 0.1857 \\ 0.8268 & 0.1857 & 4.0158 \end{bmatrix}, \quad Y = \begin{bmatrix} 0.0759 \\ -0.0530 \\ 0.0786 \end{bmatrix}.$$

Then, according to (6.50), the desired estimator gain is designed as

$$G = \begin{bmatrix} 0.0140 \\ -0.0120 \\ 0.0173 \end{bmatrix}.$$

In the simulation, the initial values of the states are taken as $x_1(s) = 0.3$, $x_2(s) = 0.2$, and $x_3(s) = -0.15$ for $s \in [0, 2]$ and the variances of Gaussian white-noises are set as $q_{11} = q_{22} = 0.2$, $q_{33} = 0.3$, $q_{ij} = 0$ $(i \neq j)$, and $Q_v = 2$. Simulation results are shown in Figs. 6.1–6.4. In Fig. 6.1, the above picture plots the real state $x_1(k)$ and its estimate $\hat{x}_1(k)$ and the state estimation error $e_1(k)$ is depicted in the picture below. Same results are shown in Figs. 6.2 and 6.3 for states $x_2(k)$ and $x_3(k)$, respectively. Fig. 6.4 draws the event-based release instants and the corresponding release intervals. The simulation results have demonstrated that the state estimator designed in Section 6.1 performs very well.

6.3.2 Example 2

In this example, we would reveal the effectiveness of the ET state estimation scheme proposed in Section 6.2 for GRNs with MJPs and TVDs.

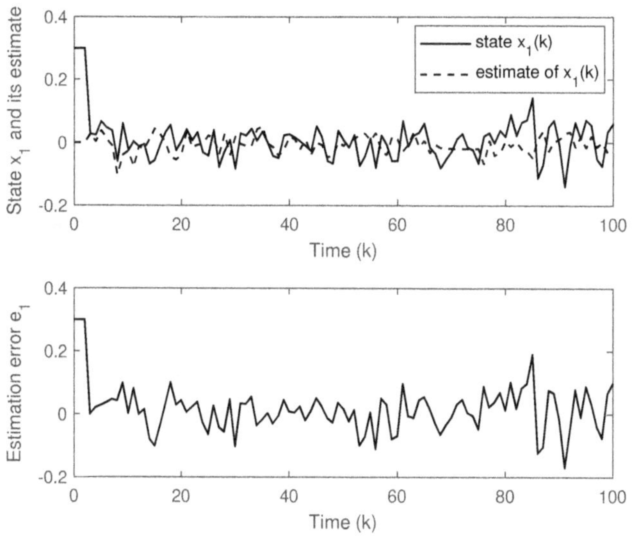

FIGURE 6.1
State $x_1(k)$, its estimate and estimation error.

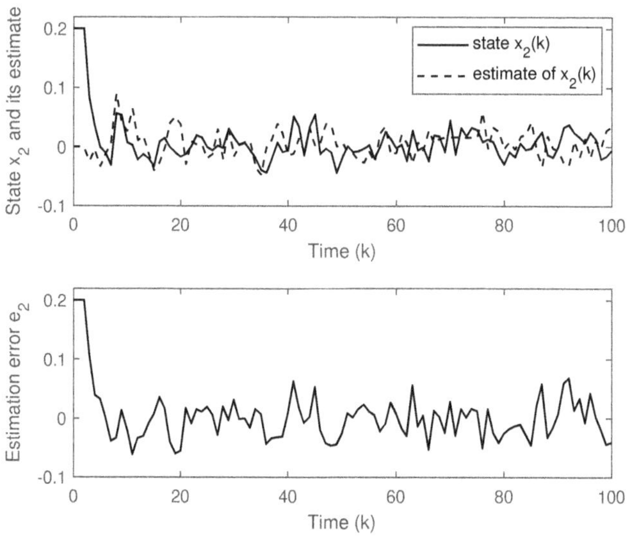

FIGURE 6.2
State $x_2(k)$, its estimate and estimation error.

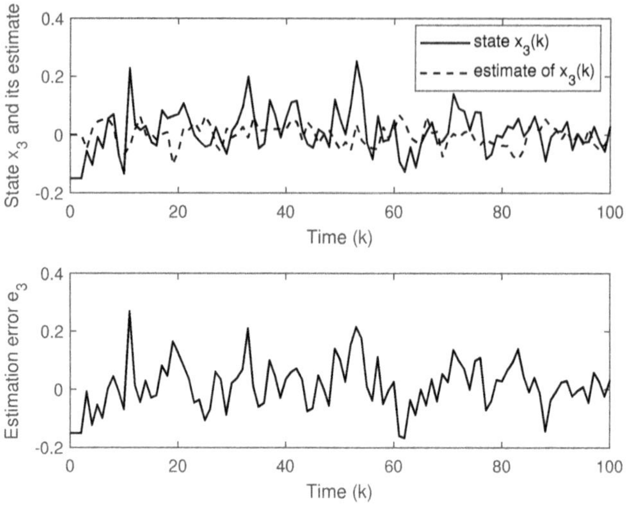

FIGURE 6.3
State $x_3(k)$, its estimate and estimation error.

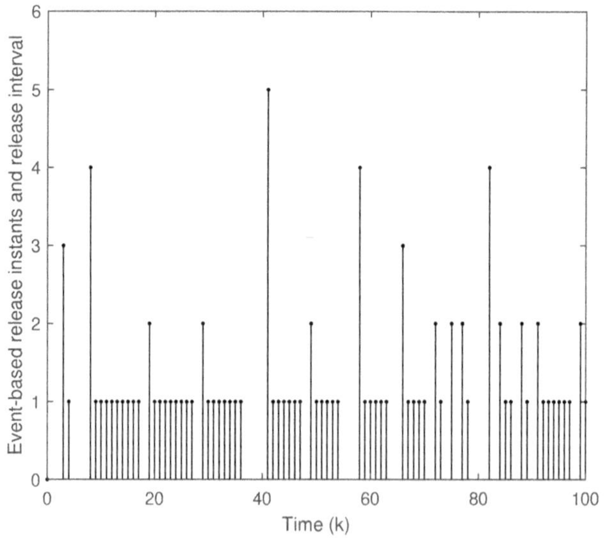

FIGURE 6.4
Event-based release instants and release intervals.

Consider a Markovian chain taking values in a finite state space $\mathcal{S} = \{1, 2\}$ with the transition probability matrix $\Lambda = [\pi_{ij}]_{p\times p}$ given by

$$\Lambda = \begin{bmatrix} \pi_{11} & \pi_{12} \\ \pi_{21} & \pi_{22} \end{bmatrix} = \begin{bmatrix} 0.4 & 0.6 \\ 0.7 & 0.3 \end{bmatrix}.$$

The GRN is assumed to have two genes with the following parameters:

$$\begin{aligned}
&A_1 = \text{diag}\{0.3, 0.4\}, \quad A_2 = \text{diag}\{0.2, 0.1\}, \quad C_1 = \text{diag}\{0.1, 0.12\}, \\
&C_2 = \text{diag}\{0.15, 0.1\}, \quad D_1 = \text{diag}\{0.2, 0.12\}, \quad D_2 = \text{diag}\{0.1, 0.14\}, \\
&B_1 = \begin{bmatrix} 0 & -0.2 \\ -0.2 & 0 \end{bmatrix}, \quad B_2 = \begin{bmatrix} 0 & -0.3 \\ -0.3 & 0 \end{bmatrix}, \quad E_{m1} = \begin{bmatrix} -0.1 \\ 0.12 \end{bmatrix}, \\
&E_{m2} = \begin{bmatrix} 0.18 \\ 0.15 \end{bmatrix}, \quad E_{p1} = \begin{bmatrix} 0.13 \\ 0.1 \end{bmatrix}, \quad E_{p2} = \begin{bmatrix} 0.2 \\ 0.14 \end{bmatrix},
\end{aligned}$$

and the intensity functions of intrinsic perturbations are selected as

$$\begin{aligned}
g_m(\bar{m}(k), \bar{p}(k)) &= \frac{1}{4}(\bar{m}(k) + \bar{p}(k)), \\
g_p(\bar{m}(k), \bar{p}(k)) &= \frac{1}{4}(\bar{m}(k) - \bar{p}(k)),
\end{aligned}$$

which imply $G_{m1} = 0.125I$, $G_{m2} = 0.125I$, $G_{p1} = 0.125I$ and $G_{p2} = 0.125I$.

The nonlinear regulatory function is taken as $f_j(p_j) = p_j^2/(1 + p_j^2)$ $(j = 1, 2)$. It can be easily verified that $f_j(p_j)$ $(j = 1, 2)$ satisfy Assumption 6.1 with $u_1 = u_2 = 0.65$. The TVDs are chosen as $\tau(k) = 1 + \sin^2(k\pi)$ and $\theta(k) = 1 + \cos(k\pi)$ from which, we have $\tau_m = 1$, $\tau_M = 2$, $\theta_m = 0$ and $\theta_M = 2$.

The parameters of measurement outputs are given by

$$\begin{aligned}
L_{m1} &= \begin{bmatrix} -1.08 & 1.1 \end{bmatrix}, \quad L_{p1} = \begin{bmatrix} 1 & 1.5 \end{bmatrix}, \\
L_{m2} &= \begin{bmatrix} 1.1 & 1 \end{bmatrix}, \quad L_{p2} = \begin{bmatrix} 0.9 & 1.3 \end{bmatrix}.
\end{aligned}$$

We set the H_∞ performance level and the threshold of the triggering condition as $\gamma = 0.8$ and $\sigma = 0.2$, respectively. By using the Matlab LMI Toolbox, LMIs (6.88) and (6.89) are solved and a set of feasible solutions is obtained as follows:

$$X_1 = \begin{bmatrix} 0.1960 & -0.0246 & 0.0319 & -0.0131 \\ -0.0246 & 0.2026 & -0.0231 & -0.1249 \\ 0.0319 & -0.0231 & 0.1149 & -0.0788 \\ -0.0131 & -0.1249 & -0.0788 & 0.1451 \end{bmatrix},$$

$$\tilde{P}_{21} = \begin{bmatrix} 1.3124 & -0.1268 & -0.0569 & -0.0345 \\ -0.1268 & 1.4645 & -0.1587 & -0.1670 \\ -0.0569 & -0.1587 & 1.5395 & -0.2351 \\ -0.0345 & -0.1670 & -0.2351 & 1.5257 \end{bmatrix},$$

$$Y_1 = \begin{bmatrix} -0.0277 \\ 0.2095 \\ 0.1079 \\ 0.0766 \end{bmatrix}, \quad Y_2 = \begin{bmatrix} 0.1204 \\ 0.0982 \\ 0.0935 \\ 0.0969 \end{bmatrix}.$$

Then, according to (6.90), the desired estimator gain matrices can be designed as

$$\hat{A}_1 = \begin{bmatrix} 0.1503 & -0.0087 & 0.0251 & -0.0181 \\ -0.0011 & 0.1282 & -0.0110 & -0.0832 \\ 0.0259 & -0.0128 & 0.0680 & -0.0485 \\ -0.0013 & -0.0700 & -0.0418 & 0.0781 \end{bmatrix},$$

$$\hat{A}_2 = \begin{bmatrix} 0.1652 & -0.0461 & -0.0110 & -0.0595 \\ -0.0404 & 0.0633 & -0.0126 & -0.0315 \\ -0.0172 & -0.0161 & 0.0798 & -0.0280 \\ -0.0546 & -0.0238 & -0.0191 & 0.0450 \end{bmatrix},$$

$$\hat{G}_1 = \begin{bmatrix} 0.0012 \\ 0.1635 \\ 0.0997 \\ 0.0835 \end{bmatrix}, \quad \hat{G}_2 = \begin{bmatrix} 0.1071 \\ 0.0955 \\ 0.0883 \\ 0.0903 \end{bmatrix}.$$

In the simulation, the exogenous disturbance input is selected as $v(k) = \sin(k)\exp(-0.1k)$ and the initial values of the GRN (6.57) are set as $\bar{m}(-2) = \begin{bmatrix}0.1 & -0.1\end{bmatrix}^T$, $\bar{m}(-1) = \begin{bmatrix}-0.1 & 0.1\end{bmatrix}^T$, $\bar{m}(0) = \begin{bmatrix}0.1 & 0.12\end{bmatrix}^T$, $\bar{p}(-2) = \begin{bmatrix}0.1 & 0.1\end{bmatrix}^T$, $\bar{p}(-1) = \begin{bmatrix}0 & -0.12\end{bmatrix}^T$, and $\bar{p}(0) = \begin{bmatrix}0.12 & -0.13\end{bmatrix}^T$. Simulation results are shown in Figs. 6.5–6.9. Figs. 6.5 and 6.6 depict the true concentration of mRNA and its estimate while the true concentration of protein and its estimate are shown in Figs. 6.7 and 6.8. Fig. 6.9 depicts the

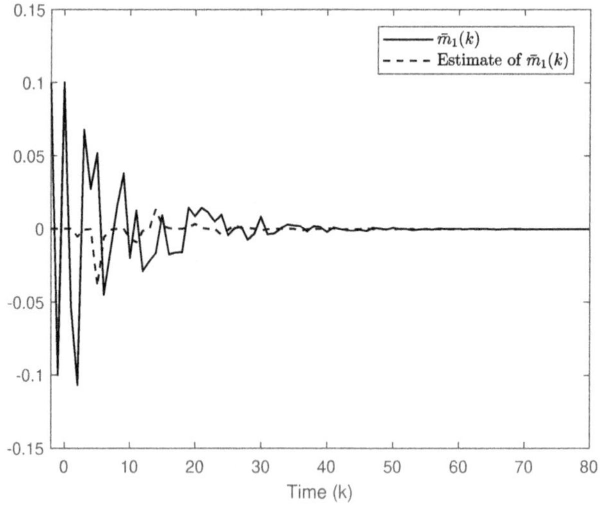

FIGURE 6.5
mRNA concentration $\bar{m}_1(k)$ and its estimate.

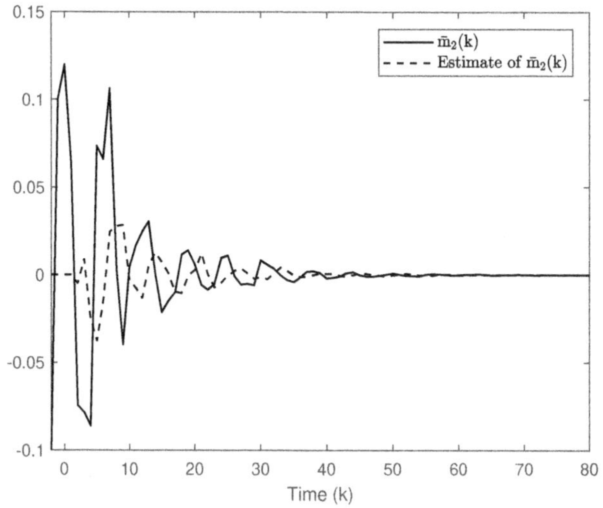

FIGURE 6.6
mRNA concentration $\bar{m}_2(k)$ and its estimate.

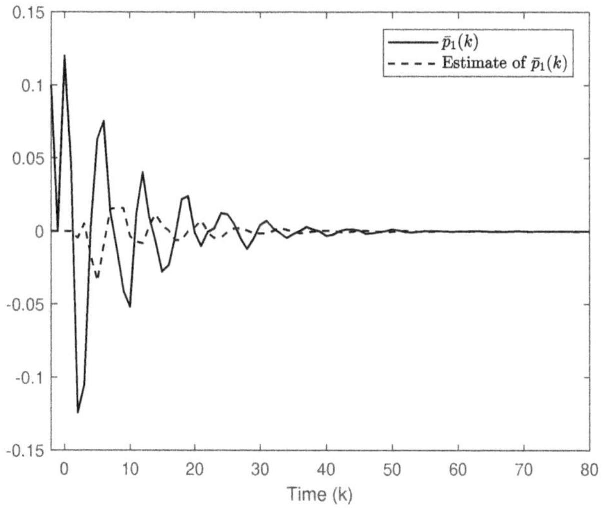

FIGURE 6.7
Protein concentration $\bar{p}_1(k)$ and its estimate.

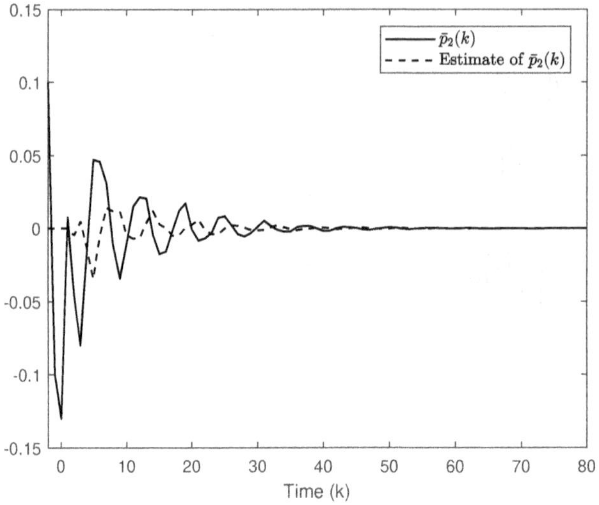

FIGURE 6.8
Protein concentration $\bar{p}_2(k)$ and its estimate.

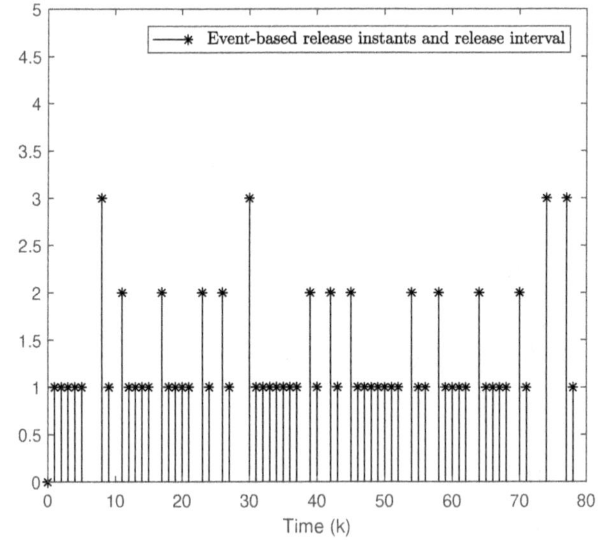

FIGURE 6.9
Event-based release instants and release intervals.

event-based release instants and release interval. In order to determine if the H_∞ performance requirement is satisfied, the experiment is repeated 80 times and the l_2-norm of the estimation error $e(k)$ and the external disturbance $v(k)$ are computed as 0.4803 and 1.1573, respectively. Accordingly, the actual l_2-gain from the exogenous disturbance to the estimation error can be obtained as 0.415, which is lower than the given H_∞ performance index $\gamma = 0.8$. The simulation results have verified that the H_∞ estimator designed in Section 6.2 performs very well and the proposed ET strategy reduces the transmission times effectively.

6.4 Summary

In this chapter, the ET state estimation problem has been solved for discrete-time neural networks. For the purpose of energy saving, the ET mechanism has been adopted and the measurement outputs are only transmitted to the estimator when a certain triggered condition is met. Firstly, a new ET estimation technique has been developed for the multi-delayed neural networks with SPs and IMs. By using the Lyapunov functional approach, a sufficient condition has been given under which the estimation error dynamics is exponentially ultimately bounded in the mean square and the ultimate boundedness of the error dynamics has been given. Also, in this chapter, we have investigated the ET H_∞ state estimation problem for a class of discrete-time stochastic GRNs with MJPs and TVDs. By recurring to the similar methods, sufficient conditions have been derived to guarantee that the estimation error dynamics is exponentially stable and the H_∞ performance requirement is satisfied. At last, some numerical simulation examples are presented to illustrate the effectiveness of the proposed ET state estimation scheme.

7

Event-Triggered Fusion Estimation for Multi-Rate Systems

This chapter is concerned with the ET fusion estimation problem for multi-rate systems. The multi-rate systems under consideration include several sensor nodes with different sampling rates. The ET robust fusion estimation problem is firstly considered for a class of uncertain multi-rate sampled-data systems with stochastic nonlinearities and coloured measurement noises. A new augmentation approach is proposed by which the multi-rate system is transformed into the single-rate system. The main purpose is to design a set of ET local filters for each sensor node such that the upper bound of each local FEC is guaranteed and minimized at each sampling instant. Sufficient conditions are established for the existence of upper bounds of the local FECs, and then the local filter gains are parameterized according to certain matrix recursions. Subsequently, for the local state estimates, a new fusion estimation scheme is proposed with the help of CI method and the consistency of the proposed CI-based fusion estimation scheme is shown. Moreover, the ET fusion estimation problem is also studied for a class of multi-rate systems subject to sensor degradations. A set of random variables obeying known probability distributions are used to characterize the phenomenon of the sensor degradations. By using similar analysis techniques, the corresponding ET filters are designed. Finally, some simulation examples are exploited to illustrate the main results of this chapter.

7.1 Event-Triggered Fusion Estimation with Coloured Measurement Noises

7.1.1 Problem Formulation

Consider a class of nonlinear uncertain stochastic systems as follows:

$$x_{(k+1)h_0} = (A + \Delta A_{kh_0})x_{kh_0} + g(x_{kh_0}, \eta_{kh_0}) + Bw_{kh_0} \tag{7.1}$$

where h_0 is the sampling period, $x_{kh_0} \in \mathbb{R}^{n_x}$ and $w_{kh_0} \in \mathbb{R}^{n_w}$ are the state vector and the process noises at the sampling instant kh_0, $g(\cdot,\cdot)$ is

DOI: 10.1201/9781003307648-7

a stochastic nonlinear function, A and B are the known matrices with appropriate dimensions, ΔA_{kh_0} is an unknown matrix representing the PU.

The process noise w_{kh_0} is represented by a zero-mean Gaussian white noise with $\mathbb{E}\{w_i w_j^T\} = \delta_{ij} R_i$, where $R_i > 0$ and $\delta_{ij} = 1$ if $i = j$, otherwise, $\delta_{ij} = 0$. The uncertain parameter matrix ΔA_{kh_0} satisfies

$$\Delta A_{kh_0} = H F_{kh_0} E \quad \text{with} \quad F_{kh_0} F_{kh_0}^T \le I, \quad \text{for all} \quad k = 1, 2, \cdots . \tag{7.2}$$

The stochastic nonlinear function $g(\cdot, \cdot)$ satisfies

$$\begin{aligned} & g(0, \eta_k) = 0, \ \mathbb{E}\{g(x_k, \eta_k)|x_k\} = 0, \\ & \mathbb{E}\{g(x_k, \eta_k) g(x_l, \eta_l)^T |x_l\} = 0, \ k \neq l, \\ & \mathbb{E}\{g(x_k, \eta_k) g(x_k, \eta_k)^T |x_k\} = \sum_{i=1}^{r} \Pi_k^i x_k^T \Gamma_k^i x_k \end{aligned} \tag{7.3}$$

where r is a known positive integer, Π_k^i and Γ_k^i are known matrices with appropriate dimensions.

In this section, M sensors with different sampling periods are used to measure the information from the sampling-data system (7.1) and the measurement model of the ith sensor is described by

$$\begin{aligned} y_{kn_i h_0, i} &= C_i x_{kn_i h_0} + v_{kn_i h_0, i}, \\ v_{kn_i h_0, i} &= D_i v_{(k-1)n_i h_0, i} + e_{(k-1)n_i h_0, i} \end{aligned} \tag{7.4}$$

where n_i is a positive integer, $y_{kn_i h_0, i} \in \mathbb{R}^{n_y}$ is the measurement output from the ith sensor at sampling instant $kn_i h_0$, $v_{kn_i h_0, i}$ is the coloured measurement noise, $e_{kn_i h_0, i}$ is a zero-mean Gaussian white noise, and C_i and D_i are known matrices with appropriate dimensions. The Gaussian white noise $e_{kn_i h_0, i}$ satisfies

$$\mathbb{E}\{e_{m,i} e_{n,j}^T\} = \delta_{mn}(\delta_{ij} Q_{m,i}) \tag{7.5}$$

where $Q_{m,i} > 0$.

The initial state x_0 is a stochastic variable with $\mathbb{E}\{x_0\} = \bar{x}_0$ and $\mathbb{E}\{(x_0 - \bar{x}_0)(x_0 - \bar{x}_0)^T\} = P_{0|0}$. All the stochastic variables aforementioned are assumed to be uncorrelated mutually.

Note that the sampling period of systems described by (7.1) is h_0 while the sampling periods of measurements received from sensors are $n_i h_0$ ($i = 1, 2, \cdots, M$). Therefore, the system to be considered is actually a multi-rate system (such as signal processing system of software-defined radios) which complicates the system analysis and synthesis. An alternative way is to transform the multi-rate system into a single-rate system. For the purpose of notation simplicity, in the sequel, the sampling instant $kn_i h_0$ is simply denoted by kn_i.

For each interval $[kn_i+1,(k+1)n_i]$, the dynamics of system (7.1) can be expressed by

$$\begin{aligned}x_{kn_i+l+1} =&(A+\Delta A_{kn_i+l})x_{kn_i+l}+g(x_{kn_i+l},\eta_{kn_i+l})+Bw_{kn_i+l}\\ =&A^{(l+1)}x_{kn_i}+\sum_{i=0}^{l}A^i(\Delta A_{kn_i+l-i}x_{kn_i+l-i}\\ &+g(x_{kn_i+l-i},\eta_{kn_i+l-i})+Bw_{kn_i+l-i}),\ 0\le l\le n_i-1.\end{aligned}$$

Thus, for the measurement received from the ith sensor together with the system (7.1), the single-rate system is given by

$$\begin{cases}x_{(k+1)n_i} =\tilde{A}_i x_{kn_i}+\breve{A}_i\Delta A_{kn_i}x_{kn_i}+\bar{A}_i\bar{\Delta}_{kn_i}\\ \qquad\qquad +\breve{A}_i g(x_{kn_i},\eta_{kn_i})+\bar{A}_i\bar{g}_{kn_i}+\bar{B}_i\bar{w}_{kn_i},\\ y_{kn_i,i} =C_i x_{kn_i}+v_{kn_i,i}\end{cases} \tag{7.6}$$

for all $i=1,2,\cdots,M$, where

$$\begin{aligned}&\tilde{A}_i=A^{n_i},\ \breve{A}_i=A^{n_i-1},\ \bar{A}_i=\begin{bmatrix}A^{n_i-2} & A^{n_i-3} & \cdots & I\end{bmatrix},\\ &\bar{B}_i=\begin{bmatrix}A^{n_i-1}B & A^{n_i-2}B & \cdots & B\end{bmatrix},\ \bar{w}_{kn_i}=\begin{bmatrix}w_{kn_i}^T & \cdots & w_{(k+1)n_i-1}^T\end{bmatrix}^T,\\ &\bar{\Delta}_{kn_i}=\begin{bmatrix}(\Delta A_{kn_i+1}x_{kn_i+1})^T & \cdots & (\Delta A_{(k+1)n_i-1}x_{(k+1)n_i-1})^T\end{bmatrix}^T,\\ &\bar{g}_{kn_i}=\begin{bmatrix}g(x_{kn_i+1},\eta_{kn_i+1})^T & \cdots & g(x_{(k+1)n_i-1},\eta_{(k+1)n_i-1})^T\end{bmatrix}^T.\end{aligned}$$

Noting that the measurement noise $v_{kn_i,i}$ is coloured, we construct new measurements as follows

$$\begin{aligned}z_{kn_i,i} =&y_{kn_i,i}-D_i y_{(k-1)n_i,i}\\ =&C_i x_{kn_i}+v_{kn_i,i}-D_i(C_i x_{(k-1)n_i}+v_{(k-1)n_i,i})\\ =&C_i x_{kn_i}-D_i C_i x_{(k-1)n_i}+e_{(k-1)n_i,i}\end{aligned} \tag{7.7}$$

where $z_{kn_i,i}$ is the actual measurement from the ith sensor used to estimate the states. Then, it follows from (7.6) and (7.7) that

$$\begin{cases}x_{(k+1)n_i} =\tilde{A}_i x_{kn_i}+\breve{A}_i\Delta A_{kn_i}x_{kn_i}+\bar{A}_i\bar{\Delta}_{kn_i}+\breve{A}_i g(x_{kn_i},\eta_{kn_i})\\ \qquad\qquad +\bar{A}_i\bar{g}_{kn_i}+\bar{B}_i\bar{w}_{kn_i},\\ z_{kn_i,i} =C_i x_{kn_i}-D_i C_i x_{(k-1)n_i}+e_{(k-1)n_i,i}\end{cases} \tag{7.8}$$

for all $i=1,2,\cdots,M$.

In this section, an event-triggering mechanism is adopted to determine whether the actual measurements are transmitted to the filters. We represent the transmission instants $l_m^i (0\le l_0^i\le l_1^i\le l_2^i\le\cdots)$ as follows:

$$l_{m+1}^i=\min\{kn_i|kn_i>l_m^i,f(\varrho_{kn_i,i},\delta_i)>0\} \tag{7.9}$$

where

$$\varrho_{kn_i,i} = z_{kn_i,i} - z_{l^i_m,i}, \; f(\varrho_{kn_i,i}, \delta_i) = \varrho^T_{kn_i,i}\varrho_{kn_i,i} - \delta_i \tag{7.10}$$

with constants $\delta_i > 0$ $(i = 1, 2, \cdots, M)$.

Based on the measurement from the ith sensor $z_{l^i_m}$, the ith local filter is constructed as follows

$$\begin{cases} \hat{x}_{(k+1)n_i|kn_i} = \tilde{A}_i \hat{x}_{kn_i|kn_i}, \\ \hat{x}_{(k+1)n_i|(k+1)n_i} = \hat{x}_{(k+1)n_i|kn_i} + K_{(k+1)n_i}(z_{l^i_m,i} - C_i \hat{x}_{(k+1)n_i|kn_i} \\ \qquad + D_i C_i \hat{x}_{kn_i|kn_i}), \quad \hat{x}_{0|0} = \bar{x}_0 \end{cases} \tag{7.11}$$

where $\hat{x}_{kn_i|kn_i}$ is the estimate of x_{kn_i} based on the observations $\{z_{j,i}\}_{j=l^i_0}^{l^i_m}$ and $K_{(k+1)n_i}$ is the filter gain matrix to be designed. The design objective of the ith local filter (7.11) is to find the filter gain matrix $K_{(k+1)n_i}$ such that the upper bounds of the FECs at each sampling instant are minimized.

It should be pointed out that the estimates are generated from the ith local filter (7.11) at time instants $\{0, n_i h_0, 2n_i h_0, \cdots\}$. As such, the estimates obtained from some filters may be generated at same time instants. Therefore, our another objective is to develop an appropriate fusion estimation scheme to derive the fusion state estimate with the local estimates generated by all the filters described by (7.11).

In this section, the local ET filters are first designed such that the upper bounds of all the local FECs at each sampling instant are minimized. Then, based on the derived local estimates, a fusion estimation scheme is proposed with the help of the CI fusion method.

7.1.2 Design of Local Filters

The following lemmas will be used in the derivation of our main results.

Lemma 7.1 *[102] For arbitrary matrices A, B and $X = X^T \geq 0$ with appropriate dimensions and a positive scalar ε, the following inequality holds:*

$$AXB^T + BXA^T \leq \varepsilon AXA^T + \varepsilon^{-1}BXB^T.$$

Lemma 7.2 *[189] Assume that the matrices A, H, F, E with appropriate dimensions such that $FF^T \leq I$ are given. Let X be a symmetric positive definite matrix and $\gamma > 0$ be an arbitrary positive constant such that $\gamma^{-1}I - EXE^T > 0$. Then, the following inequality holds:*

$$(A + HFE)X(A + HFE)^T \leq A(X^{-1} - \gamma E^T E)^{-1}A^T + \gamma^{-1}HH^T.$$

Define the filtering error and prediction error by $\tilde{x}_{kn_i|kn_i} = x_{kn_i} - \hat{x}_{kn_i|kn_i}$ and $\tilde{x}_{(k+1)n_i|kn_i} = x_{(k+1)n_i} - \hat{x}_{(k+1)n_i|kn_i}$, respectively, and it is easily known from (7.8) and (7.11) that

$$\begin{aligned}
\tilde{x}_{(k+1)n_i|kn_i} &= \tilde{A}_i\tilde{x}_{kn_i|kn_i} + \check{A}_i\Delta A_{kn_i}x_{kn_i} + \bar{A}_i\bar{\Delta}_{kn_i} + \check{A}_i g(x_{kn_i}, \eta_{kn_i}) \\
&\quad + \bar{A}_i\bar{g}_{kn_i} + \bar{B}_i\bar{w}_{kn_i} \\
&= (\tilde{A}_i + \check{A}_i\Delta A_{kn_i})\tilde{x}_{kn_i|kn_i} + \check{A}_i\Delta A_{kn_i}\hat{x}_{kn_i|kn_i} + \bar{A}_i\bar{\Delta}_{kn_i} \\
&\quad + \check{A}_i g(x_{kn_i}, \eta_{kn_i}) + \bar{A}_i\bar{g}_{kn_i} + \bar{B}_i\bar{w}_{kn_i}, \qquad (7.12) \\
\tilde{x}_{(k+1)n_i|(k+1)n_i} &= (I - K_{(k+1)n_i}C_i)\tilde{x}_{(k+1)n_i|kn_i} + K_{(k+1)n_i}D_iC_i\tilde{x}_{kn_i|kn_i} \\
&\quad + K_{(k+1)n_i}\varrho_{(k+1)n_i,i} - K_{(k+1)n_i}e_{kn_i,i}. \qquad (7.13)
\end{aligned}$$

From (7.13), the FEC can be obtained as follows

$$\begin{aligned}
&P_{(k+1)n_i|(k+1)n_i} \\
&\triangleq \mathbb{E}\{\tilde{x}_{(k+1)n_i|(k+1)n_i}\tilde{x}^T_{(k+1)n_i|(k+1)n_i}\} \\
&= \Omega_{(k+1)n_i|kn_i} + K_{(k+1)n_i}D_iC_iP_{kn_i|kn_i}C_i^TD_i^TK^T_{(k+1)n_i} \\
&\quad + K_{(k+1)n_i}\varrho_{(k+1)n_i,i}\varrho^T_{(k+1)n_i,i}K^T_{(k+1)n_i} + K_{(k+1)n_i}Q_{k,k}K^T_{(k+1)n_i} \\
&\quad + (I - K_{(k+1)n_i}C_i)\mathbb{E}\{\tilde{x}_{(k+1)n_i|kn_i}\tilde{x}^T_{kn_i|kn_i}\}C_i^TD_i^TK^T_{(k+1)n_i} \\
&\quad + K_{(k+1)n_i}D_iC_i\mathbb{E}\{\tilde{x}_{kn_i|kn_i}\tilde{x}^T_{(k+1)n_i|kn_i}\}(I - K_{(k+1)n_i}C_i)^T \\
&\quad + (I - K_{(k+1)n_i}C_i)\mathbb{E}\{\tilde{x}_{(k+1)n_i|kn_i}\varrho^T_{(k+1)n_i,i}\}K^T_{(k+1)n_i} \\
&\quad + K_{(k+1)n_i}\mathbb{E}\{\varrho_{(k+1)n_i,i}\tilde{x}^T_{(k+1)n_i|kn_i}\}(I - K_{(k+1)n_i}C_i)^T \\
&\quad + K_{(k+1)n_i}\mathbb{E}\{\varrho_{(k+1)n_i,i}\tilde{x}^T_{kn_i|kn_i}\}C_i^TD_i^TK^T_{(k+1)n_i} \\
&\quad + K_{(k+1)n_i}D_iC_i\mathbb{E}\{\tilde{x}_{kn_i|kn_i}\varrho^T_{(k+1)n_i,i}\}K^T_{(k+1)n_i} \qquad (7.14)
\end{aligned}$$

where $\Omega_{(k+1)n_i|kn_i} = (I - K_{(k+1)n_i}C_i)P_{(k+1)n_i|kn_i}(I - K_{(k+1)n_i}C_i)^T$.

In (7.14), the cross term $\mathbb{E}\{\tilde{x}_{(k+1)n_i|kn_i}\tilde{x}^T_{kn_i|kn_i}\}$ is computed by

$$\begin{aligned}
\mathbb{E}\{\tilde{x}_{(k+1)n_i|kn_i}\tilde{x}^T_{kn_i|kn_i}\} &= (\tilde{A}_i + \check{A}_i\Delta A_{kn_i})P_{kn_i|kn_i} + \check{A}_i\Delta A_{kn_i}\mathbb{E}\{\hat{x}_{kn_i|kn_i} \\
&\quad \times\tilde{x}^T_{kn_i|kn_i}\} + \bar{A}_i\mathbb{E}\{\bar{\Delta}_{kn_i}\tilde{x}^T_{kn_i|kn_i}\} \qquad (7.15)
\end{aligned}$$

and, from (7.3) and (7.12), the second moment of the prediction error can be obtained as follows

$$\begin{aligned}
P_{(k+1)n_i|kn_i} \triangleq &\mathbb{E}\{\tilde{x}_{(k+1)n_i|kn_i}\tilde{x}^T_{(k+1)n_i|kn_i}\} \\
= &(\tilde{A}_i + \check{A}_i\Delta A_{kn_i})P_{kn_i|kn_i}(\tilde{A}_i + \check{A}_i\Delta A_{kn_i})^T \\
&+ \check{A}_i\Delta A_{kn_i}\mathbb{E}\{\hat{x}_{kn_i|kn_i}\hat{x}^T_{kn_i|kn_i}\}\Delta A^T_{kn_i}\check{A}_i^T \\
&+ \check{A}_i\Big(\sum_{i=1}^{r}\Pi_k^i\mathrm{tr}\big(\mathbb{E}\{x_{kn_i}x^T_{kn_i}\}\Gamma_k^i\big)\Big)\check{A}_i^T + \bar{B}_i\bar{R}_{k,i}\bar{B}_i^T \\
&+ (\tilde{A}_i + \check{A}_i\Delta A_{kn_i})\mathbb{E}\{\tilde{x}_{kn_i|kn_i}\hat{x}^T_{kn_i|kn_i}\}\Delta A^T_{kn_i}\check{A}_i^T + \Theta^1_{k,i} \\
&+ \check{A}_i\Delta A_{kn_i}\mathbb{E}\{\hat{x}_{kn_i|kn_i}\tilde{x}^T_{kn_i|kn_i}\}(\tilde{A}_i + \check{A}_i\Delta A_{kn_i})^T + \Theta^2_{k,i} \\
&\qquad (7.16)
\end{aligned}$$

where

$$
\begin{aligned}
\Theta^1_{k,i} =& (\tilde{A}_i + \breve{A}_i\Delta A_{kn_i})\mathbb{E}\{\tilde{x}_{kn_i|kn_i}\bar{\Delta}^T_{kn_i})\}\bar{A}^T_i + \bar{A}_i\mathbb{E}\{\bar{\Delta}_{kn_i}\tilde{x}^T_{kn_i|kn_i}\} \\
&\times(\tilde{A}_i + \breve{A}_i\Delta A_{kn_i})^T + (\tilde{A}_i + \breve{A}_i\Delta A_{kn_i})\mathbb{E}\{\tilde{x}_{kn_i|kn_i}\bar{g}^T_{kn_i})\}\bar{A}^T_i \\
&+\bar{A}_i\mathbb{E}\{\bar{g}_{kn_i}\tilde{x}^T_{kn_i|kn_i}\}(\tilde{A}_i + \breve{A}_i\Delta A_{kn_i})^T + \bar{A}_i\mathbb{E}\{\bar{\Delta}_{kn_i}\bar{g}^T_{kn_i}\}\bar{A}^T_i \\
&+\bar{A}_i\mathbb{E}\{\bar{g}_{kn_i}\bar{\Delta}^T_{kn_i})\}\bar{A}^T_i + \bar{A}_i\mathbb{E}\{\bar{\Delta}_{kn_i}\bar{\Delta}^T_{kn_i}\}\bar{A}^T_i \\
&+\bar{A}_i\mathbb{E}\{\bar{g}_{kn_i}\bar{g}^T_{kn_i}\}\bar{A}^T_i + \bar{A}_i\mathbb{E}\{\bar{g}_{kn_i}\bar{w}^T_{kn_i}\}\bar{B}^T_i + \bar{B}_i\mathbb{E}\{\bar{w}_{kn_i}\bar{g}^T_{kn_i}\}\bar{A}^T_i \\
&+\breve{A}_i\Delta A_{kn_i}\mathbb{E}\{\hat{x}_{kn_i|kn_i}\bar{\Delta}^T_{kn_i}\}\bar{A}^T_i + \bar{A}_i\mathbb{E}\{\bar{\Delta}_{kn_i}\hat{x}^T_{kn_i|kn_i}\}\Delta A^T_{kn_i}\breve{A}^T_i \\
&+\breve{A}_i\Delta A_{kn_i}\mathbb{E}\{\hat{x}_{kn_i|kn_i}\bar{g}^T_{kn_i}\}\bar{A}^T_i + \bar{A}_i\mathbb{E}\{\bar{g}_{kn_i}\hat{x}^T_{kn_i|kn_i})\}\Delta A^T_{kn_i}\breve{A}^T_i \\
&+\bar{A}_i\mathbb{E}\{\bar{\Delta}_{kn_i}g^T(x_{kn_i},\eta_{kn_i})\}\breve{A}^T_i + \breve{A}_i\mathbb{E}\{g(x_{kn_i},\eta_{kn_i})\bar{\Delta}^T_{kn_i})\}\bar{A}^T_i \\
&+\bar{A}_i\mathbb{E}\{\bar{g}_{kn_i}g^T(x_{kn_i},\eta_{kn_i})\}\breve{A}^T_i + \breve{A}_i\mathbb{E}\{g(x_{kn_i},\eta_{kn_i})\bar{g}^T_{kn_i}\}\bar{A}^T_i \\
&+\bar{A}_i\mathbb{E}\{\bar{\Delta}_{kn_i}\bar{w}^T_{kn_i}\}\bar{B}^T_i + \bar{B}_i\mathbb{E}\{\bar{w}_{kn_i}\bar{\Delta}^T_{kn_i}\}\bar{A}^T_i, \\
\Theta^2_{k,i} =& (\tilde{A}_i + \breve{A}_i\Delta A_{kn_i})\mathbb{E}\{\tilde{x}_{kn_i|kn_i}g^T(x_{kn_i},\eta_{kn_i})\}\breve{A}^T_i + \breve{A}_i\mathbb{E}\{g(x_{kn_i},\eta_{kn_i}) \\
&\times\tilde{x}^T_{kn_i|kn_i}\}(\tilde{A}_{k,i} + \breve{A}_i\Delta A_{kn_i})^T + \breve{A}_i\Delta A_{kn_i}\mathbb{E}\{\hat{x}_{kn_i|kn_i}g^T(x_{kn_i},\eta_{kn_i})\}\breve{A}^T_i \\
&+\breve{A}_i\mathbb{E}\{g(x_{kn_i},\eta_{kn_i})\hat{x}^T_{kn_i|kn_i}\}\Delta A^T_{kn_i}\breve{A}^T_i.
\end{aligned}
$$

As the filter design method proposed in [70], the following condition is required

$$
(I - K_{(k+1)n_i}C_i)\bar{A}_i = 0. \tag{7.17}
$$

Remark 7.1 *In this section, the uncertainties and nonlinearities are considered which result in redundant terms in the derived single-rate system and the condition (7.17) is employed to eliminate the influence of redundant terms. Actually, if the uncertainties and nonlinearities on interval* $[kn_i+1,(k+1)n_i-1]$ *are viewed as the unknown input, the condition (7.17) is just the decoupling condition described in [70].*

Then, it can be derived that

$$
\begin{aligned}
&\Omega_{(k+1)n_i|kn_i} \\
=&(I - K_{(k+1)n_i}C_i)(\tilde{A}_i + \breve{A}_i\Delta A_{kn_i})P_{kn_i|kn_i}(\tilde{A}_i + \breve{A}_i\Delta A_{kn_i})^T \\
&\times(I - K_{(k+1)n_i}C_i)^T + (I - K_{(k+1)n_i}C_i)\breve{A}_i\Delta A_{kn_i} \\
&\times\mathbb{E}\{\hat{x}_{kn_i|kn_i}\hat{x}^T_{kn_i|kn_i}\}\Delta A^T_{kn_i}\breve{A}^T_i(I - K_{(k+1)n_i}C_i)^T \\
&+(I - K_{(k+1)n_i}C_i)\breve{A}_i\left(\sum_{i=1}^{r}\Pi^i_k\mathrm{tr}\big(\mathbb{E}\{x_{kn_i}x^T_{kn_i}\}\Gamma^i_k\big)\right) \\
&\times\breve{A}^T_i(I - K_{(k+1)n_i}C_i)^T + (I - K_{(k+1)n_i}C_i)\bar{B}_i\bar{R}_{k,i}\bar{B}^T_i \\
&\times(I - K_{(k+1)n_i}C_i)^T + (I - K_{(k+1)n_i}C_i)(\tilde{A}_i + \breve{A}_i\Delta A_{kn_i}) \\
&\times\mathbb{E}\{\tilde{x}_{kn_i|kn_i}\hat{x}^T_{kn_i|kn_i}\}\Delta A^T_{kn_i}\breve{A}^T_i(I - K_{(k+1)n_i}C_i)^T \\
&+(I - K_{(k+1)n_i}C_i)\breve{A}_i\Delta A_{kn_i}\mathbb{E}\{\hat{x}_{kn_i|kn_i}\tilde{x}^T_{kn_i|kn_i}\} \\
&\times(\tilde{A}_i + \breve{A}_i\Delta A_{kn_i})^T(I - K_{(k+1)n_i}C_i)^T.
\end{aligned} \tag{7.18}
$$

Consequently, it follows from (7.14)–(7.18) that

$$\begin{aligned}
&P_{(k+1)n_i|(k+1)n_i}\\
&= \Omega_{(k+1)n_i|kn_i} + K_{(k+1)n_i}D_iC_iP_{kn_i|kn_i}C_i^TD_i^TK_{(k+1)n_i}^T\\
&\quad +K_{(k+1)n_i}\varrho_{(k+1)n_i,i}\varrho_{(k+1)n_i,i}^TK_{(k+1)n_i}^T + K_{(k+1)n_i}Q_{k,k}K_{(k+1)n_i}^T\\
&\quad +(I-K_{(k+1)n_i}C_i)(\tilde{A}_i+\breve{A}_i\Delta A_{kn_i})P_{kn_i|kn_i}C_i^TD_i^TK_{(k+1)n_i}^T\\
&\quad +K_{(k+1)n_i}D_iC_iP_{kn_i|kn_i}(\tilde{A}_i+\breve{A}_i\Delta A_{kn_i})^T(I-K_{(k+1)n_i}C_i)^T\\
&\quad +(I-K_{(k+1)n_i}C_i)\breve{A}_i\Delta A_{kn_i}\mathbb{E}\{\hat{x}_{kn_i|kn_i}\tilde{x}_{kn_i|kn_i}^T\}C_i^TD_i^TK_{(k+1)n_i}^T\\
&\quad +K_{(k+1)n_i}D_iC_i\mathbb{E}\{\tilde{x}_{kn_i|kn_i}\hat{x}_{kn_i|kn_i}^T\}\Delta A_{kn_i}^T\breve{A}_i^T(I-K_{(k+1)n_i}C_i)^T\\
&\quad +(I-K_{(k+1)n_i}C_i)\mathbb{E}\{\tilde{x}_{(k+1)n_i|kn_i}\varrho_{(k+1)n_i,i}^T\}K_{(k+1)n_i}^T\\
&\quad +K_{(k+1)n_i}\mathbb{E}\{\varrho_{(k+1)n_i,i}\tilde{x}_{(k+1)n_i|kn_i}^T\}(I-K_{(k+1)n_i}C_i)^T\\
&\quad +K_{(k+1)n_i}\mathbb{E}\{\varrho_{(k+1)n_i,i}\tilde{x}_{kn_i|kn_i}^T\}C_i^TD_i^TK_{(k+1)n_i}^T\\
&\quad +K_{(k+1)n_i}D_iC_i\mathbb{E}\{\tilde{x}_{kn_i|kn_i}\varrho_{(k+1)n_i,i}^T\}K_{(k+1)n_i}^T. \qquad (7.19)
\end{aligned}$$

We define a new recursion as follows:

$$\begin{aligned}
\Xi_{(k+1)n_i,i} =&(I-K_{(k+1)n_i}C_i)\Psi_1(\Xi_{kn_i,i})(I-K_{(k+1)n_i}C_i)^T\\
&+K_{(k+1)n_i}\Psi_2(\Xi_{kn_i,i})K_{(k+1)n_i}^T, \qquad (7.20)\\
\Xi_{0,i} =&P_{0|0}>0
\end{aligned}$$

where

$$\begin{aligned}
\Psi_1(\Xi_{kn_i,i}) =&\alpha_{2,kn_i}(\tilde{A}_i(\Xi_{kn_i,i}^{-1}-\gamma_{kn_i}E^TE)^{-1}\tilde{A}_i^T+\gamma_{kn_i}^{-1}\breve{A}_iHH^T\breve{A}_i^T)+\alpha_{1,kn_i}\breve{A}_i\\
&\times\Big(\sum_{i=1}^{r}\Pi_k^i\mathrm{tr}\big(((1+\varepsilon_{6,kn_i})\Xi_{kn_i,i}+(1+\varepsilon_{6,kn_i}^{-1})\hat{x}_{kn_i|kn_i}\hat{x}_{kn_i|kn_i}^T\big)\\
&\times\Gamma_k^i)\Big)\breve{A}_i^T+\alpha_{3,kn_i}\lambda_{kn_i}\breve{A}_iHH^T\breve{A}_i^T+\alpha_{1,kn_i}\bar{B}_i\bar{R}_{k,i}\bar{B}_i^T,\\
\Psi_2(\Xi_{kn_i,i}) =&\alpha_4D_iC_i\Xi_{kn_i,i}C_i^TD_i^T+Q_{k,k}+\alpha_{5,kn_i}I
\end{aligned}$$

and

$$\begin{aligned}
\lambda_{kn_i} &= \lambda_{\max}\big(E\hat{x}_{kn_i|kn_i}\hat{x}_{kn_i|kn_i}^TE^T\big),\ \alpha_{1,kn_i}=1+\varepsilon_{1,kn_i},\\
\alpha_{2,kn_i} &= (1+\varepsilon_{4,kn_i})\alpha_{1,kn_i}+\varepsilon_{3,kn_i}^{-1},\ \alpha_{3,kn_i}=(1+\varepsilon_{4,kn_i}^{-1})\alpha_{1,kn_i}+\varepsilon_{5,kn_i}^{-1},\\
\alpha_{4,kn_i} &= 1+\varepsilon_{3,kn_i}+\varepsilon_{5,kn_i}+\varepsilon_{2,kn_i},\ \alpha_{5,kn_i}=(1+\varepsilon_{1,kn_i}^{-1}+\varepsilon_{2,kn_i}^{-1})\delta_i, \qquad (7.21)
\end{aligned}$$

ε_{1,kn_i}, ε_{2,kn_i}, ε_{3,kn_i}, ε_{4,kn_i}, ε_{5,kn_i}, ε_{6,kn_i} and γ_{kn_i} are positive constants and $\Xi_{kn_i,i}$ satisfies

$$\gamma_{kn_i}^{-1}I-E\Xi_{kn_i,i}E^T>0. \qquad (7.22)$$

The following theorem provides an upper bound of the FEC.

Theorem 7.1 *Under the condition (7.17), the solution to the recursion (7.20) with (7.22) is an upper bound of the solution to the recursion (7.19), i.e.,* $P_{kn_i|kn_i} \leq \Xi_{kn_i,i}$.

Proof *In this proof, we choose the mathematical induction approach. Firstly, it can be obtained from the initial condition that* $\Xi_{0,i} \geq P_{0|0}$ *holds. Next, we assume that* $\Xi_{kn_i,i}$ *satisfies* $P_{kn_i|kn_i} \leq \Xi_{kn_i,i}$ *and then we aim to show* $P_{(k+1)n_i|(k+1)n_i} \leq \Xi_{(k+1)n_i,i}$.

Firstly, it can be obtained that

$$\varrho_{(k+1)n_i,i}\varrho_{(k+1)n_i,i}^T \leq \varrho_{(k+1)n_i,i}^T\varrho_{(k+1)n_i,i}I \leq \delta_i I. \tag{7.23}$$

Following by Lemma 7.1 and $P_{kn_i|kn_i} \leq \Xi_{kn_i,i}$, *we have*

$$\begin{aligned}&(I - K_{(k+1)n_i}C_i)\mathbb{E}\{\tilde{x}_{(k+1)n_i|kn_i}\varrho_{(k+1)n_i,i}^T\}K_{(k+1)n_i}^T\\&+ K_{(k+1)n_i}\mathbb{E}\{\varrho_{(k+1)n_i,i}\tilde{x}_{(k+1)n_i|kn_i}^T\}(I - K_{(k+1)n_i}C_i)^T\\ \leq&\varepsilon_{1,kn_i}\Omega_{(k+1)n_i|kn_i} + \varepsilon_{1,kn_i}^{-1}\delta_i K_{(k+1)n_i}K_{(k+1)n_i}^T\end{aligned} \tag{7.24}$$

and

$$\begin{aligned}&K_{(k+1)n_i}\mathbb{E}\{\varrho_{(k+1)n_i,i}\tilde{x}_{kn_i|kn_i}^T\}C_i^T D_i^T K_{(k+1)n_i}^T\\&+ K_{(k+1)n_i}D_iC_i\mathbb{E}\{\tilde{x}_{kn_i|kn_i}\varrho_{(k+1)n_i,i}^T\}K_{(k+1)n_i}^T\\ \leq&\varepsilon_{2,kn_i}K_{(k+1)n_i}D_iC_iP_{kn_i|kn_i}C_i^T D_i^T K_{(k+1)n_i}^T + \varepsilon_{2,kn_i}^{-1}\delta_i K_{(k+1)n_i}K_{(k+1)n_i}^T\\ \leq&\varepsilon_{2,kn_i}K_{(k+1)n_i}D_iC_i\Xi_{kn_i,i}C_i^T D_i^T K_{(k+1)n_i}^T + \varepsilon_{2,kn_i}^{-1}\delta_i K_{(k+1)n_i}K_{(k+1)n_i}^T.\end{aligned} \tag{7.25}$$

From (7.22), it is known that

$$\gamma_{kn_i}^{-1}I - EP_{kn_i|kn_i}E^T > 0, \tag{7.26}$$

by which and (7.2), it can be obtained from Lemma 7.2 that

$$\begin{aligned}&(\tilde{A}_i + \breve{A}_i\Delta A_{kn_i})P_{kn_i|kn_i}(\tilde{A}_i + \breve{A}_i\Delta A_{kn_i})^T\\ \leq&\tilde{A}_i(P_{kn_i|kn_i}^{-1} - \gamma_{kn_i}E^T E)^{-1}\tilde{A}_i^T + \gamma_{kn_i}^{-1}\breve{A}_i HH^T\breve{A}_i^T\\ \leq&\tilde{A}_i(\Xi_{kn_i,i}^{-1} - \gamma_{kn_i}E^T E)^{-1}\tilde{A}_i^T + \gamma_{kn_i}^{-1}\breve{A}_i HH^T\breve{A}_i^T.\end{aligned} \tag{7.27}$$

By using Lemma 7.1, it follows from (7.27) that

$$\begin{aligned}&(I - K_{(k+1)n_i}C_i)(\tilde{A}_i + \breve{A}_i\Delta A_{kn_i})P_{kn_i|kn_i}C_i^T D_i^T K_{(k+1)n_i}^T\\&+ K_{(k+1)n_i}D_iC_iP_{kn_i|kn_i}(\tilde{A}_i + \breve{A}_i\Delta A_{kn_i})^T(I - K_{(k+1)n_i}C_i)^T\\ \leq&\varepsilon_{3,kn_i}^{-1}(I - K_{(k+1)n_i}C_i)(\tilde{A}_i + \breve{A}_i\Delta A_{kn_i})P_{kn_i|kn_i}(\tilde{A}_i + \breve{A}_i\Delta A_{kn_i})^T\\&\times(I - K_{(k+1)n_i}C_i)^T + \varepsilon_{3,kn_i}K_{(k+1)n_i}D_iC_iP_{kn_i|kn_i}C_i^T D_i^T K_{(k+1)n_i}^T\\ \leq&\varepsilon_{3,kn_i}^{-1}(I - K_{(k+1)n_i}C_i)(\tilde{A}_i(P_{kn_i|kn_i}^{-1} - \gamma_{kn_i}E^T E)^{-1}\tilde{A}_i^T + \gamma_{kn_i}^{-1}\breve{A}_i HH^T\breve{A}_i^T)\\&\times(I - K_{(k+1)n_i}C_i)^T + \varepsilon_{3,kn_i}K_{(k+1)n_i}D_iC_iP_{kn_i|kn_i}C_i^T D_i^T K_{(k+1)n_i}^T,\\ \leq&\varepsilon_{3,kn_i}^{-1}(I - K_{(k+1)n_i}C_i)(\tilde{A}_i(\Xi_{kn_i|kn_i}^{-1} - \gamma_{kn_i}E^T E)^{-1}\tilde{A}_i^T + \gamma_{kn_i}^{-1}\breve{A}_i HH^T\breve{A}_i^T)\\&\times(I - K_{(k+1)n_i}C_i)^T + \varepsilon_{3,kn_i}K_{(k+1)n_i}D_iC_i\Xi_{kn_i|kn_i}C_i^T D_i^T K_{(k+1)n_i}^T.\end{aligned} \tag{7.28}$$

Noting (7.2) again, we have

$$\breve{A}_i \Delta A_{kn_i} \mathbb{E}\{\hat{x}_{kn_i|kn_i} \hat{x}^T_{kn_i|kn_i}\} \Delta A^T_{kn_i} \breve{A}^T_i \leq \lambda_{kn_i} \breve{A}_i H H^T \breve{A}^T_i. \tag{7.29}$$

From (7.27), (7.29), and Lemma 7.1, we further obtain

$$\begin{aligned}
&(\tilde{A}_i + \breve{A}_i \Delta A_{kn_i}) \mathbb{E}\{\tilde{x}_{kn_i|kn_i} \hat{x}^T_{kn_i|kn_i}\} \Delta A^T_{kn_i} \breve{A}^T_i \\
&\quad + \breve{A}_i \Delta A_{kn_i} \mathbb{E}\{\hat{x}_{kn_i|kn_i} \tilde{x}^T_{kn_i|kn_i}\} (\tilde{A}_i + \breve{A}_i \Delta A_{kn_i})^T \\
&\leq \varepsilon_{4,kn_i} (\tilde{A}_i + \breve{A}_i \Delta A_{kn_i}) P_{kn_i|kn_i} (\tilde{A}_i + \breve{A}_i \Delta A_{kn_i})^T \\
&\quad + \varepsilon^{-1}_{4,kn_i} \breve{A}_i \Delta A_{kn_i} \mathbb{E}\{\hat{x}_{kn_i|kn_i} \hat{x}^T_{kn_i|kn_i}\} \Delta A^T_{kn_i} \breve{A}^T_i \\
&\leq \varepsilon_{4,kn_i} (\tilde{A}_i (\Xi^{-1}_{kn_i,i} - \gamma_{kn_i} E^T E)^{-1} \tilde{A}^T_i + \gamma^{-1}_{kn_i} \breve{A}_i H H^T \breve{A}^T_i) \\
&\quad + \varepsilon^{-1}_{4,kn_i} \lambda_{kn_i} \breve{A}_i H H^T \breve{A}^T_i
\end{aligned} \tag{7.30}$$

and

$$\begin{aligned}
&(I - K_{(k+1)n_i} C_i) \breve{A}_i \Delta A_{kn_i} \mathbb{E}\{\hat{x}_{kn_i|kn_i} \tilde{x}^T_{kn_i|kn_i}\} C^T_i D^T_i K^T_{(k+1)n_i} \\
&\quad + K_{(k+1)n_i} D_i C_i \mathbb{E}\{\tilde{x}_{kn_i|kn_i} \hat{x}^T_{kn_i|kn_i}\} \Delta A^T_{kn_i} \breve{A}^T_i (I - K_{(k+1)n_i} C_i)^T \\
&\leq \varepsilon^{-1}_{5,kn_i} (I - K_{(k+1)n_i} C_i) \breve{A}_i \Delta A_{kn_i} \mathbb{E}\{\hat{x}_{kn_i|kn_i} \hat{x}^T_{kn_i|kn_i}\} \Delta A^T_{kn_i} \breve{A}^T_i \\
&\quad \times (I - K_{(k+1)n_i} C_i)^T + \varepsilon_{5,kn_i} K_{(k+1)n_i} D_i C_i P_{kn_i|kn_i} C^T_i D^T_i K^T_{(k+1)n_i} \\
&\leq \varepsilon^{-1}_{5,kn_i} (I - K_{(k+1)n_i} C_i) \lambda_{kn_i} \breve{A}_i H H^T \breve{A}^T_i (I - K_{(k+1)n_i} C_i)^T \\
&\quad + \varepsilon_{5,kn_i} K_{(k+1)n_i} D_i C_i \Xi_{kn_i|kn_i} C^T_i D^T_i K^T_{(k+1)n_i}.
\end{aligned} \tag{7.31}$$

For the term $\mathbb{E}\{x_{kn_i} x^T_{kn_i}\}$, we have

$$\begin{aligned}
\mathbb{E}\{x_{kn_i} x^T_{kn_i}\} &= \mathbb{E}\{(\hat{x}_{kn_i|kn_i} + \tilde{x}_{kn_i|kn_i})(\hat{x}_{kn_i|kn_i} + \tilde{x}_{kn_i|kn_i})^T \\
&\leq (1 + \varepsilon_{6,kn_i}) P_{kn_i|kn_i} + (1 + \varepsilon^{-1}_{6,kn_i}) \hat{x}_{kn_i|kn_i} \hat{x}^T_{kn_i|kn_i} \\
&\leq (1 + \varepsilon_{6,kn_i}) \Xi_{kn_i|kn_i} + (1 + \varepsilon^{-1}_{6,kn_i}) \hat{x}_{kn_i|kn_i} \hat{x}^T_{kn_i|kn_i}.
\end{aligned} \tag{7.32}$$

Now, it readily follows from (7.18), (7.19), (7.23)-(7.32) that

$$P_{(k+1)n_i|(k+1)n_i} \leq \Xi_{(k+1)n_i,i}.$$

Therefore, in virtue of mathematical induction, it is shown that $P_{kn_i|kn_i} \leq \Xi_{kn_i,i}$ for $k = 0, 1, 2, \cdots$. The proof of Theorem 7.1 is complete.

Remark 7.2 *Due to the influence of the PUs, stochastic nonlinearities as well as the ET mechanism, it is usually difficult to obtain the exact value of the FECs and an alternative way is to look for their upper bound instead. In Theorem 7.1, an upper bound of the FEC is given.*

Remark 7.3 *Note that, when deriving the upper bound, we have used Lemmas 7.1 and 7.2 at each time instant and this may give rise to some conservativeness. Such conservativeness can be reduced effectively by adjusting the parameters ε and γ.*

In the following theorem, the local filter gain is designed such that the upper bound $\Xi_{(k+1)n_i,i}$ is minimized with the constraint (7.17).

Theorem 7.2 *The upper bound given by the recursion (7.20) with (7.22) is minimized with the constraint (7.17) by the following filter gain*

$$\begin{aligned} K_{(k+1)n_i} =& \Big[\Psi_1(\Xi_{kn_i,i})C_i^T - (\Psi_1(\Xi_{kn_i,i})C_i^T\Theta_k^{-1}C_i\bar{A}_i - \bar{A}_i) \\ & (\bar{A}_i^TC_i^T\Theta_k^{-1}C_i\bar{A}_i)^{-1}\bar{A}_i^TC_i^T\Big]\Theta_k^{-1} \end{aligned} \tag{7.33}$$

where

$$\Theta_k = C_i\Psi_1(\Xi_{kn_i,i})C_i^T + \Psi_2(\Xi_{kn_i,i}). \tag{7.34}$$

Proof *By means of Lagrange multiplier method, we introduce the auxiliary function*

$$\begin{aligned} \Upsilon(K_{(k+1)n_i}, \Lambda_{(k+1)n_i}) =& \Xi_{(k+1)n_i,i} + (K_{(k+1)n_i}C_i\bar{A}_i - \bar{A}_i)\Lambda_{(k+1)n_i}^T \\ & + \Lambda_{(k+1)n_i}(K_{(k+1)n_i}C_i\bar{A}_i - \bar{A}_i)^T \end{aligned} \tag{7.35}$$

where $\Lambda_{(k+1)n_i}$ is the Lagrange factor.

By using "completing the square", it can be obtained that

$$\begin{aligned} & \Upsilon(K_{(k+1)n_i}, \Lambda_{(k+1)n_i}) \\ =& \left[K_{(k+1)n_i} - \left(\Psi_1(\Xi_{kn_i,i})C_i^T - \Lambda_{(k+1)n_i}\bar{A}_i^TC_i^T\right)\Theta_k^{-1}\right]\Theta_k^{-1} \\ & \times\left[K_{(k+1)n_i} - \left(\Psi_1(\Xi_{kn_i,i})C_i^T - \Lambda_{(k+1)n_i}\bar{A}_i^TC_i^T\right)\Theta_k^{-1}\right]^T \\ & -\left(\Psi_1(\Xi_{kn_i,i})C_i^T - \Lambda_{(k+1)n_i}\bar{A}_i^TC_i^T\right)\Theta_k^{-1} \\ & \times\left(\Psi_1(\Xi_{kn_i,i})C_i^T - \Lambda_{(k+1)n_i}\bar{A}_i^TC_i^T\right)^T \\ & -\Lambda_{(k+1)n_i}\bar{A}_i^T - \bar{A}_i\Lambda_{(k+1)n_i}^T + \Psi_1(\Xi_{kn_i,i}). \end{aligned} \tag{7.36}$$

Therefore, under the constraint (7.17), the filter gain given by

$$K_{(k+1)n_i} = \left(\Psi_1(\Xi_{kn_i,i})C_i^T - \Lambda_{(k+1)n_i}\bar{A}_i^TC_i^T\right)\Theta_k^{-1} \tag{7.37}$$

minimizes the $\Xi_{(k+1)n_i,i}$.

Substitution of $K_{(k+1)n_i}$ given by (7.37) into constraint condition (7.17) yields

$$\Lambda_{(k+1)n_i} = \Big(\Psi_1(\Xi_{kn_i,i})C_i^T\Theta_k^{-1}C_i\bar{A}_i - \bar{A}_i\Big)(\bar{A}_i^T C_i^T\Theta_k^{-1}C_i\bar{A}_i)^{-1}. \quad (7.38)$$

It can be easily seen that the filter gain given by (7.33) with (7.34) is exactly the same as the one given by (7.37) with (7.38). Therefore, the proof of Theorem 7.2 is accomplished.

The upper bound of the local FEC has been obtained and the filter gain matrix $K_{(k+1)n_i}$ has been designed such that the upper bound of the local FEC at each sampling instant is minimized. Next, based on the local estimates derived, a fusion estimation scheme is proposed by employing the CI fusion method.

7.1.3 Fusion Estimation

Note that the local estimates are generated in different sampling rate, and hence the cross-covariance of estimates at same time instant is usually not available. Therefore, we adopt the CI fusion estimation scheme, which can be stated as follows.

Set $\Omega_k = \{i|k = m_{k,i}n_i, m_{k,i}$ is a positive integer, $i = 1, 2, \cdots, M\}$. In this subsection, the local estimates are only fused at those time instant kh_0 when the set Ω_k is not empty. Denote by $\hat{x}_k^0$ and Ξ_k^0 the fusion estimate and fusion "covariance", respectively. The fusion scheme is adopted by

$$\begin{aligned} \Xi_k^0 &= \left(\sum_{i\in\Omega_k} \omega_i \Xi_{m_{k,i}n_i,i}^{-1}\right)^{-1}, \\ \hat{x}_k^0 &= \Xi_k^0 \sum_{i\in\Omega_k} \omega_i \Xi_{m_{k,i}n_i,i}^{-1}\hat{x}_{m_{k,i}n_i|m_{k,i}n_i} \end{aligned} \quad (7.39)$$

where $\omega_i \geq 0$ satisfying

$$\sum_{i\in\Omega_k} \omega_i = 1. \quad (7.40)$$

The fusion objective is formulated as the following optimization problem

$$\begin{aligned} &\min_{\omega_i}\{\mathrm{tr}(\Xi_k^0)\}, \\ &\text{s.t.} \sum_{i\in\Omega_k} \omega_i = 1, \omega_i \geq 0, i \in \Omega_k. \end{aligned} \quad (7.41)$$

In the following theorem, the CI-based fusion scheme expressed by (7.39) is consistent.

Theorem 7.3 *For the multi-rate system (7.8) with local filter (7.11), the CI-based fusion scheme (7.39)–(7.41) is consistent, i.e.*

$$\bar{P}_{CL} = \mathbb{E}\{(x_k - \hat{x}_k^0)(x_k - \hat{x}_k^0)^T\} \leq \Xi_k^0. \tag{7.42}$$

Proof *Noticing the fact that $P_{m_{k,i}n_i|m_{k,i}n_i} \leq \Xi_{m_{k,i}n_i,i}$, the proof of Theorem 7.3 can be directly obtained by following the proof provided in [21,67].*

The above design process can be summarised as the following algorithm.

Fusion estimation algorithm	
Step 1.	Give the parameters $\varepsilon_{1,0n_i}$, $\varepsilon_{2,0n_i}$, $\varepsilon_{3,0n_i}$, $\varepsilon_{4,0n_i}$, $\varepsilon_{5,0n_i}$, $\varepsilon_{6,0n_i}$, δ_i and γ_{0n_i} $(i = 1, 2, \ldots, M)$, and set the initial conditions as $\hat{x}_{0n_i\|0n_i} = \bar{x}_0$ and $\Xi_{0n_i,i} = P_{0\|0}$ and $k = 0$, generate fusion estimation $\hat{x}_0^0$;
Step 2.	If $\Omega_k \neq \emptyset$, go to Step 3, else, go to Step 6;
Step 3.	Calculate $K_{m_{k,i}n_i}$ $(i \in \Omega_k)$ from (7.33), and compute $\hat{x}_{m_{k,i}n_i\|m_{k,i}n_i}$ from local filter (7.11);
Step 4.	Compute the fusion estimation $\hat{x}_k^0$ and Ξ_k^0 from (7.39) and (7.41) on the basis of local estimate generated by node i $(i \in \Omega_k)$;
Step 5.	Give parameters $\varepsilon_{1,m_{k,i}n_i}$, $\varepsilon_{2,m_{k,i}n_i}$, $\varepsilon_{3,m_{k,i}n_i}$, $\varepsilon_{4,m_{k,i}n_i}$, $\varepsilon_{5,m_{k,i}n_i}$, $\varepsilon_{6,m_{k,i}n_i}$ and $\gamma_{m_{k,i}n_i}$, and calculate $\Xi_{(m_{k,i}+1)n_i,i}$ from (7.20);
Step 6.	Set $k = k + 1$ and go to Step 2;
Step 7.	Output $\hat{x}_{m_{k,i}n_i\|m_{k,i}n_i}$, $\hat{x}_k^0$, $\Xi_{m_{k,i}n_i,i}$ and Ξ_k^0;
Step 8.	Stop.

Until now, we have dealt with the filtering problem for multi-rate uncertain systems with stochastic nonlinearities and coloured measurement noises. In the design process, an upper bound of the FEC has been derived and the designed local filters minimize the upper bound of their FECs separately. Based on the obtained local filters, a consistent CI-based fusion scheme has been proposed.

7.2 Event-Triggered Fusion Estimation with Sensor Degradations

7.2.1 Problem Formulation

The considered discrete system can be formulated as follows:

$$x_{(k+1)T} = Ax_{kT} + Bw_{kT} \tag{7.43}$$

where $T > 0$ is a system sampling period, $x_{kT} \in \mathbb{R}^{n_x}$ is a state vector at sampling instant kT, $w_{kT} \in \mathbb{R}^{n_x}$ is the zero-mean process noise with variance $R_{kT} > 0$ and A, B are known time-invariant matrices having appropriate dimensions.

The output of the ith ($i = 1, 2, \ldots, M$) sensor node with sensor degradations and stochastic nonlinearities, is described by

$$y_{kl_iT} = \Theta_{kl_iT} C_i x_{kl_iT} + s(x_{kl_iT}, \varsigma_{kl_iT}) + v_{kl_iT} \tag{7.44}$$

where $l_i > 0$ is an integer, $y_{kl_iT} \in \mathbb{R}^{n_y}$ is a measurement vector at sampling instant kl_iT, C_i is an appropriate-dimensional matrix, $v_{kl_iT} \in \mathbb{R}^{n_y}$ is the ith zero-mean measurement noise having variance $Q_{kl_iT} > 0$, and $\Theta_{kl_iT} = \mathrm{diag}\{\theta_{1,kl_iT}, \theta_{2,kl_iT}, \ldots, \theta_{n_y,kl_iT}\}$, where θ_{m,kl_iT} ($m = 1, 2, \ldots, n_y$) is a random variable obeying the uniform distribution over the given interval $[\underline{\kappa}_m,\ \bar{\kappa}_m]$ ($0 \le \underline{\kappa}_m < \bar{\kappa}_m \le 1$) with the mathematical expectation $\alpha_m = \frac{\bar{\kappa}_m + \underline{\kappa}_m}{2}$ and variance $\sigma_m^2 = \frac{(\bar{\kappa}_m - \underline{\kappa}_m)^2}{12}$.

Assumption 7.1 *x_0, w_{kl_iT}, v_{kl_iT} and θ_{m,kl_iT} are mutually independent.*

The stochastic nonlinearities function $s(\cdot,\cdot)$ satisfies

$$\begin{aligned} &s(0, \varsigma_{kl_iT}) = 0,\ \mathbb{E}\{s(x_{kl_iT}, \varsigma_{kl_iT})|x_{kl_iT}\} = 0, \\ &\mathbb{E}\{s(x_{kl_iT}, \varsigma_{kl_iT}) s(x_{jl_iT}, \varsigma_{jl_iT})^T |x_{jl_iT}\} = 0, \quad k \ne j, \\ &\mathbb{E}\{s(x_{kl_iT}, \varsigma_{kl_iT}) s(x_{kl_iT}, \varsigma_{kl_iT})^T |x_{kl_iT}\} = \sum_{i=1}^{r} \Pi^i_{kl_iT} x^T_{kl_T} \Gamma^i_{kl_iT} x_{kl_iT} \end{aligned} \tag{7.45}$$

where $r > 0$ is a given integer and $\Pi^i_{kl_iT}$, $\Gamma^i_{kl_iT}$ are known matrices.

To facilitate the filter design, we'd like to transform the system (7.43) and ith measurement (7.44) into an single-rate system. For each interval $[kl_i + 1\ (k+1)l_i]$, the system dynamics in (1) can be expressed by (T is omitted for brevity)

$$\begin{aligned} x_{kl_i+l+1} =& A x_{kl_i+l} + B w_{kl_i+l} \\ =& A^{l+1} x_{kl_i} + \sum_{n=0}^{l} B w_{kl_i+l-n},\ 0 \le l \le l_i - 1. \end{aligned}$$

Therefore, the single-rate system with ith measurement is given by

$$\begin{cases} x_{(k+1)l_i} =& \mathscr{A}_i x_{kl_i} + \mathscr{B}_i \bar{w}_{kl_i} \\ y_{kl_i} =& \Theta_{kl_i} C_i x_{kl_i} + s(x_{kl_i}, \varsigma_{kl_i}) + v_{kl_i} \end{cases} \tag{7.46}$$

where

$$\begin{aligned} \mathscr{A}_i =& A^{l_i},\ \mathscr{B}_i = \begin{bmatrix} A^{l_i-1}B & A^{l_i-2}B & \cdots & AB & B \end{bmatrix}, \\ \bar{w}_{kl_i} =& \begin{bmatrix} w_{kl_i} & w_{kl_i+1} & \cdots & w_{kl_i+l_i-1} \end{bmatrix}^T. \end{aligned} \tag{7.47}$$

Remark 7.4 *Note that, the system under consideration is actually an multi-rate system since the sampling period of the considered system (7.43) is kT while the sampling period of the i-th measurement (7.44) is kl_iT. As such, it is difficult to extend the existing filtering methods to the multi-rate system. As such, we transform the multi-rate systems into an single-rate system with the aid of augmentation technique.*

In order to effectively reduce the measurement transmission frequency, the ET mechanism is employed. We set the transmission instants $0 = k_0^i < k_1^i < k_2^i < \cdots < k_l^i < \cdots$, which is computed by

$$k_{l+1}^i = \min\left\{kl_i \in \mathbb{N} | kl_i > k_l^i, f(\varrho_{kl_i}, \delta_i) > 0\right\} \tag{7.48}$$

with

$$\begin{aligned} f(\varrho_{kl_i}, \delta_i) &= \varrho_{kl_i}^T \varrho_{kl_i} - \delta_i, \\ \varrho_{kl_i} &= y_{k_l^i} - y_{kl_i} \end{aligned} \tag{7.49}$$

where $\delta_i > 0$ is a given scalar and $y_{k_l^i}$ represents the ith transmission measurement at the latest event time.

Remark 7.5 *In networked control systems, control/measurement signal is often transmitted via sensors. It should be pointed out that the communication channel resources and the energy possessed by each sensor are limited. However, the traditional time-triggering mechanism usually transmits signals in a fixed time period which results in a waste of channel resources and sensor energies. Different from the time-triggering mechanism, the ET mechanism can reduce the signal transmission, and thus achieve the purpose of resource saving.*

Based on the received measurement $y_{k_l^i}$ from the ith sensor node, the corresponding local recursive filter is constructed as

$$\begin{cases} \hat{x}_{kl_i|(k-1)l_i} = \mathscr{A}_i \hat{x}_{(k-1)l_i|(k-1)l_i} \\ \hat{x}_{kl_i|kl_i} = \hat{x}_{kl_i|(k-1)l_i} + K_{kl_i}\left(y_{k_l^i} - \bar{\Theta}_{kl_i}\hat{x}_{kl_i|(k-1)l_i}\right) \end{cases} \tag{7.50}$$

where $\hat{x}_{kl_i|(k-1)l_i}$ and $\hat{x}_{kl_i|kl_i}$ are, respectively, the prediction and the estimate of the state x_{kl_i}, $\bar{\Theta}_{kl_i} = \mathbb{E}\{\Theta_{kl_i}\} = \mathrm{diag}\{\alpha_1, \alpha_2, \ldots, \alpha_{n_y}\}$, and K_{kl_i} is the filter parameter to be designed.

Let $\tilde{x}_{kl_i|kl_i} = x_{kl_i} - \hat{x}_{kl_i|kl_i}$ be the filtering error and $\tilde{x}_{kl_i|(k-1)l_i} = x_{kl_i} - \hat{x}_{kl_i|(k-1)l_i}$ be the prediction error. It follows from (7.46) and (7.50) that

$$\tilde{x}_{kl_i|(k-1)l_i} = \mathscr{A}_i \tilde{x}_{(k-1)l_i|(k-1)l_i} + \mathscr{B}_i \bar{w}_{(k-1)l_i}, \tag{7.51}$$

$$\begin{aligned} \tilde{x}_{kl_i|kl_i} = &\left(I - K_{kl_i}\bar{\Theta}_{kl_i}C_i\right)\tilde{x}_{kl_i|(k-1)l_i} - K_{kl_i}\varrho_{kl_i} - K_{kl_i}\tilde{\Theta}_{kl_i}C_i x_{kl_i} \\ &- K_{kl_i}s(x_{kl_i}, \varsigma_{kl_i}) - K_{kl_i}v_{kl_i} \end{aligned} \tag{7.52}$$

where $\tilde{\Theta}_{kl_i} = \Theta_{kl_i} - \bar{\Theta}_{kl_i}$.

Then, we define the PEC by

$$P_{kl_i|(k-1)l_i} = \mathbb{E}\left\{\tilde{x}_{kl_i|(k-1)l_i}\tilde{x}^T_{kl_i|(k-1)l_i}\right\}, \tag{7.53}$$

and the FEC is

$$P_{kl_i|kl_i} = \mathbb{E}\left\{\tilde{x}_{kl_i|kl_i}\tilde{x}^T_{kl_i|kl_i}\right\}. \tag{7.54}$$

In this section, our main purpose can be summarized as follows: 1) for each local recursive filter of the form (7.50), we desire to find an upper bound on each local FEC $\bar{P}_{kl_i|kl_i}$, and then choose the appropriate parameter K_{kl_i}, such that the obtained upper bound is minimized; 2) a multi-rate fusion estimation algorithm is subsequently conducted based on the received local estimates $\hat{x}_{kl_i|kl_i}$ and the minimal upper bounds $\bar{P}_{kl_i|kl_i}$ at each step.

7.2.2 Design of Local Filters

Lemma 7.3 *The PEC $P_{kl_i|(k-1)l_i} = \mathbb{E}\{\tilde{x}_{kl_i|(k-1)l_i}\tilde{x}^T_{kl_i|(k-1)l_i}\}$ can be derived as follows*

$$P_{kl_i|(k-1)l_i} = \mathscr{A}_i P_{(k-1)l_i|(k-1)l_i}\mathscr{A}_i^T + \mathscr{B}_i R_{(k-1)l_i}\mathscr{B}_i^T. \tag{7.55}$$

Proof *This proof can be accomplished immediately according to (7.51) and (7.53), and the rest of proof is omitted.*

Lemma 7.4 *The FEC $P_{kl_i|kl_i} = \mathbb{E}\{\tilde{x}_{kl_i|kl_i}\tilde{x}^T_{kl_i|kl_i}\}$ can be derived as*

$$\begin{aligned}
P_{kl_i|kl_i} =&(I - K_{kl_i}\bar{\Theta}_{kl_i}C_i)P_{kl_i|(k-1)l_i}(I - K_{kl_i}\bar{\Theta}_{kl_i}C_i)^T \\
&+ K_{kl_i}\mathbb{E}\{\varrho_{kl_i}\varrho^T_{kl_i}\}K^T_{kl_i} + K_{kl_i}Q_{kl_i}K^T_{kl_i} \\
&+ K_{kl_i}\left(\sum_{i=1}^{r}\Pi^i_{kl_i}\operatorname{tr}\left(\mathbb{E}\{x_{kl_i}x^T_{kl_i}\}\Gamma^i_{kl_i}\right)\right)K^T_{kl_i} \\
&+ K_{kl_i}\left(\check{\Theta}_{kl_i}\circ\mathbb{E}\{C_i x_{kl_i}x^T_{kl_i}C_i^T\}\right)K^T_{kl_i} \\
&- (I - K_{kl_i}\bar{\Theta}_{kl_i}C_i)\mathbb{E}\{\tilde{x}_{kl_i|(k-1)l_i}\varrho^T_{kl_i}\}K^T_{kl_i} \\
&- K_{kl_i}\mathbb{E}\{\varrho_{kl_i}\tilde{x}^T_{kl_i|(k-1)l_i}\}(I - K_{kl_i}\bar{\Theta}_{kl_i}C_i)^T \\
&- K_{kl_i}\mathbb{E}\{\tilde{\Theta}_{kl_i}C_i x_{kl_i}\tilde{x}^T_{kl_i|(k-1)l_i}\}(I - K_{kl_i}\bar{\Theta}_{kl_i}C_i)^T \\
&- (I - K_{kl_i}\bar{\Theta}_{kl_i}C_i)\mathbb{E}\{\tilde{x}_{kl_i|(k-1)l_i}x^T_{kl_i}C_i^T\tilde{\Theta}^T_{kl_i}\}K^T_{kl_i} \\
&+ K_{kl_i}\mathbb{E}\{\varrho_{kl_i}v^T_{kl_i}\}K^T_{kl_i} + K_{kl_i}\mathbb{E}\{v_{kl_i}\varrho^T_{kl_i}\}K^T_{kl_i} \\
&- K_{kl_i}\mathbb{E}\{\varrho_{kl_i}x^T_{kl_i}C_i^T\tilde{\Theta}^T_{kl_i}\}K^T_{kl_i} - K_{kl_i}\mathbb{E}\{x_{kl_i}C_i\tilde{\Theta}_{kl_i}\varrho^T_{kl_i}\}K^T_{kl_i}
\end{aligned} \tag{7.56}$$

where $\check{\Theta}_{kl_i} = \{\sigma^2_{1,kl_i}, \sigma^2_{2,kl_i}, \ldots, \sigma^2_{n_y,kl_i}\}$ and $R_{(k-1)l_i}$ is the covariance of $\bar{w}_{(k-1)l_i}$.

Proof *From the definition of $P_{kl_i|kl_i}$ in (7.54), we have*

$$\begin{aligned}P_{kl_i|kl_i} =&\mathbb{E}\Big\{\tilde{x}_{kl_i|kl_i}\tilde{x}^T_{kl_i|kl_i}\Big\}\\
=&(I-K_{kl_i}\bar{\Theta}_{kl_i}C_i)P_{kl_i|(k-1)l_i}(I-K_{kl_i}\bar{\Theta}_{kl_i}C_i)^T\\
&+K_{kl_i}\mathbb{E}\{\varrho_{kl_i}\varrho^T_{kl_i}\}K^T_{kl_i}+K_{kl_i}Q_{kl_i}K^T_{kl_i}\\
&+K_{kl_i}\left(\sum_{i=1}^{r}\Pi^i_{kl_i}\mathrm{tr}\Big(\mathbb{E}\{x_{kl_i}x^T_{kl_i}\}\Gamma^i_{kl_i}\Big)\right)K^T_{kl_i}\\
&+K_{kl_i}\Big(\check{\Theta}_{kl_i}\circ\mathbb{E}\{C_ix_{kl_i}x^T_{kl_i}C_i^T\}\Big)K^T_{kl_i}-\chi_{1,k}-\chi^T_{1,k}\\
&-\chi_{2,k}-\chi^T_{2,k}-\chi_{3,k}-\chi^T_{3,k}-\chi_{4,k}-\chi^T_{4,k}\\
&+\chi_{5,k}+\chi^T_{5,k}-\chi_{6,k}-\chi^T_{6,k}+\chi_{7,k}+\chi^T_{7,k}\\
&+\chi_{8,k}+\chi^T_{8,k}+\chi_{9,k}+\chi^T_{9,k}+\chi_{10,k}+\chi^T_{10,k}\end{aligned}\tag{7.57}$$

where

$$\begin{aligned}\chi_{1,k}=&K_{kl_i}\mathbb{E}\{\varrho_{kl_i}\tilde{x}^T_{kl_i|(k-1)l_i}\}(I-K_{kl_i}\bar{\Theta}_{kl_i}C_i)^T,\\
\chi_{2,k}=&K_{kl_i}\mathbb{E}\{\tilde{\Theta}_{kl_i}C_ix_{kl_i}\tilde{x}^T_{kl_i|(k-1)l_i}\}(I-K_{kl_i}\bar{\Theta}_{kl_i}C_i)^T,\\
\chi_{3,k}=&K_{kl_i}\mathbb{E}\{s(x_{kl_i},\varsigma_{kl_i})\tilde{x}^T_{kl_i|(k-1)l_i}\}(I-K_{kl_i}\bar{\Theta}_{kl_i}C_i)^T,\\
\chi_{4,k}=&K_{kl_i}\mathbb{E}\{v_{kl_i}\tilde{x}^T_{kl_i|(k-1)l_i}\}(I-K_{kl_i}\bar{\Theta}_{kl_i}C_i)^T,\\
\chi_{5,k}=&K_{kl_i}\mathbb{E}\{s(x_{kl_i},\varsigma_{kl_i})x^T_{kl_i}C_i^T\tilde{\Theta}^T_{kl_i}\}K^T_{kl_i},\ \chi_{10,k}=K_{kl_i}\mathbb{E}\{\varrho_{kl_i}v^T_{kl_i}\}K^T_{kl_i},\\
\chi_{6,k}=&K_{kl_i}\mathbb{E}\{\varrho_{kl_i}x^T_{kl_i}C_i^T\tilde{\Theta}^T_{kl_i}\}K^T_{kl_i},\ \chi_{7,k}=K_{kl_i}\mathbb{E}\{v_{kl_i}x^T_{kl_i}C_i\tilde{\Theta}^T_{kl_i}\}K^T_{kl_i},\\
\chi_{8,k}=&K_{kl_i}\mathbb{E}\{s(x_{kl_i},\varsigma_{kl_i})v^T_{kl_i}\}K^T_{kl_i},\ \chi_{9,k}=K_{kl_i}\mathbb{E}\{s(x_{kl_i},\varsigma_{kl_i})\varrho^T_{kl_i}\}K^T_{kl_i}.\end{aligned}$$

By noting Assumption 7.1, it is not difficult to show that the terms

$$\chi_{3,k}=0,\ \chi_{4,k}=0,\ \chi_{5,k}=0,\ \chi_{7,k}=0,\ \chi_{8,k}=0,\ \chi_{9,k}=0.\tag{7.58}$$

Then, the FEC can be written as

$$\begin{aligned}P_{kl_i|kl_i} =&\mathbb{E}\Big\{\tilde{x}_{kl_i|kl_i}\tilde{x}^T_{kl_i|kl_i}\Big\}\\
=&(I-K_{kl_i}\bar{\Theta}_{kl_i}C_i)P_{kl_i|(k-1)l_i}(I-K_{kl_i}\bar{\Theta}_{kl_i}C_i)^T\\
&+K_{kl_i}\mathbb{E}\{\varrho_{kl_i}\varrho^T_{kl_i}\}K^T_{kl_i}+K_{kl_i}Q_{kl_i}K^T_{kl_i}\\
&+K_{kl_i}\left(\sum_{i=1}^{r}\Pi^i_{kl_i}\mathrm{tr}\Big(\mathbb{E}\{x_{kl_i}x^T_{kl_i}\}\Gamma^i_{kl_i}\Big)\right)K^T_{kl_i}\\
&+K_{kl_i}\Big(\check{\Theta}_{kl_i}\circ\mathbb{E}\{C_ix_{kl_i}x^T_{kl_i}C_i^T\}\Big)K^T_{kl_i}-\chi_{1,k}-\chi^T_{1,k}\\
&-\chi_{2,k}-\chi^T_{2,k}-\chi_{6,k}-\chi^T_{6,k}+\chi_{10,k}+\chi^T_{10,k}.\end{aligned}\tag{7.59}$$

The proof is now complete.

Lemma 7.5 *For arbitrary positive scalars $\varepsilon_{i,k}$ $(i=1,2,3,4,5)$, the PEC and FEC satisfy the following relationship*

$$\begin{aligned}P_{kl|(k-1)l_i} \triangleq&\Lambda(P_{(k-1)l_i|(k-1)l_i})\\ \triangleq&\mathscr{A}_i P_{(k-1)l_i|(k-1)l_i}\mathscr{A}_i^T+\mathscr{B}_i R_{(k-1)l_i}\mathscr{B}_i^T\end{aligned} \tag{7.60}$$

and

$$\begin{aligned}P_{kl_i|kl_i} \leq&\bar{\Lambda}(P_{kl_i|(k-1)l_i})\\ \triangleq&\zeta_{1,k}(I-K_{kl_i}\bar{\Theta}_{kl_i}C_i)P_{kl_i|(k-1)l_i}(I-K_{kl_i}\bar{\Theta}_{kl_i}C_i)^T\\ &+(1+\varepsilon_{2,k}^{-1}+\varepsilon_{4,k}^{-1})K_{kl_i}\Big\{\check{\Theta}_{kl_i}\circ C_i\big((1+\varepsilon_{5,k})P_{kl_i|(k-1)l_i}\\ &+(1+\varepsilon_{5,k}^{-1})\hat{x}_{kl_i|(k-1)l_i}\hat{x}_{kl_i|(k-1)l_i}^T\big)C_i^T\Big\}K_{kl_i}^T\\ &+K_{kl_i}\bigg(\sum_{i=1}^{r}\Pi_{kl_i}^i\operatorname{tr}\Big(\big((1+\varepsilon_{5,k})P_{kl_i|(k-1)l_i}\\ &+(1+\varepsilon_{5,k}^{-1})\hat{x}_{kl_i|(k-1)l_i}\hat{x}_{kl_i|(k-1)l_i}^T\big)\Gamma_{kl_i}^i\Big)\bigg)K_{kl_i}^T\\ &+\zeta_{2,k}K_{kl_i}K_{kl_i}^T\delta_i+(1+\varepsilon_{3,k}^{-1})K_{kl_i}Q_{kl_i}K_{kl_i}^T\end{aligned} \tag{7.61}$$

where $\zeta_{1,k}=1+\varepsilon_{1,k}+\varepsilon_{2,k}$ *and* $\zeta_{2,k}=1+\varepsilon_{1,k}^{-1}+\varepsilon_{3,k}+\varepsilon_{4,k}$.

Proof *By virtue of the triggering condition (7.48), one has*

$$\varrho_{kl_i}\varrho_{kl_i}^T\leq\varrho_{kl_i}^T\varrho_{kl_i}I\leq\delta_i I. \tag{7.62}$$

By applying Lemma 7.1 and noting the sixth to thirteenth terms in (7.56), we have

$$\begin{aligned}&-(I-K_{kl_i}\bar{\Theta}_{kl_i}C_i)\mathbb{E}\{\tilde{x}_{kl_i|(k-1)l_i}\varrho_{kl_i}^T\}K_{kl_i}^T\\ &\quad-K_{kl_i}\mathbb{E}\{\varrho_{kl_i}\tilde{x}_{kl_i|(k-1)l_i}^T\}(I-K_{kl_i}\bar{\Theta}_{kl_i}C_i)^T\\ \leq&\varepsilon_{1,k}(I-K_{kl_i}\bar{\Theta}_{kl_i}C_i)P_{kl_i|(k-1)l_i}(I-K_{kl_i}\bar{\Theta}_{kl_i}C_i)^T\\ &+\varepsilon_{1,k}^{-1}K_{kl_i}K_{kl_i}^T\delta_i,\end{aligned} \tag{7.63}$$

$$\begin{aligned}&-K_{kl_i}\mathbb{E}\{\tilde{\Theta}_{kl_i}C_i x_{kl_i}\tilde{x}_{kl_i|(k-1)l_i}^T\}(I-K_{kl_i}\bar{\Theta}_{kl_i}C_i)^T\\ &\quad-(I-K_{kl_i}\bar{\Theta}_{kl_i}C_i)\mathbb{E}\{\tilde{x}_{kl_i|(k-1)l_i}x_{kl_i}^T C_i^T\tilde{\Theta}_{kl_i}^T\}K_{kl_i}^T\\ \leq&\varepsilon_{2,k}(I-K_{kl_i}\bar{\Theta}_{kl_i}C_i)P_{kl_k|(k-1)l_i}(I-K_{kl_i}\bar{\Theta}_{kl_i}C_i)^T\\ &+\varepsilon_{2,k}^{-1}K_{kl_i}\left(\check{\Theta}_{kl_i}\circ\mathbb{E}\{C_i x_{kl_i}x_{kl_i}^T C_i^T\}\right)K_{kl_i}^T,\end{aligned} \tag{7.64}$$

$$\begin{aligned}&K_{kl_i}\mathbb{E}\{\varrho_{kl_i}v_{kl_i}^T\}K_{kl_i}^T+K_{kl_i}\mathbb{E}\{v_{kl_i}\varrho_{kl_i}^T\}K_{kl_i}^T\\ \leq&\varepsilon_{3,k}K_{kl_i}K_{kl_i}^T\delta_i+\varepsilon_{3,k}^{-1}K_{kl_i}Q_{kl_i}K_{kl_i}^T,\end{aligned} \tag{7.65}$$

and

$$\begin{aligned}&-K_{kl_i}\mathbb{E}\{\varrho_{kl_i}x_{kl_i}^T C_i^T\tilde{\Theta}_{kl_i}^T\}K_{kl_i}^T-K_{kl_i}\mathbb{E}\{x_{kl_i}C_i\tilde{\Theta}_{kl_i}\varrho_{kl_i}^T\}K_{kl_i}^T\\ \leq&\varepsilon_{4,k}^{-1}K_{kl_i}\left(\check{\Theta}_{kl_i}\circ\mathbb{E}\{C_i x_{kl_i}x_{kl_i}^T C_i^T\}\right)K_{kl_i}^T+\varepsilon_{4,k}K_{kl_i}K_{kl_i}^T\delta_i.\end{aligned} \tag{7.66}$$

Using Lemma 7.1 again, we have

$$\begin{aligned}\mathbb{E}\{x_{kl_i}x_{kl_i}^T\} =&\mathbb{E}\Big\{(\tilde{x}_{kl_i|(k-1)l_i}+\hat{x}_{kl_i|(k-1)l_i})(\tilde{x}_{kl_i|(k-1)l_i}+\hat{x}_{kl_i|(k-1)l_i})^T\Big\}\\ \leq&(1+\varepsilon_{5,k})P_{kl_i|(k-1)l_i}+(1+\varepsilon_{5,k}^{-1})\hat{x}_{kl_i|(k-1)l_i}\hat{x}_{kl_i|(k-1)l_i}^T.\end{aligned} \tag{7.67}$$

Similarly,

$$\begin{aligned}&K_{kl_i}\left(\sum_{i=1}^{r}\Pi_{kl_i}^i\text{tr}(\mathbb{E}\{x_{kl_i}x_{kl_i}^T\}\Gamma_{kl_i}^i)\right)K_{kl_i}^T\\ \leq&K_{kl_i}\left(\sum_{i=1}^{r}\Pi_{kl_i}^i\text{tr}\Big(((1+\varepsilon_{5,k})P_{kl_i|(k-1)l_i}(1+\varepsilon_{5,k}^{-1})\right.\\ &\left.\times\hat{x}_{kl_i|(k-1)l_i}\hat{x}_{kl_i|(k-1)l_i}^T)\Gamma_{kl_i}^i\Big)\right)K_{kl_i}^T.\end{aligned} \tag{7.68}$$

It follows from (7.56), (7.62)–(7.68) that

$$\begin{aligned}P_{kl_i|kl_i}\leq&\zeta_{1,k}(I-K_{kl_i}\bar{\Theta}_{kl_i}C_i)P_{kl_i|(k-1)l_i}(I-K_{kl_i}\bar{\Theta}_{kl_i}C_i)^T\\ &+(1+\varepsilon_{2,k}^{-1}+\varepsilon_{4,k}^{-1})K_{kl_i}\Big\{\check{\Theta}_{kl_i}\circ C_i((1+\varepsilon_{5,k})P_{kl_i|(k-1)l_i}\\ &+(1+\varepsilon_{5,k}^{-1})\hat{x}_{kl_i|(k-1)l_i}\hat{x}_{kl_i|(k-1)l_i}^T)C_i^T\Big\}K_{kl_i}^T\\ &+K_{kl_i}\left(\sum_{i=1}^{r}\Pi_{kl_i}^i\text{tr}\Big(((1+\varepsilon_{5,k})P_{kl_i|(k-1)l_i}\right.\\ &\left.+(1+\varepsilon_{5,k}^{-1})\hat{x}_{kl_i|(k-1)l_i}\hat{x}_{kl_i|(k-1)l_i}^T)\Gamma_{kl_i}^i\Big)\right)K_{kl_i}^T\\ &+\zeta_{2,k}K_{kl_i}K_{kl_i}^T\delta_i+(1+\varepsilon_{4,k}^{-1})K_{kl_i}Q_{kl_i}K_{kl_i}^T.\end{aligned} \tag{7.69}$$

The proof is now complete.

The following theorem provides the upper bound on each local FEC.

Theorem 7.4 *Consider the multi-rate system (7.43) with the filter (7.50). The PEC $P_{kl_i|(k-1)l_i}$ and the FEC $P_{kl_i|kl_i}$ are given in (7.55) and (7.56), respectively. Let the parameters $\varepsilon_{i,k}$ $(i=1,2,3,4,5)$ be positive scalars. If the matrices $\bar{P}_{kl|(k-1)l_i}$ and $\bar{P}_{kl_i|kl_i}$ satisfy*

$$\begin{aligned}\bar{P}_{kl|(k-1)l_i}\triangleq&\Lambda(\bar{P}_{(k-1)l_i|(k-1)l_i})\\ \triangleq&\mathscr{A}_i\bar{P}_{(k-1)l_i|(k-1)l_i}\mathscr{A}_i^T+\mathscr{B}_iR_{(k-1)l_i}\mathscr{B}_i^T\end{aligned} \tag{7.70}$$

and

$$\bar{P}_{kl_i|kl_i}\triangleq\bar{\Lambda}(P_{kl_i|(k-1)l_i})$$

$$
\begin{aligned}
&\triangleq \zeta_{1,k}(I - K_{kl_i}\bar{\Theta}_{kl_i}C_i)\bar{P}_{kl_i|(k-1)l_i}(I - K_{kl_i}\bar{\Theta}_{kl_i}C_i)^T \\
&\quad + (1+\varepsilon_{2,k}^{-1}+\varepsilon_{4,k}^{-1})K_{kl_i}\Big\{\check{\Theta}_{kl_i} \circ C_i\big((1+\varepsilon_{5,k})\bar{P}_{kl_i|(k-1)l_i} \\
&\quad + (1+\varepsilon_{5,k}^{-1})\hat{x}_{kl_i|(k-1)l_i}\hat{x}_{kl_i|(k-1)l_i}^T\big)C_i^T\Big\}K_{kl_i}^T \\
&\quad + K_{kl_i}\Bigg(\sum_{i=1}^{r}\Pi_{kl_i}^i \operatorname{tr}\Big(\big((1+\varepsilon_{5,k})\bar{P}_{kl_i|(k-1)l_i} \\
&\quad + (1+\varepsilon_{5,k}^{-1})\hat{x}_{kl_i|(k-1)l_i}\hat{x}_{kl_i|(k-1)l_i}^T\big)\Gamma_{kl_i}^i\Big)\Bigg)K_{kl_i}^T \\
&\quad + \zeta_{2,k}K_{kl_i}K_{kl_i}^T\delta_i + (1+\varepsilon_{3,k}^{-1})K_{kl_i}Q_{kl_i}K_{kl_i}^T
\end{aligned} \tag{7.71}
$$

with initial value $P_{0|0} \le \bar{P}_{0|0}$, then we conclude that $\bar{P}_{kl_i|kl_i}$ is the upper bound on the FEC $P_{kl_i|kl_i}$, that is, $P_{kl_i|kl_i} \le \bar{P}_{kl_i|kl_i}$.

Proof *We intend to confirm this theorem by using mathematical induction. It can be obtained from the initial condition that $P_{0|0} \le \Sigma_{0|0}$. Suppose that $P_{(k-1)l_i|(k-1)l_i} \le \bar{P}_{(k-1)l_i|(k-1)l_i}$ is valid. Then, we need to show that $P_{kl_i|kl_i} \le \bar{P}_{kl_i|kl_i}$.*

From (7.61) and (7.71), one has

$$
\bar{P}_{kl_i|kl_i} - P_{kl_i|kl_i} \ge \bar{\Lambda}(\bar{P}_{kl_i|(k-1)l_i}) - \bar{\Lambda}(P_{kl_i|(k-1)l_i}). \tag{7.72}
$$

Moreover, by noting (7.60) and (7.70), we further have

$$
\bar{P}_{kl_i|(k-1)l_i} - P_{kl_i|(k-1)l_i} = \Lambda(\bar{P}_{(k-1)l_i|(k-1)l_i}) - \Lambda(P_{(k-1)l_i|(k-1)l_i}). \tag{7.73}
$$

By noting the assumption $P_{(k-1)l_i|(k-1)l_i} \le \bar{P}_{(k-1)l_i|(k-1)l_i}$, we have $P_{kl_i|(k-1)l_i} \le \bar{P}_{kl_i|(k-1)l_i}$, which further implies that $P_{kl_i|kl_i} \le \bar{P}_{kl_i|kl_i}$, and the proof is now complete.

The following theorem shows that the designed filter parameter K_{kl_i} for each local filter is able to minimize the upper bound on the FEC at each step.

Theorem 7.5 *The upper bound $\bar{P}_{kl_i|kl_i}$ given in* (7.71) *is minimized by the following filter parameter*

$$
K_{kl_i} = \Phi_{kl_i}\Psi_{kl_i}^{-1} \tag{7.74}
$$

where

$$
\Phi_{kl_i} = \zeta_{1,k}\bar{P}_{kl_i|(k-1)l_i}C_i^T\bar{\Theta}_{kl_i}^T
$$

and

$$
\begin{aligned}
\Psi_{kl_i} =& \zeta_{1,k}\bar{\Theta}_{kl_i}C_i\bar{P}_{kl_i|(k-1)l_i}C_i^T\bar{\Theta}_{kl_i}^T + (1+\varepsilon_{3,k}^{-1})Q_{kl_i} \\
&+ \sum_{i=1}^{r}\Pi_{kl_i}^i \operatorname{tr}\Big(\big((1+\varepsilon_{5,k})\bar{P}_{kl_i|(k-1)l_i}
\end{aligned}
$$

$$+(1+\varepsilon_{5,k}^{-1})\hat{x}_{kl_i|(k-1)l_i}\hat{x}_{kl_i|(k-1)l_i}^T)\Gamma_{kl_i}^i\Big)I$$
$$+(1+\varepsilon_{2,k}^{-1}+\varepsilon_{4,k}^{-1})\check{\Theta}_{kl_i}\circ C_i\big((1+\varepsilon_{5,k})\bar{P}_{kl_i|(k-1)l_i}$$
$$+(1+\varepsilon_{5,k}^{-1})\hat{x}_{kl_i|(k-1)l_i}\hat{x}_{kl_i|(k-1)l_i}^T\big)C_i^T+\zeta_{2,k}\delta_i I.$$

Then, the minimal upper bound $\bar{P}_{kl_i|kl_i}$ can be obtained by the following recursion

$$\bar{P}_{kl_i|kl_i}=-\Phi_{kl_i}\Psi_{kl_i}^{-1}\Phi_{kl_i}^T+\zeta_{1,k}\bar{P}_{kl_i|(k-1)l_i}. \tag{7.75}$$

Proof *By virtue of the completing-square technique, the upper bound in (7.71) can be rewritten as*

$$\begin{aligned}\bar{P}_{kl_i|kl_i}=&\zeta_{1,k}\bar{P}_{kl_i|(k-1)l_i}-\zeta_{1,k}\bar{P}_{kl_i|(k-1)l_i}C_i^T\bar{\Theta}_{kl_i}^T K_{kl_i}^T\\&-\zeta_{1,k}K_{kl_i}\bar{\Theta}_{kl_i}C_i\bar{P}_{kl_i|(k-1)l_i}^T\\&+\zeta_{1,k}K_{kl_i}\bar{\Theta}_{kl_i}C_i\bar{P}_{kl_i|(k-1)l_i}C_i^T\bar{\Theta}_{kl_i}^T K_{kl_i}^T\\&+(1+\varepsilon_{2,k}^{-1}+\varepsilon_{4,k}^{-1})K_{kl_i}\Big\{\check{\Theta}_{kl_i}\circ C_i\big((1+\varepsilon_{5,k})\bar{P}_{kl_i|(k-1)l_i}\\&+(1+\varepsilon_{5,k}^{-1})\hat{x}_{kl_i|(k-1)l_i}\hat{x}_{kl_i|(k-1)l_i}^T\big)C_i^T\Big\}K_{kl_i}^T\\&+K_{kl_i}\Bigg(\sum_{i=1}^{r}\Pi_{kl_i}^i\operatorname{tr}\Big(\big((1+\varepsilon_{5,k})\bar{P}_{kl_i|(k-1)l_i}\\&+(1+\varepsilon_{5,k}^{-1})\hat{x}_{kl_i|(k-1)l_i}\hat{x}_{kl_i|(k-1)l_i}^T\big)\Gamma_{kl_i}^i\Big)\Bigg)K_{kl_i}^T\\&+\zeta_{2,k}K_{kl_i}K_{kl_i}^T\delta_i+(1+\varepsilon_{3,k}^{-1})K_{kl_i}Q_{kl_i}K_{kl_i}^T\\=&K_{kl_i}\Psi_{kl_i}K_{kl_i}^T-\Phi_{kl_i}K_{kl_i}^T-K_{kl_i}^T\Phi_{kl_i}+\zeta_{1,k}\bar{P}_{kl_i|(k-1)l_i}\\=&\big(K_{kl_i}-\Phi_{kl_i}\Psi_{kl_i}^{-1}\big)\Psi_{kl_i}\big(K_{kl_i}-\Phi_{kl_i}\Psi_{kl_i}^{-1}\big)^T\\&-\Phi_{kl_i}\Psi_{kl_i}^{-1}\Phi_{kl_i}^T+\zeta_{1,k}\bar{P}_{kl_i|(k-1)l_i}.\end{aligned} \tag{7.76}$$

Then, from $\Psi_{kl_i}>0$, it is easy to derive the minimized upper bound $\bar{P}_{kl_i|kl_i}$ by choosing the filter parameter $K_{kl_i}=\Phi_{kl_i}\Psi_{kl_i}^{-1}$, and the minimum of $\bar{P}_{kl_i|kl_i}$ is given by (7.75). The proof is now complete.

Remark 7.6 *Due to the influence of the sensor degradations, stochastic nonlinearities as well as the ET mechanism, it is usually difficult to obtain the exact value of the FEC and an alternative way is to look for its upper bound instead. In Theorems 7.4 and 7.5, an upper bound of the FEC is given and minimized by the designed filter parameter.*

7.2.3 Fusion Estimation

Set $\Pi_k=\{i|k=m_{i,k}l_i,\ m_{i,k}$ is a positive integer, $i\in(1,2,\ldots,M)\}$. The strategy of the fusion estimation is that all the local estimates are fused only when the set is not empty. Suppose that $\hat{x}_k$ and Σ_k are, respectively, the fusion estimate and fusion covariance. The fusion estimation approach is adopted by

$$\begin{aligned}\Sigma_k =&\left(\sum_{i\in\Pi_k}\mu_i\bar{P}^{-1}_{m_{i,k}l_i|m_{i,k}l_i}\right)^{-1},\\ \hat{x}_k =&\Sigma_k\sum_{i\in\Pi_k}\mu_i\bar{P}^{-1}_{m_{i,k}l_i|m_{i,k}l_i}\hat{x}_{m_{i,k}l_i|m_{i,k}l_i}\end{aligned}\tag{7.77}$$

where $\mu_i \geq 0$ satisfies

$$\sum_{i\in\Pi_k}\mu_i = 1. \tag{7.78}$$

Then, the issue of fusion estimation can be formulated as the following convex optimization problem

$$\begin{aligned}&\min_{\mu_i}\{\text{tr}(\Sigma_k)\},\\ &\text{s.t.}\quad \sum_{i\in\Pi_k}\mu_i = 1,\ \mu_i \geq 0.\end{aligned}\tag{7.79}$$

Theorem 7.6 *For the given multi-rate system (7.43) with (7.44), the proposed multi-rate fusion estimation approach (7.77)–(7.79) is consistent, i.e.,*

$$P_k \leq \Sigma_k \tag{7.80}$$

where

$$P_k = \mathbb{E}\{(x_k - \hat{x}_k)(x_k - \hat{x}_k)^T\}. \tag{7.81}$$

Proof *From Theorem 7.4, we have* $P_{kl_i|kl_i} \leq \bar{P}_{kl_i|kl_i}$*. Combining the results in [21], thus we can draw the conclusion that the equation (7.80) is valid. The proof is omitted here for conciseness.*

Remark 7.7 *Due to the existence of stochastic nonlinearities, sensor degradation and ET mechanisms, it is difficult to derive the cross-covariance among the sensor nodes. An alternative way (CI fusion) to solve this problem has been presented in [21] and the detailed proof for the consistency of CI-based fusion estimation has also been demonstrated in [21] in the sense of minimal mean-squared error.*

Overall, the above designing process can be summarized as the following Algorithm.

Remark 7.8 *To date, we have studied the fusion estimation problems for multi-rate systems with stochastic nonlinearities subject to sensor degradations under the ET mechanisms. The local filters have been designed where each upper bound on the FEC has been guaranteed, and then such an upper bound has been minimized by the designed filter parameter in Theorems 7.4 and 7.5. The primary difficulties caused by the multi-rates, sensor degradations as well as ET mechanisms have all been tackled.*

Multi-rate fusion estimation algorithm	
Step 1.	Let parameters $\varepsilon_{i,k}$, $(i = 1,2,3,4,5)$ and δ_i be given positive scalars. Set initial values $\hat{x}_{0\|0} = \bar{x}_0$, $\bar{P}_{0\|0} = P_{0\|0}$, $\hat{x}_0 = \bar{x}_0$ and $\Sigma_0 = P_{0\|0}$, the length of time horizon N and $k = 0$;
Step 2.	Calculate the filter parameter K_{kl_i} from (7.74), local estimate $\hat{x}_{kl_i\|kl_i}$ via (7.50), and local upper bound $\bar{P}_{kl_i\|kl_i}$ according to (7.71);
Step 3.	If $\Pi_k \neq \emptyset$, go to Step 4, else go to Step 5;
Step 4.	Calculate the fusion estimate $\hat{x}_k$ and its covariance Σ_k by (7.77);
Step 5.	If $k \leq N$, set $k = k+1$ and go to Step 2, else go to Step 6;
Step 6.	Stop.

7.3 Illustrative Examples

In this section, some numerical examples are presented to verify the effectiveness of the ET fusion method proposed in this chapter.

7.3.1 Example 1

Consider a nonlinear uncertain stochastic systems described by (7.1) with parameters

$$A = \begin{bmatrix} 0.1 & -0.3 \\ 0.5 & -0.1 \end{bmatrix}, \quad B = \begin{bmatrix} 1 & 0 \\ 0 & 1 \end{bmatrix},$$

uncertainty

$$\Delta A_{kh_0} = \begin{bmatrix} -0.2 & -0.06 \\ -0.01 & -0.1 \end{bmatrix} \begin{bmatrix} 0.5\sin(kh_0) & 0 \\ 0 & -0.4\cos(kh_0) \end{bmatrix} \begin{bmatrix} -0.3 & -0.02 \\ 0.02 & 0.2 \end{bmatrix},$$

and nonlinear function

$$g(x_{kh_0}, \eta_{kh_0}) = \begin{bmatrix} 0.2 \\ 0.1 \end{bmatrix} \left[0.1\text{sign}(x^1_{kh_0})x^1_k\eta^1_{kh_0} + 0.1\text{sign}(x^2_{kh_0})x^2_{kh_0}\eta^2_{kh_0} \right].$$

It is easily checked that the nonlinear function (7.3.1) satisfies (7.3) with

$$\Pi^1_k = \begin{bmatrix} 0.04 & 0.02 \\ 0.02 & 0.01 \end{bmatrix}, \quad \Gamma^1_k = \begin{bmatrix} 0.01 & 0 \\ 0 & 0.01 \end{bmatrix}.$$

The mathematical expectation and variance of the initial value are assumed to be $\mathbb{E}\{x_0\} = \begin{bmatrix} 0.3 & -0.5 \end{bmatrix}^T$ and $\text{Var}(x_0) = 0.3I_2$, respectively. The variance of the process noise w_{kh_0} is $R_{kh_0} = \begin{bmatrix} 2.5 & 0 \\ 0 & 2 \end{bmatrix}$.

Suppose that there are two sensor nodes whose sampling rates are $2h_0$ and $3h_0$ and other measurement parameters are given by

$$C_1 = \begin{bmatrix} 0.5 & -0.6 \\ 0.8 & -0.1 \end{bmatrix}, \quad C_2 = \begin{bmatrix} -0.6 & -0.2 \\ -0.6 & 0.3 \end{bmatrix},$$
$$D_1 = \begin{bmatrix} -0.2 & -0.3 \\ -0.5 & -0.2 \end{bmatrix}, \quad D_2 = \begin{bmatrix} 0.6 & -1 \\ 0.2 & -1.2 \end{bmatrix}.$$

The variances of the measurement noises $e_{kn_i h_0, i}$ $(i = 1, 2)$ are

$$Q_{kn_1 h_0, 1} = Q_{kn_2 h_0, 2} = \begin{bmatrix} 0.18 & 0 \\ 0 & 0.2 \end{bmatrix}.$$

In the design of filters, other parameters are taken as $\varepsilon_{1,kn_1} = 0.2$, $\varepsilon_{2,kn_1} = 0.6$, $\varepsilon_{3,kn_1} = 0.002$, $\varepsilon_{4,kn_1} = 0.1$, $\varepsilon_{5,kn_1} = 0.002$, $\varepsilon_{6,kn_1} = 0.1$, $\varepsilon_{1,kn_2} = 0.2$, $\varepsilon_{2,kn_2} = 0.2$, $\varepsilon_{3,kn_2} = 0.002$, $\varepsilon_{4,kn_2} = 0.2$, $\varepsilon_{5,kn_2} = 0.002$, $\varepsilon_{6,kn_2} = 0.4$, $\delta_1 = 3$, $\delta_2 = 8$, $\gamma_{kn_1} = 0.001$, and $\gamma_{kn_2} = 0.002$. Letting $\Xi_{0,1} = \Xi_{0,2} = \text{Var}(x_0) = 0.5I_2$, all the upper bounds $\Xi_{kn_i,i}$ can be recursively computed in terms of (7.20) and, from (7.33) with (7.34), the local filter gains K_{kn_i} can be obtained.

In the simulation, we have implemented the simulation 300 times and the simulation results are shown in Figs. 7.1–7.5. In Fig. 7.1, the local estimates

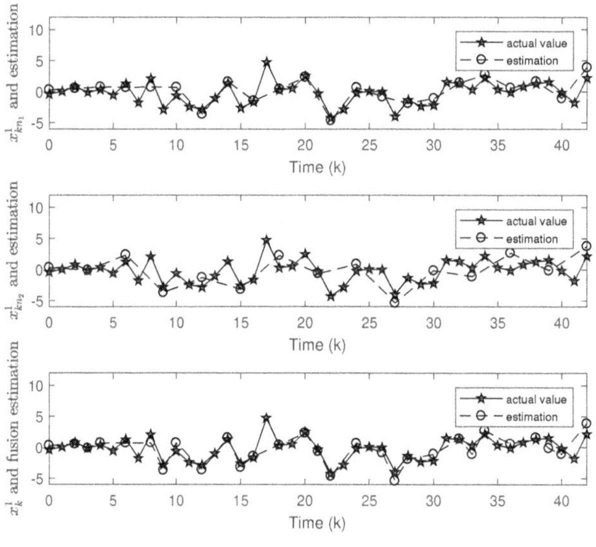

FIGURE 7.1
State x_k^1 and its estimate.

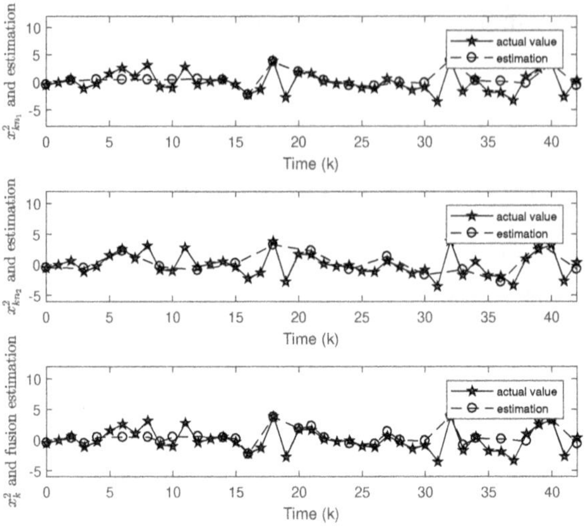

FIGURE 7.2
State x_k^2 and its estimate.

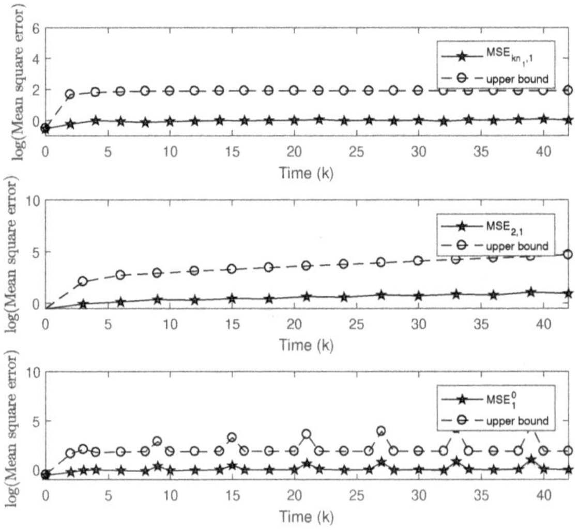

FIGURE 7.3
MSEs of $\hat{x}_k^1$ and its upper bound.

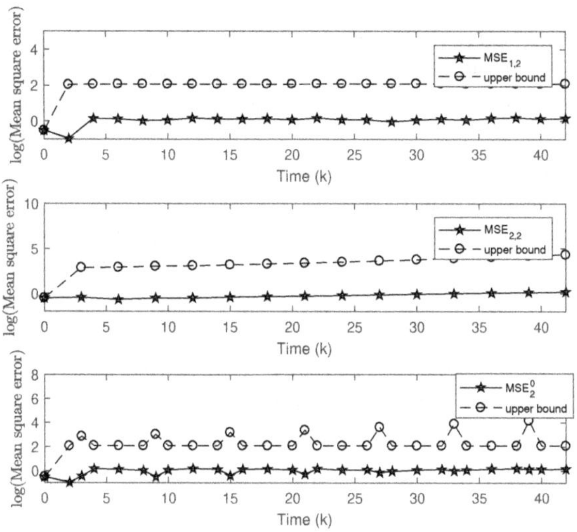

FIGURE 7.4
MSEs of $\hat{x}_k^2$ and its upper bound.

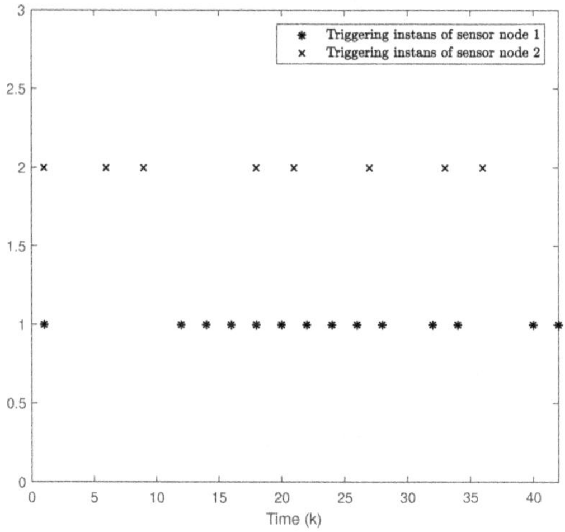

FIGURE 7.5
Triggering instants of each sensor node.

of the first element of the state from the 1st and 2nd filters are plotted in the first and second pictures, respectively, and the third picture depicts the fusion estimate of the first element of the state. The mean-square-errors (MSEs) (see [58] for its definition) of estimation and the average of their upper bounds are shown in Fig. 7.3, where the above two pictures are for the case of local estimation for the first element of the state and the third picture gives the results for the fusion estimation error of the first element of the state. The corresponding results for the second element of the state are shown in Figs. 7.2 and 7.4. It can be seen from the simulation results that the local estimate with faster sampling rate is closer to the actual state value, and the fusion estimate outperforms the local estimates. Moreover, the MSEs of state estimation stay below their upper bounds which conforms well to our theoretical analysis. In Fig. 7.5, the triggering instants are presented.

7.3.2 Example 2

We consider that there are two sensor nodes with sampling rates being $2T$ and $3T$, respectively. The parameters for the system (7.43) under consideration are given by

$$A = \begin{bmatrix} -0.6 & -0.2 \\ 0.5 & -0.2 \end{bmatrix}, \ B = \begin{bmatrix} -0.15 & -0.2 \\ -0.2 & 0.3 \end{bmatrix},$$
$$C_1 = [0.5 \ -0.6], \ C_2 = [-0.6 \ 0.2],$$

and stochastic nonlinear function is described by

$$s(x_{kl_i}, \varsigma_{kl_i}) = 0.1(0.1\text{sign}(x^1_{kl_i})x^1_{kl_i}\varsigma^1_{kl_i} + 0.1\text{sign}(x^2_{kl_i})x^2_{kl_i}\varsigma^2_{kl_i}) \tag{7.82}$$

where ς_{kl_i} stands for Gaussian noise sequence with unity covariance and zero mean.

It is not difficult to get that the equation (7.82) caters for (7.45) with

$$\Pi_{kl_i} = 0.01, \quad \Gamma_{kl_i} = \begin{bmatrix} 0.01 & 0 \\ 0 & 0.01 \end{bmatrix}.$$

The degradation coefficient θ_{m,kl_i} ($m = 1, 2, \ldots, n_y$) is assumed to follow the uniform distribution over the interval $[0.8, 1]$. In addition, the covariance of the plant noise w_k and measurement noise v_k are assumed to be $R_{kT} = \begin{bmatrix} 3 & 0 \\ 0 & 2.5 \end{bmatrix}$ and $Q_{2kT} = Q_{3kT} = 0.1$, respectively. The thresholds of the triggering condition are chosen as $\delta_1 = \delta_2 = 0.2$. In this example, we set $\varepsilon_{1,k} = \varepsilon_{2,k} = 10$, $\varepsilon_{3,k} = 0.1$, and $\varepsilon_{4,k} = \varepsilon_{5,k} = 1$. According to the above parameters, the local filter parameter K_{kl_i} can be calculated by (7.74) at each sampling instant.

The experimental results are listed in Figs. 7.6–7.10. In Fig. 7.6, the first subplot plots the actual state x^1_k and its estimate with the rate of $2T$, the second subplot depicts the actual state x^1_k and its estimate with the rate of $3T$, and the third subplot shows the actual state x^1_k and its fusion estimate.

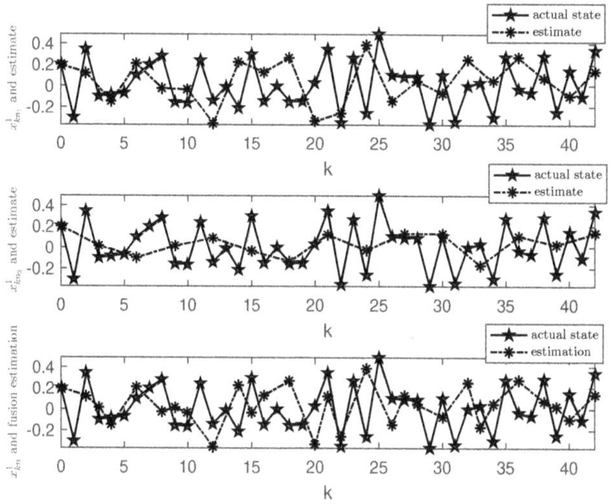

FIGURE 7.6
State x_k^1 and its estimate.

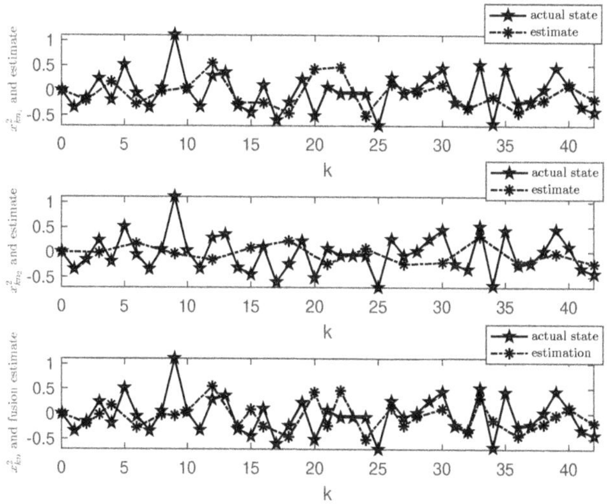

FIGURE 7.7
State x_k^2 and its estimate.

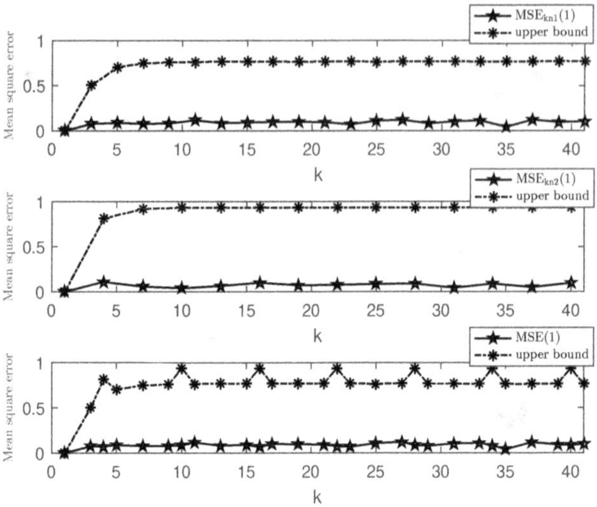

FIGURE 7.8
MSE of $\hat{x}_k^1$ and its upper bound.

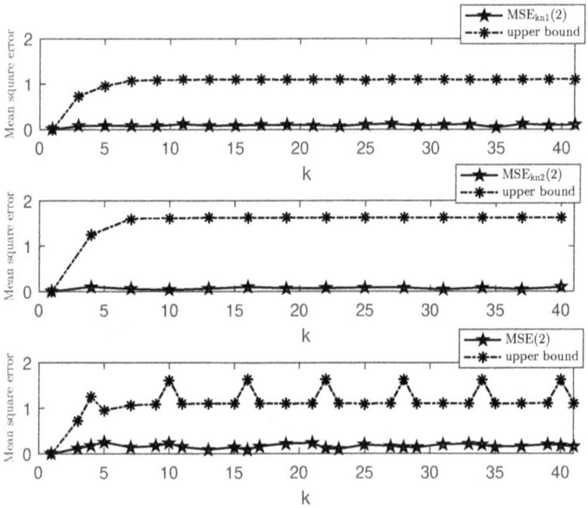

FIGURE 7.9
MSE of $\hat{x}_k^2$ and its upper bound.

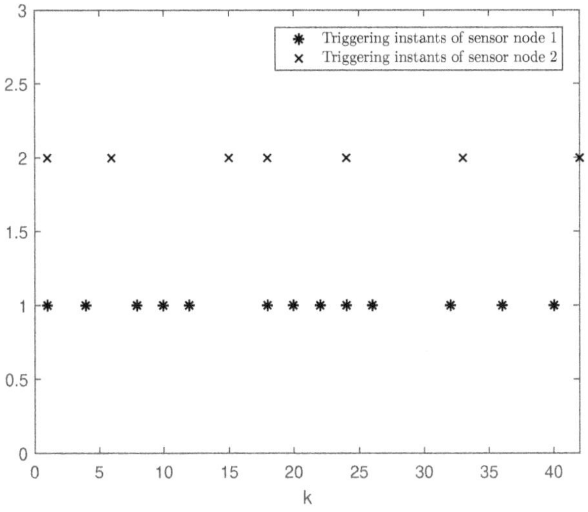

FIGURE 7.10
Triggering instants.

The corresponding results for x_k^2 are depicted in Fig. 7.7. The mean square errors (MSEs) (defined by $\text{MSE}_{i,k} \triangleq \frac{1}{300}\sum_{t=1}^{300}(x_k^i - \hat{x}_{k|k}^i)^2$, $i = 1, 2$), and corresponding minimal upper bound are plotted for x_k^1 with the rate of $2T$ and $3T$ in the first and the second subplots in Fig. 7.8. The third subplot in Fig. 7.8 depicts the MSEs and the upper bounds on the MSEs of the fusion estimates for x_k^1. The corresponding results for x_k^2 are shown in Fig. 7.9. Moreover, Fig. 7.10 displays the triggering instants. From Figs. 7.6–7.9, it is observed that: 1) the local estimates with faster sampling rates are closer to the actual states; 2) the fusion estimates outperform the local ones; and 3) the estimation performance index is satisfied since the MSEs are below the corresponding upper bounds.

Due mainly to the effects brought by the ET mechanism, we have discussed the upper bound $\bar{P}_{kl_i|kl_i}$ with different triggering thresholds. Fig. 7.11 shows the upper bound $\bar{P}_{kl_i|kl_i}$ with different triggering thresholds of $\delta_1 = 0.1$, $\delta_2 = 0.15$, and $\delta_3 = 0.2$. It is trivial to see that the upper bound $\bar{P}_{kl_i|kl_i}$ monotonically increases as δ_i increases. It can be concluded that with the increase of the threshold δ_i, the number of event-triggering decreases, but the cost is to sacrifice certain estimation accuracy.

Remark 7.9 *Due mainly to the effects of the sensor degradations, stochastic nonlinearities as well as ET mechanisms, the corresponding local upper bound of the FEC is treated as the performance index of the filtering. Such a method*

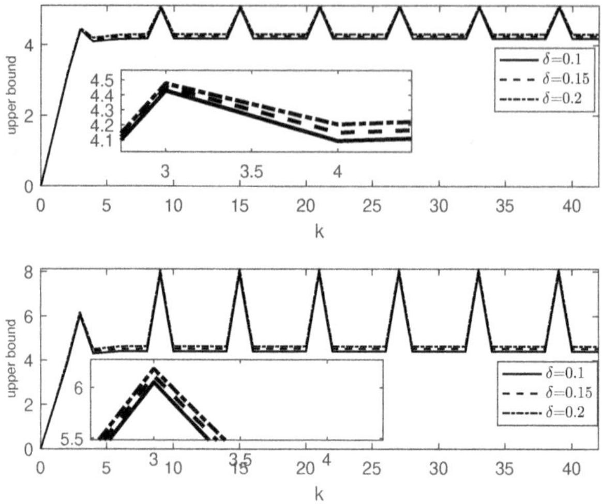

FIGURE 7.11
Upper bounds with different thresholds.

will inevitably produce conservatism and affect the real filtering accuracy. Although the filter parameter is designed to minimize the upper bound at each iteration, actually the minimized upper bound is not tight. It is noted from Theorems 7.4 and 7.5 that the minimum upper bound is closely related to the parameters ε_i $(i = 1, 2, \ldots, 5)$. *Therefore, these parameters should be carefully selected in the experiment to further improve the experimental results.*

7.4 Summary

In this chapter, we have firstly investigated the ET robust fusion estimation problem for a class of uncertain multi-rate sampled-data systems. Both the stochastic nonlinearities and the coloured measurement noises have been taken into account. A new augmentation approach has been proposed to transform the multi-rate sampled-data system into a single-rate system. The local filters have been constructed where the recursions of the upper bounds on the FECs have been guaranteed. By properly designing the filter parameters, such upper bounds have been then minimized at each iteration. Subsequently, by virtue of the modified CI method, the received local estimates have been fused together in the fusion centre. Furthermore, the ET fusion estimation has also been considered for multi-rate systems with sensor degradations. The phenomenon of sensor degradations has been characterized by a set of random variables with known probability distributions. The ET filter design problem has also been discussed. Finally, some simulation examples have been provided to verify the effectiveness of the fusion estimation scheme proposed.

8

Synchronization Control under Dynamic Event-Triggered Mechanisms

In this chapter, the synchronization control problem is considered for a class of discrete time-delay complex dynamical networks under a dynamic ET mechanism. For the efficiency of energy utilization, we make the first attempt to introduce a dynamic ET strategy into the design of synchronization controllers for complex dynamical networks. A new discrete-time version of the dynamic ET mechanism is proposed in terms of the absolute errors between control input updates. By constructing an appropriate Lyapunov functional, the dynamics of each network node combined with the introduced ET mechanism are first analyzed, and a sufficient condition is then provided under which the synchronization error dynamics is exponentially ultimately bounded. Subsequently, a set of the desired synchronization controllers is designed by solving a matrix inequality. Finally, a simulation example is provided to verify the effectiveness of the proposed dynamic ET synchronization control scheme.

8.1 Problem Formulation

Consider the following class of complex dynamical networks consisting of N coupled nodes:

$$x_i(k+1) = f(x_i(k)) + \sum_{j=1}^{N} w_{ij}\Gamma x_j(k-\tau(k)) + u_i(k) \tag{8.1}$$

for $i = 1, 2, \ldots, N$, where $x_i(k) \in \mathbb{R}^n$ is the state vector and $u_i(k) \in \mathbb{R}^n$ is the control input of the ith node; $W = (w_{ij})_{N\times N}$ is the coupled configuration matrix of the network with $w_{ij} \geq 0 (i \neq j)$ but not all zeros and is assumed to be symmetric and satisfy $w_{ii} = -\sum_{j=1,j\neq i}^{N} w_{ij}$; $\Gamma = \text{diag}\{r_1, r_2, \ldots, r_n\}$ is an inner-coupling matrix; and $\tau(k)$ denotes the time-varying coupling delay satisfying $0 < \tau_m \leq \tau(k) \leq \tau_M$.

DOI: 10.1201/9781003307648-8

The nonlinear function $f : \mathbb{R}^n \to \mathbb{R}^n$ is continuous and satisfy $f(0) = 0$ and the following sector-bounded condition:

$$[f(x) - f(y) - U_1(x - y)]^T [f(x) - f(y) - U_2(x - y)] \leq 0, \tag{8.2}$$

for all $x, y \in \mathbb{R}^n$, where U_1 and U_2 are known matrices with $U_1 - U_2 \geq 0$.

Suppose that $s(k) \in \mathbb{R}^n$ is the solution of the unforced isolate node system satisfying

$$s(k+1) = f(s(k)). \tag{8.3}$$

Letting the synchronization error be $e_i(k) = x_i(k) - s(k)$, we have the following error dynamics:

$$e_i(k+1) = \tilde{f}(e_i(k)) + \sum_{j=1}^{N} w_{ij}\Gamma e_j(k - \tau(k)) + u_i(k), \tag{8.4}$$

for $i = 1, 2, \ldots, N$, where $\tilde{f}(e_i(k)) = f(x_i(k)) - f(s(k))$.

For the ith node, a state feedback controller is chosen as follows:

$$u_i(k) = u_i(r_l^i) = K_i e_i(r_l^i), \quad r_l^i \leq k < r_{l+1}^i, \tag{8.5}$$

where K_i is the feedback parameter to be designed and $\{r_l^i\}_{0 \leq l < +\infty}$ are the execution time instants determined based on the following triggering condition:

$$r_{l+1}^i = \min\left\{k \in \mathbb{N} | k > r_l^i, \ \frac{1}{\theta_i}\eta_i(k) + \sigma_i - \varepsilon_i^T(k)\varepsilon_i(k) \leq 0\right\} \tag{8.6}$$

with $r_0^i = 0$, where σ_i and θ_i are given positive scalars, $\varepsilon_i(k)$ is defined by $\varepsilon_i(k) = e_i(k) - e_i(r_l^i)$, and $\eta_i(k)$ is an internal dynamical variable satisfying

$$\eta_i(k+1) = \lambda_i \eta_i(k) + \sigma_i - \varepsilon_i^T(k)\varepsilon_i(k), \quad \eta_i(0) = \eta_0^i \tag{8.7}$$

with $\lambda_i \in (0, 1)$ being a given constant and $\eta_0^i \geq 0$ being the initial condition.

Remark 8.1 *It is seen from (8.6) that, when the parameter θ_i approaches infinity, the dynamic triggering condition (8.6) becomes the static triggering condition proposed in [101]. Moreover, the dynamic triggering condition (8.6) is constructed by means of absolute errors between control input updates, and such a mechanism is different from that in [60].*

With the dynamic ET mechanism given by (8.6) with (8.7), the error dynamics (8.4) can be written as

$$e_i(k+1) = \tilde{f}(e_i(k)) + \sum_{j=1}^{N} w_{ij}\Gamma e_j(k - \tau(k)) + K_i e_i(r_l^i), \tag{8.8}$$

when $k \in [r_l^i, r_{l+1}^i)$ for $i = 1, 2, \ldots, N$.

Note that the definition $\varepsilon_i(k) = e_i(k) - e_i(r_l^i)$ on the local time interval $k \in [r_l^i, r_{l+1}^i)$ is still valid on the global time interval $k \in [0, +\infty)$ as l takes from 0 to ∞. Then, we further have

$$e_i(k+1) = \tilde{f}(e_i(k)) + \sum_{j=1}^{N} w_{ij}\Gamma e_j(k - \tau(k)) + K_i(e_i(k) - \varepsilon_i(k)) \tag{8.9}$$

for $i = 1, 2, \ldots, N$.

By Setting

$$\begin{aligned}
\varepsilon(k) &= \begin{bmatrix} \varepsilon_1^T(k) & \varepsilon_2^T(k) & \cdots & \varepsilon_N^T(k) \end{bmatrix}^T, \\
F(k) &= \begin{bmatrix} \tilde{f}^T(e_1(k)) & \tilde{f}^T(e_2(k)) & \cdots & \tilde{f}^T(e_N(k)) \end{bmatrix}^T, \\
e(k) &= \begin{bmatrix} e_1^T(k) & e_2^T(k) & \cdots & e_N^T(k) \end{bmatrix}^T, \; K = \operatorname{diag}\{K_1, K_2, \ldots, K_N\},
\end{aligned}$$

the error dynamics (8.9) can be expressed by

$$e(k+1) = F(k) + (W \otimes \Gamma)e(k - \tau(k)) + Ke(k) - K\varepsilon(k). \tag{8.10}$$

Our aim in this chapter is to, under the dynamic ET mechanism by (8.6) with (8.7), find the controller gain matrix $K = \operatorname{diag}\{K_1, K_2, \ldots, K_N\}$ such that the complex dynamical network (8.1) is exponentially bounded synchronized.

Definition 8.1 *The complex dynamical network (8.1) is said to achieve exponentially ultimately bounded synchronization if the error dynamics (8.10) is exponentially ultimately bounded.*

8.2 Main Results

In this section, a sufficient condition is established to ensure the exponentially ultimately bounded synchronization of the dynamical network (8.1), and the corresponding synchronization controllers are designed.

Lemma 8.1 *For the dynamic ET mechanism given by (8.6) and (8.7) with the initial value $\eta_0^i \geq 0$, the internal dynamic variable satisfies $\eta_i(k) \geq 0$ for all $k \geq 0$ if the parameters λ_i $(0 < \lambda_i < 1)$ and θ_i $(\theta_i > 0)$ satisfy $\lambda_i \theta_i \geq 1$.*

Proof *According to the triggering condition (8.6), one has*

$$\frac{1}{\theta_i}\eta_i(k) + \sigma_i - \varepsilon_i^T(k)\varepsilon_i(k) \geq 0$$

for all $k \geq 0$. Then, it follows from (8.7) that

$$\eta_i(k+1) \geq (\lambda_i - \frac{1}{\theta_i})\eta_i(k) \geq \ldots \geq (\lambda_i - \frac{1}{\theta_i})^{k+1}\eta_0^i,$$

from which it is easily seen that $\eta_i(k) \geq 0$ for all $k \geq 0$ under the condition $\lambda_i\theta_i \geq 1$ and $\eta_0^i \geq 0$. The proof is complete.

Remark 8.2 *Lemma 8.1 guarantees that the internal dynamical variable $\eta_i(k)$ is non-negative for all $k \geq 0$. Therefore, the value of $\sigma_i - \varepsilon_i^T(k)\varepsilon_i(k)$ is not required to be nonnegative always, which means that triggering times under the dynamic triggering condition (8.6) are reduced, and thus the efficiency of energy utilization can be improved potentially.*

Theorem 8.1 *For all $i = 1, 2, \ldots, N$, suppose that λ_i $(0 < \lambda_i < 1)$ and θ_i $(\theta_i > 0)$ satisfy $\lambda_i\theta_i \geq 1$ and let the controller gain K be given. The error dynamics (8.10) is exponentially ultimately bounded if there exist positive definite matrices $P \in \mathbb{R}^{nN\times nN}$, $Q \in \mathbb{R}^{nN\times nN}$ and positive scalars γ_1 and γ_2 satisfying*

$$\Pi = \begin{bmatrix} \Pi_1 & \Pi_2 \\ * & -P \end{bmatrix} < 0 \tag{8.11}$$

where

$$\begin{aligned}
\Pi_1 &= \begin{bmatrix} \Phi_{11} & -\gamma_1\tilde{U}_2 & 0 & 0 & 0 \\ * & -\gamma_1 I & 0 & 0 & 0 \\ * & * & -Q & 0 & 0 \\ * & * & * & \Phi_{44} & 0 \\ * & * & * & * & \Phi_{55} \end{bmatrix}, \\
\Pi_2 &= \begin{bmatrix} PK & P & P(W\otimes\Gamma) & -PK & 0 \end{bmatrix}^T, \\
\Phi_{11} &= -P + (1+\tau_M-\tau_m)Q - \gamma_1\tilde{U}_1, \\
\tilde{U}_1 &= \frac{\hat{U}_1^T\hat{U}_2 + \hat{U}_2^T\hat{U}_1}{2}, \quad \tilde{U}_2 = -\frac{\hat{U}_1^T + \hat{U}_2^T}{2}, \\
\hat{U}_1 &= \mathrm{diag}_N\{U_1\}, \quad \hat{U}_2 = \mathrm{diag}_N\{U_2\}, \\
\Phi_{44} &= \mathrm{diag}\{(\frac{1}{\theta_1}+\gamma_2)I, \ldots, (\frac{1}{\theta_N}+\gamma_2)I\}, \\
\Phi_{55} &= \mathrm{diag}\{\frac{\lambda_1 - 1 + \gamma_2}{\theta_1}, \ldots, \frac{\lambda_N - 1 + \gamma_2}{\theta_N}\}.
\end{aligned} \tag{8.12}$$

Furthermore, if the inequality (8.11) is solvable, the ultimate bound of the synchronization error dynamics (8.10) is given by

$$\bar{\kappa} = \frac{u_0}{c(u_0-1)}\sum_{i=1}^{N}\left(\frac{\sigma_i}{\theta_i} + \gamma_2\sigma_i\right) \tag{8.13}$$

where $c = \min\{\lambda_{\min}(P), \frac{1}{\theta_1}, \ldots, \frac{1}{\theta_N}\}$ *and* $u_0 > 1$ *satisfies*

$$\varrho(u_0) + \tau_M u_0^{\tau_M} \pi(u_0) = 0 \tag{8.14}$$

with

$$\begin{aligned} &\varrho(u_0) = u_0 a + (u_0 - 1)b,\ \pi(u_0) = (u_0 - 1)(\tau_M - \tau_m + 1)\lambda_{\max}(Q). \\ &a = \lambda_{\max}(\Pi_1 + \Pi_2 P^{-1}\Pi_2^T),\ b = \max\{\lambda_{\max}(P), \frac{1}{\theta_1}, \cdots, \frac{1}{\theta_N}\}. \end{aligned} \tag{8.15}$$

Proof *Choose the following Lyapunov functional candidate:*

$$M(k) = V(k) + \eta(k) \tag{8.16}$$

where

$$\begin{aligned} V(k) =& e^T(k)Pe(k) + \sum_{i=k-\tau(k)}^{k-1} e^T(i)Qe(i) \\ &+ \sum_{j=k-\tau_M+1}^{k-\tau_m} \sum_{i=j}^{k-1} e^T(i)Qe(i), \\ \eta(k) =& \sum_{i=1}^{N} \frac{1}{\theta_i}\eta_i(k). \end{aligned}$$

The difference of $V(k)$ *along the systems (8.10) is calculated as follows*

$$\begin{aligned} \Delta V(k) =& V(k+1) - V(k) \\ \leq& \Big[F(k) + (W \otimes \Gamma)e(k - \tau(k)) + Ke(k) - K\varepsilon(k)\Big]^T P\Big[F(k) \\ &+ Ke(k) + (W \otimes \Gamma)e(k - \tau(k)) - K\varepsilon(k)\Big] + (1 + \tau_M - \tau_m) \\ &\times e^T(k)Qe(k) - e^T(k)Pe(k) - e^T(k - \tau(k))Qe(k - \tau(k)) \end{aligned} \tag{8.17}$$

and, noting (8.7), the difference of $\eta(k)$ *can be obtained as follows*

$$\begin{aligned} \Delta\eta(k) =& \eta(k+1) - \eta(k) \\ =& \sum_{i=1}^{N} \frac{1}{\theta_i}(\eta_i(k+1) - \eta_i(k)) \\ =& \sum_{i=1}^{N} \frac{1}{\theta_i}[(\lambda_i - 1)\eta_i(k) + \sigma_i - \varepsilon_i^T(k)\varepsilon_i(k)] \\ =& \sum_{i=1}^{N} \frac{\lambda_i - 1}{\theta_i}\eta_i(k) + \sum_{i=1}^{N} \frac{\sigma_i}{\theta_i} - \sum_{i=1}^{N} \frac{1}{\theta_i}\varepsilon_i^T(k)\varepsilon_i(k). \end{aligned} \tag{8.18}$$

On the other hand, by using the notations in (8.12), the inequality (8.2) can be rewritten as

$$\begin{bmatrix} e(k) \\ F(k) \end{bmatrix}^T \begin{bmatrix} \tilde{U}_1 & \tilde{U}_2 \\ * & I \end{bmatrix} \begin{bmatrix} e(k) \\ F(k) \end{bmatrix} \leq 0 \tag{8.19}$$

and, from the triggering condition (8.6), we have

$$\sum_{i=1}^{N} \frac{1}{\theta_i} \eta_i(k) + \sum_{i=1}^{N} \sigma_i - \sum_{i=1}^{N} \varepsilon_i^T(k)\varepsilon_i(k) \geq 0. \tag{8.20}$$

By taking (8.19) and (8.20) into account, it follows from (8.17) and (8.18) that

$$\begin{aligned} \Delta M(k) =& \Delta V(k) + \Delta\eta(k) \\ \leq& \Big[F(k) + (W \otimes \Gamma)e(k-\tau(k)) + Ke(k) - K\varepsilon(k)\Big]^T P\Big[F(k) \\ &+ Ke(k) + (W \otimes \Gamma)e(k-\tau(k)) - K\varepsilon(k)\Big] - e^T(k-\tau(k)) \\ &\times Qe(k-\tau(k)) - \gamma_1 \begin{bmatrix} e(k) \\ F(k) \end{bmatrix}^T \begin{bmatrix} \tilde{U}_1 & \tilde{U}_2 \\ * & I \end{bmatrix} \begin{bmatrix} e(k) \\ F(k) \end{bmatrix} \\ &- \varepsilon^T(k)\Phi_{44}\varepsilon(k) + (1 + \tau_M - \tau_m)e^T(k)Qe(k) - e^T(k)Pe(k) \\ &+ \sum_{i=1}^{N}(\gamma_2\sigma_i + \frac{\sigma_i}{\theta_i}) + \sum_{i=1}^{N} \frac{\lambda_i - 1 + \gamma_2}{\theta_i}\eta_i(k) \\ =& \zeta^T(k)(\Pi_1 + \Pi_2 P^{-1}\Pi_2^T)\zeta(k) + \sum_{i=1}^{N}(\frac{\sigma_i}{\theta_i} + \gamma_2\sigma_i) \end{aligned} \tag{8.21}$$

where

$$\begin{aligned} \zeta(k) &= \begin{bmatrix} e^T(k) & F^T(k) & e^T(k-\tau(k)) & \varepsilon^T(k) & \bar{\eta}^T(k) \end{bmatrix}^T, \\ \bar{\eta}(k) &= \begin{bmatrix} \eta_1^{\frac{1}{2}}(k) & \eta_2^{\frac{1}{2}}(k) & \dots & \eta_N^{\frac{1}{2}}(k) \end{bmatrix}^T. \end{aligned}$$

By using the Schur Complement Lemma, it is seen that inequality (8.11) holds if and only if the following is true:

$$\Pi_1 + \Pi_2 P^{-1}\Pi_2^T < 0, \tag{8.22}$$

from which we further have

$$\Delta M(k) \leq a\|\psi(k)\|^2 + \sum_{i=1}^{N}(\frac{\sigma_i}{\theta_i} + \gamma_2\sigma_i) \tag{8.23}$$

where $\psi(k) = \begin{bmatrix} e^T(k) & \bar{\eta}^T(k) \end{bmatrix}^T$ *and* a *is defined in (8.15).*

From the definition of $M(k)$*, it can be obtained that*

$$\begin{aligned} M(k) \leq &(1+\tau_M-\tau_m)\lambda_{\max}(Q) \sum_{i=k-\tau_M}^{k-1} \|e(i)\|^2 \\ &+ \lambda_{\max}(P)\|e(k)\|^2 + \sum_{i=1}^{N} \frac{1}{\theta_i}\eta_i(k) \\ \leq &(1+\tau_M-\tau_m)\lambda_{\max}(Q) \sum_{i=k-\tau_M}^{k-1} \|\psi(i)\|^2 + b\|\psi(k)\|^2. \end{aligned} \tag{8.24}$$

For a given scalar $u > 1$*, we calculate*

$$\begin{aligned} &u^{k+1}M(k+1) - u^k M(k) \\ =&u^{k+1}(M(k+1) - M(k)) + u^k(u-1)M(k) \\ \leq &u^k \varrho(u)\|\psi(k)\|^2 + u^k \pi(u) \sum_{i=k-\tau_M}^{k-1} \|\psi(i)\|^2 \\ &+ u^{k+1}\sum_{i=1}^{N}(\frac{\sigma_i}{\theta_i} + \gamma_2\sigma_i). \end{aligned} \tag{8.25}$$

where $\varrho(u)$ *and* $\pi(u)$ *are defined in (8.15).*

Let $r \geq \tau_M + 1$ *be a positive integer. Summing up both sides of (8.25) from* 0 *to* $r-1$ *with respect to* k *yields*

$$\begin{aligned} &u^r M(r) - M(0) \\ \leq &\varrho(u)\sum_{k=0}^{r-1} u^k\|\psi(k)\|^2 + \pi(u)\sum_{k=0}^{r-1}\sum_{i=k-\tau_M}^{k-1} u^k\|\psi(i)\|^2 \\ &+ \frac{u(1-u^r)}{1-u}\sum_{i=1}^{N}(\frac{\sigma_i}{\theta_i} + \gamma_2\sigma_i), \end{aligned} \tag{8.26}$$

in which the term $\pi(u)\sum_{k=0}^{r-1}\sum_{i=k-\tau_M}^{k-1} u^k\|\psi(i)\|^2$ *can be calculated as follows:*

$$\begin{aligned} &\pi(u)\sum_{k=0}^{r-1}\sum_{i=k-\tau_M}^{k-1} u^k\|\psi(i)\|^2 \\ \leq &\pi(u)\Bigg(\sum_{i=-\tau_M}^{-1}\sum_{k=0}^{i+\tau_M} + \sum_{i=0}^{r-1-\tau_M}\sum_{k=i+1}^{i+\tau_M} + \sum_{i=r-\tau_M}^{r-1}\sum_{k=i+1}^{r-1}\Bigg)u^k\|\psi(i)\|^2 \\ \leq &\pi(u)\tau_M\sum_{i=-\tau_M}^{-1} u^{i+\tau_M}\|\psi(i)\|^2 + \pi(u)\tau_M\sum_{i=0}^{r-1-\tau_M} u^{i+\tau_M}\|\psi(i)\|^2 \end{aligned}$$

$$+\pi(u)\tau_M \sum_{i=r-\tau_M}^{r-1} u^{i+\tau_M}\|\psi(i)\|^2$$

$$\leq \pi(u)\frac{\tau_M u^{\tau_M}}{u-1}\max_{-\tau_M\leq i\leq 0}\|\psi(i)\|^2 + \pi(u)\tau_M u^{\tau_M}\sum_{i=0}^{r-1} u^i\|\psi(i)\|^2. \tag{8.27}$$

Then, it is obtained that

$$\begin{aligned}
&u^r M(r) - M(0) \\
\leq\ & \varrho(u)\sum_{k=0}^{r-1} u^k\|\psi(k)\|^2 + \tau_M u^{\tau_M}\pi(u)\sum_{i=0}^{r-1} u^i\|\psi(i)\|^2 \\
&+\frac{\tau_M u^{\tau_M}\pi(u)}{u-1}\max_{-\tau_M\leq i\leq 0}\|\psi(i)\|^2 + \frac{u(1-u^r)}{1-u}\sum_{i=1}^{N}(\frac{\sigma_i}{\theta_i}+\gamma_2\sigma_i) \\
=\ & \frac{\tau_M u^{\tau_M}\pi(u)}{u-1}\max_{-\tau_M\leq i\leq 0}\|\psi(i)\|^2 + \frac{u(1-u^r)}{1-u}\sum_{i=1}^{N}(\frac{\sigma_i}{\theta_i}+\gamma_2\sigma_i) \\
&+\beta(u)\sum_{k=0}^{r-1} u^k\|\psi(k)\|^2
\end{aligned} \tag{8.28}$$

where $\beta(u) = \varrho(u) + \tau_M u^{\tau_M}\pi(u)$.

Since $\beta(1) = a < 0$ *and* $\beta(+\infty) = +\infty$, *we can find a scalar* $u_0 > 1$ *such that* $\beta(u_0) = 0$. *Therefore, we finally arrive at*

$$\begin{aligned}
u_0^r M(r) - M(0) \leq & \frac{\tau_M u_0^{\tau_M}\pi(u_0)}{u_0-1}\max_{-\tau_M\leq i\leq 0}\|\psi(i)\|^2 \\
& + \frac{u_0(1-u_0^r)}{1-u_0}\sum_{i=1}^{N}(\frac{\sigma_i}{\theta_i}+\gamma_2\sigma_i).
\end{aligned} \tag{8.29}$$

From (8.24), it is easily seen that

$$M(0) \leq \rho\tau_M \max_{-\tau_M\leq i\leq 0}\|\psi(i)\|^2 \tag{8.30}$$

where $\rho = \max\{b, (\tau_M - \tau_m + 1)\lambda_{\max}(Q)\}$. *Moreover, it follows from the definition of the* $M(k)$ *that*

$$M(r) \geq c\|\psi(r)\|^2. \tag{8.31}$$

By substituting (8.30) and (8.31) into (8.29), one has

$$\|\psi(r)\|^2 \leq \frac{y(u_0)}{cu_0^r} + \frac{u_0^r - 1}{cu_0^{r-1}(u_0-1)}\sum_{i=1}^{N}(\frac{\sigma_i}{\theta_i}+\gamma_2\sigma_i) \tag{8.32}$$

where

$$y(u_0) = (\rho\tau_M + \frac{\tau_M u_0^{\tau_M}\pi(u_0)}{u_0 - 1}) \max_{-\tau_M \le i \le 0} \|\psi(i)\|^2.$$

Finally, we obtain

$$\begin{aligned} \|e(r)\|^2 \le & \|\psi(r)\|^2 \\ \le & \frac{y(u_0)}{cu_0^r} + \frac{u_0^r - 1}{cu_0^{r-1}(u_0 - 1)} \sum_{i=1}^{N} (\frac{\sigma_i}{\theta_i} + \gamma_2\sigma_i) \end{aligned} \tag{8.33}$$

which, according to Definition 6.1, means that the error dynamics (8.10) is exponentially ultimately bounded by setting

$$\nu = \frac{y(u_0)}{c}, \quad \mu = \frac{1}{u_0}, \quad \kappa(r) = \frac{u_0^r - 1}{cu_0^{r-1}(u_0 - 1)} \sum_{i=1}^{N} (\frac{\sigma_i}{\theta_i} + \gamma_2\sigma_i).$$

Moreover, the ultimate bound is

$$\bar{\kappa} = \lim_{r \to +\infty} \kappa(r) = \frac{u_0}{c(u_0 - 1)} \sum_{i=1}^{N} (\frac{\sigma_i}{\theta_i} + \gamma_2\sigma_i). \tag{8.34}$$

The proof of this theorem is now complete.

Remark 8.3 *Theorem 8.1 provides a synchronization criterion for the complex dynamical network (8.1), where the dynamics of the network nodes is identical. It is worth mentioning that the synchronization criterion could be derived easily for the case of nonidentical node dynamics with some trivial modifications. However, the solvability of the synchronization criterion may be reduced when the nonidentical degree of nodes increases.*

According to the performance analysis conducted in Theorem 8.1, the design approach of the desired synchronization controller is given in the following theorem.

Theorem 8.2 *For all* $i = 1, 2, \ldots, N$, *under the assumption that* $0 < \lambda_i < 1$, $\theta_i > 0$ *and* $\lambda_i\theta_i \ge 1$, *the complex dynamical network (8.1) is exponentially ultimately boundedly synchronized if there exist positive matrices* $P = \mathrm{diag}\{P_1, P_2, \ldots, P_N\} \in \mathbb{R}^{nN \times nN}$, $Q \in \mathbb{R}^{nN \times nN}$, *matrix* $X = \mathrm{diag}\{X_1, X_2, \ldots, X_N\} \in \mathbb{R}^{nN \times nN}$ *and positive scalars* γ_1, γ_2 *such that*

$$\begin{bmatrix} \Phi_{11} & -\gamma_1\tilde{U}_2 & 0 & 0 & 0 & X^T \\ * & -\gamma_1 I & 0 & 0 & 0 & P \\ * & * & -Q & 0 & 0 & (W^T \otimes \Gamma^T)P \\ * & * & * & \Phi_{44} & 0 & -X^T \\ * & * & * & * & \Phi_{55} & 0 \\ * & * & * & * & * & -P \end{bmatrix} < 0 \tag{8.35}$$

holds, where Φ_{11}, Φ_{44}, and Φ_{55} are defined in (8.12). In this case, the desired controller parameters are given by

$$K_i = P_i^{-1} X_i, \quad i = 1, 2, \ldots, N. \tag{8.36}$$

Proof *Noting the relation $P_i K_i = X_i$, the proof of Theorem 8.2 follows immediately from that of Theorem 8.1.*

It is worth mentioning that the proposed dynamic ET mechanism includes the static one as a special case, and the corresponding results for static ET scheme can be obtained immediately from Theorem 8.2 by letting $\theta_i \to +\infty$.

Corollary 8.1 *Under the following static triggering condition*

$$r_{l+1}^i = \min\{k \in \mathbb{N} | k > r_l^i, \, \sigma_i - \varepsilon_i^T(k)\varepsilon_i(k) < 0\} \tag{8.37}$$

for all $i = 1, 2, \ldots, N$, the complex dynamical network (8.1) is exponentially ultimately boundedly synchronized if there exist matrices $\tilde{P} = \text{diag}\{\tilde{P}_1, \tilde{P}_2, \ldots, \tilde{P}_N\} > 0$, $\tilde{Q} > 0$ and $\tilde{X} = \text{diag}\{\tilde{X}_1, \tilde{X}_2, \ldots, \tilde{X}_N\}$ and positive scalars $\tilde{\gamma}_1$ and $\tilde{\gamma}_2$ such that

$$\tilde{\Pi} = \begin{bmatrix} \tilde{\Pi}_1 & \tilde{\Pi}_2 \\ * & -\tilde{P} \end{bmatrix} < 0, \tag{8.38}$$

holds, where

$$\tilde{\Pi}_1 = \begin{bmatrix} \tilde{\Phi}_{11} & -\tilde{\gamma}_1 \tilde{U}_2 & 0 & 0 \\ * & -\tilde{\gamma}_1 I & 0 & 0 \\ * & * & -\tilde{Q} & 0 \\ * & * & * & -\tilde{\gamma}_2 I \end{bmatrix},$$

$$\tilde{\Pi}_2 = \begin{bmatrix} X & \tilde{P} & \tilde{P}(W \otimes \Gamma) & -X \end{bmatrix}^T,$$

$$\tilde{\Phi}_{11} = -\tilde{P} + (1 + \tau_M - \tau_m)\tilde{Q} - \tilde{\gamma}_1 \tilde{U}_1.$$

If the inequality (8.38) is feasible, the desired controller parameters are given by

$$K_i = \tilde{P}_i^{-1} \tilde{X}_i, \quad i = 1, 2, \ldots, N, \tag{8.39}$$

which achieves that the synchronization error dynamics (8.10) is exponentially ultimately bounded, and the ultimate bound is given by

$$\bar{\kappa} = \frac{\tilde{u}_0}{\tilde{c}(\tilde{u}_0 - 1)} \sum_{i=1}^{N} \tilde{\gamma}_2 \sigma_i \tag{8.40}$$

where $\tilde{c} = \lambda_{\min}(\tilde{P})$ and $\tilde{u}_0 > 1$ satisfies

$$\tilde{\varrho}(\tilde{u}_0) + \tau_M \tilde{u}_0^{\tau_M} \tilde{\pi}(\tilde{u}_0) = 0$$

with

$$\tilde{a} = \lambda_{\max}(\tilde{\Pi}_1 + \tilde{\Pi}_2\tilde{P}^{-1}\tilde{\Pi}_2^T)\ \tilde{b} = \lambda_{\max}(\tilde{P}),\ \tilde{\varrho}(\tilde{u}_0) = \tilde{u}_0\tilde{a} + (\tilde{u}_0 - 1)\tilde{b},$$
$$\tilde{\pi}(\tilde{u}_0) = (\tilde{u}_0 - 1)(\tau_M - \tau_m + 1)\lambda_{\max}(\tilde{Q}).$$

Proof *The proof of this corollary can be obtained directly from that of Theorem 8.1 by letting* $\theta_i \to +\infty$ $(i = 1, 2, \cdots, N)$.

Remark 8.4 *Until now, a set of desired controllers is designed in Theorem 8.2 such that the complex dynamical network (8.1) achieves exponentially ultimately bounded synchronization under the dynamic ET mechanism (8.6) with (8.7). As a corollary of our main results, the designed method of the static ET synchronization controllers has also been given. According to previous analysis, the dynamic ET synchronization controllers outperform the static ones in terms of the efficiency of energy utilization. In next section, the effectiveness of the developed dynamic ET control approach is verified by a numerical simulation example, and the performance comparison between dynamic and static ET mechanisms is also discussed.*

8.3 Illustrative Examples

In this section, a numerical simulation example is presented to demonstrate the effectiveness of the synchronization control approach proposed in this chapter.

8.3.1 Demonstrations of Results

Consider a dynamical network with three nodes that are connected in terms of the following coupling matrix

$$W = \begin{bmatrix} -0.2 & 0.1 & 0.1 \\ 0.1 & -0.2 & 0.1 \\ 0.1 & 0.1 & -0.2 \end{bmatrix}$$

and the inner-coupling matrix is set as $\Gamma = 0.1I$. The TVD is taken as $\tau(k) = 1 + \sin^2(k\pi)$, from which we have $\tau_m = 1$ and $\tau_M = 2$.

The nonlinear function in the dynamical network (8.1) is selected as

$$f(x_i(t)) = 0.5\big((U_1 + U_2)x_i(k) + (U_2 - U_1)\sin(k)x_i(k)\big)$$

for $i = 1, 2, 3$, where

$$U_1 = \begin{bmatrix} -1.2 & 0.1 \\ 0 & 1.2 \end{bmatrix},\quad U_2 = \begin{bmatrix} -1.4 & 0.1 \\ 0 & 1.1 \end{bmatrix}.$$

For the dynamic triggering condition (8.6) with (8.7), the thresholds are chosen as $\sigma_1 = 0.6$, $\sigma_2 = 0.6$, and $\sigma_3 = 0.8$, respectively, and the other parameters are taken as $\lambda_1 = \lambda_2 = \lambda_3 = 0.5$ and $\theta_1 = \theta_2 = \theta_3 = 6$.

Remark 8.5 *The parameters θ_i $(i = 1, 2, 3)$ are the main parameters of the dynamic ET mechanism and are usually selected according to the practical engineering requirement.*

By using the Matlab LMI Toolbox, a set of feasible solutions to the inequality (36) is obtained as follows:

$$\gamma_1 = 0.8668, \quad \gamma_2 = 0.4215,$$

$$P_1 = P_2 = P_3 = \begin{bmatrix} 0.2312 & -0.0118 \\ -0.0.0118 & 0.2859 \end{bmatrix},$$

$$X_1 = X_2 = X_3 = \begin{bmatrix} 0.3046 & -0.0122 \\ -0.0122 & -0.3032 \end{bmatrix}.$$

Then, according to (8.36), the desired controller parameters can be computed as

$$K_1 = K_2 = K_3 = \begin{bmatrix} 1.3183 & -0.1071 \\ -0.0115 & -1.0650 \end{bmatrix}.$$

In the simulation, the initial values of the complex dynamical network are set as $x_1(s) = \begin{bmatrix} 1 & 1.2 \end{bmatrix}^T$, $x_2(s) = \begin{bmatrix} 1 & 1 \end{bmatrix}^T$ and $x_3(s) = \begin{bmatrix} 1.5 & 1.5 \end{bmatrix}^T$ for $s = \{-2, -1, 0\}$, the initial values of the unforced isolate node system is chosen as $s(0) = \begin{bmatrix} 1 & 1 \end{bmatrix}^T$, and the initial value of the internal dynamic variable is taken as $\eta_0^1 = \eta_0^2 = \eta_0^3 = 1$. Simulation results are shown in Figs. 8.1–8.3. Figs. 8.1–8.2 show the state trajectories of nodes 1, 2, and 3 with and without control inputs, from which, it is seen that the desired controllers achieve the synchronization of all the states of the network nodes. The synchronization error $e(k)$ is described in Fig. 8.3.

8.3.2 Comparisons of Results

In order to compare the dynamic ET and static ET cases, the ET release instants of nodes 1, 2, and 3 for both cases are shown in Fig. 8.4, where the above picture plots the triggering instants for the dynamic case and the below picture is the one for the static case. It can be seen from Fig. 8.4 that the triggering times for the dynamic ET case are much less than the one for the static case.

To be more specific, the triggering rates and average triggering rate (ATR) of all nodes 1, 2, and 3 are given in Table 8.1 for the different selections of parameters θ_i $(i = 1, 2, 3)$. It is seen from Table 8.1 that the ATR monotonically increases when the parameters θ_i $(i = 1, 2, 3)$ increase. Note from (8.6) that the dynamic triggering case is just the static triggering

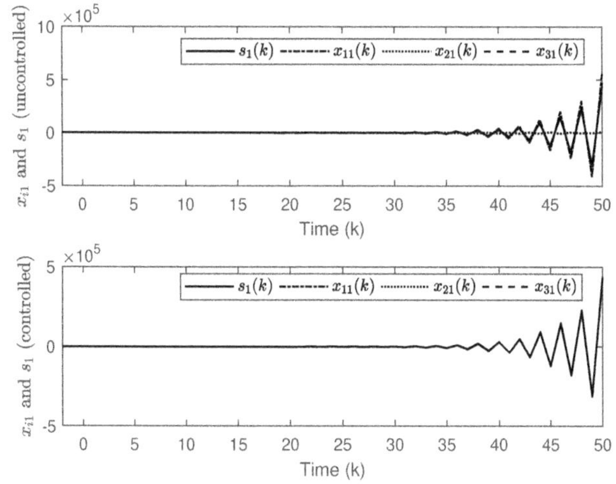

FIGURE 8.1
Uncontrolled and controlled state x_{i1} $(i = 1, 2, 3)$.

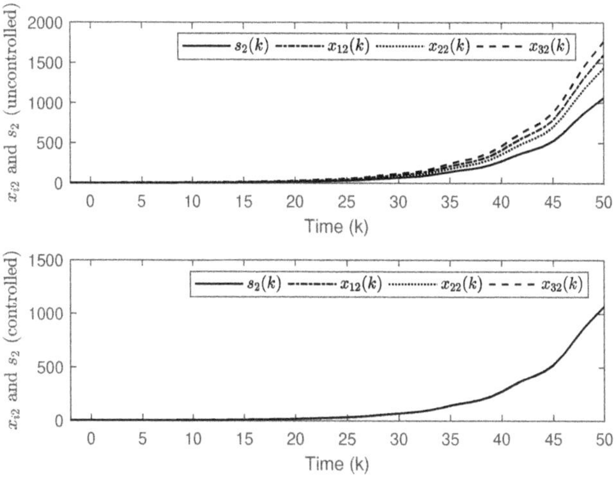

FIGURE 8.2
Uncontrolled and controlled state x_{i2} $(i = 1, 2, 3)$.

case when the parameter θ_i approaches infinity. Therefore, it is shown from Table 8.1 that the dynamic ET case reduces the triggering times effectively comparing with the static ET case.

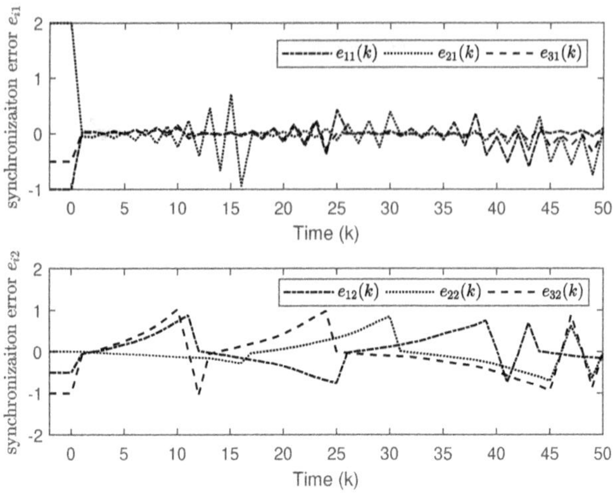

FIGURE 8.3
Synchronization error $e(k)$.

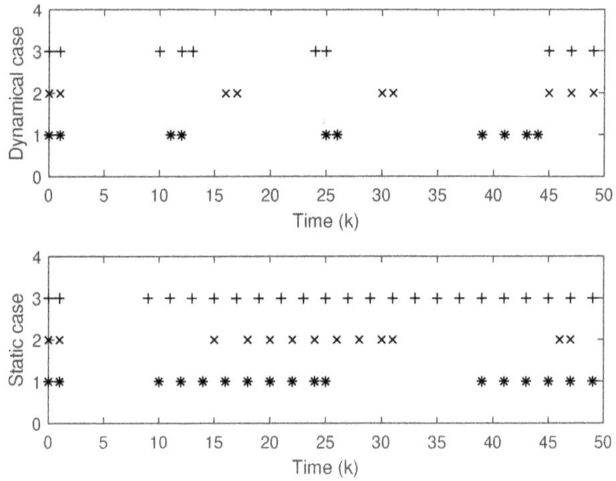

FIGURE 8.4
The dynamic and static triggering instants for controllers.

	Node 1	Node 2	Node 3	ATR
$\theta_1 = \theta_2 = \theta_3 = 6$	18%	14%	18%	16.67%
$\theta_1 = \theta_2 = \theta_3 = 30$	22%	16%	16%	18%
$\theta_1 = \theta_2 = \theta_3 = 80$	30%	16%	34%	26.67%
Static ET case	34%	20%	40%	31.33%

TABLE 8.1
Triggering rates for node i $(i = 1, 2, 3)$.

8.4 Summary

In this chapter, we have studied the dynamic ET synchronization control problem for a class of discrete time-delay complex dynamical networks. For the sake of energy saving, a new discrete-time version dynamic ET mechanism has been proposed and utilized in the design of synchronization controller. By constructing an appropriate Lyapunov functional, the dynamics of network nodes and the ET mechanism has been analyzed first in a unified way, and then a sufficient condition has been given to ensure exponentially ultimate boundedness of the synchronization error. With the obtained sufficient condition, the controller gains have been parameterized by means of the feasibility of a matrix inequality. A simulation example has been given to illustrate the effectiveness of the proposed dynamic ET synchronization control approach.

9

Filtering or Estimation under Dynamic Event-Triggered Mechanisms

This chapter deals with the filtering and state estimation problems under dynamic ET mechanisms. For the sake of energy saving, a dynamic ET mechanism is employed in the design of filters and state estimators. Firstly, the dynamic ET filtering problem is investigated for a class of discrete time-varying systems with CMs and PUs. The CMs under consideration are described by the Tobit measurement model. By means of the mathematical induction, an upper bound is derived for the FEC in terms of recursive matrix equations and such an upper bound is then minimized by designing the filter gain properly. Furthermore, the boundedness is discussed for the minimized upper bound of the FEC. Secondly, we consider the system outputs are collected through a sensor network subject to a time-varying topology that is connected via Gilbert-Elliott channels and governed by a set of Markov chains. By using similar analysis techniques, the corresponding dynamic ET distributed filtering problem is studied, and a set of parallel results is derived. Thirdly, the finite-time resilient H_∞ state estimation problem is discussed for delayed neural networks under dynamic ET mechanisms. In order to handle the possible fluctuation of the estimator gain parameters when the state estimator is implemented, a resilient state estimator is adopted. Lyapunov functional approach is carried out to obtain sufficient conditions for the existence of desired estimators ensuring both the finite-time boundedness and the H_∞ performance of the estimation error system. Finally, some numerical examples are employed to illustrate the usefulness of the proposed design techniques in this chapter.

DOI: 10.1201/9781003307648-9

9.1 Dynamic Event-Triggered Robust Filtering with Censored Measurements

9.1.1 Problem Formulation

Consider a class of linear discrete time-varying systems as follows

$$x_{k+1} = (A_k + \Delta A_k)x_k + B_k w_k, \tag{9.1}$$

$$y_k^* = C_k x_k + v_k \tag{9.2}$$

where $x_k \in \mathbb{R}^{n_x}$ is the state vector, $y_k^* \in \mathbb{R}^{n_y}$ is the uncensored measurement vector, $w_k \in \mathbb{R}^{n_x}$ and $v_k \in \mathbb{R}^{n_y}$ are zero-mean Gaussian white sequences with covariances Q_k and R_k, respectively, the initial state x_0 is a stochastic variable with $\mathbb{E}\{x_0\} = \bar{x}_0$ and $\mathbb{E}\{(x_0 - \bar{x}_0)(x_0 - \bar{x}_0)^T\} = P_{0|0}$, A_k, B_k and C_k are known parameters with appropriate dimensions, and ΔA_k is a parameter matrix representing the PU.

The parameter matrix ΔA_k is assumed to be of the following form

$$\Delta A_k = E_k F_k H_k \tag{9.3}$$

where E_k and H_k are known matrices with appropriate dimensions, F_k represents the PU satisfying $F_k F_k^T \leq I$.

The Tobit measurement model under consideration is given as follows

$$y_k = \begin{bmatrix} y_k^1 & y_k^2 & \cdots & y_k^{n_y} \end{bmatrix}^T \tag{9.4}$$

with y_k^m $(m = 1, 2, \ldots, n_y)$ defined by

$$y_k^m = \begin{cases} y_k^{*m}, & y_k^{*m} > \mathcal{I}^m \\ \mathcal{I}^m, & y_k^{*m} \leq \mathcal{I}^m \end{cases} \tag{9.5}$$

where $\mathcal{I}^m \in \mathbb{R}$ is the threshold of y_k^{*m} and y_k^{*m} is the mth element of the uncensored measurement vector y_k^*.

Remark 9.1 *In practice, as one of the measurement nonlinearities, the censoring phenomenon often occurs due to the physical causes such as the sensor saturations and limited detection effects. The major difficulty in using a standard Kalman filtering approach to dealing with the CMs is that the measurement near the censoring region is no longer Gaussian, which may lead to a biased state estimate. Therefore, in order to cater for the case of CMs, the standard Kalman filtering approach should be modified which constitutes the main motivation of this section.*

According to the analysis in [6], the measurement y_k^m can be described by

$$y_k^m = \gamma_k^m y_k^{*m} + (1 - \gamma_k^m)\mathcal{I}^m \tag{9.6}$$

where γ_k^m $(m = 1, 2, \ldots, n_y)$ is a Bernoulli random variable with the following probability distribution:

$$\text{Prob}\{\gamma_k^m = 1\} = \bar{\gamma}_k^m, \quad \text{Prob}\{\gamma_k^m = 0\} = 1 - \bar{\gamma}_k^m. \tag{9.7}$$

Remark 9.2 *As described in equation (9.6), the Bernoulli random variables are used to characterize the censoring phenomena of y_k^m. When the random variable γ_k^m is taken as $\gamma_k^m = 1$, the measurement output is $y_k^m = y_k^{*m}$, which means no censoring occurs, otherwise the measurement output is $y_k^m = \mathcal{I}^m$, which means censoring occurs. Such an alternative representation way has also been adopted in [6].*

The censoring probability $\bar{\gamma}_k^m$ is known by some statistical experiments. Alternatively, $\bar{\gamma}_k^m$ can also be approximated by (see [6])

$$\bar{\gamma}_k^m = \Phi\left(\frac{\Sigma_{l=1}^{n_x} C_k^{m,l} x_k^l - \mathcal{I}^m}{\sqrt{R_k^{m,m}}}\right) \approx \Phi\left(\frac{\Sigma_{l=1}^{n_x} C_k^{m,l} \hat{x}_{k|k-1}^l - \mathcal{I}^m}{\sqrt{R_k^{m,m}}}\right) \tag{9.8}$$

where x_k^l and $\hat{x}_{k|k-1}^l$ represent the lth element of state and prediction for x_k, respectively, $R_k^{m,m}$ and $C_k^{m,l}$ are (m,m)th and (m,l)th element of R_k and C_k, respectively, $\Phi(\cdot)$ is the cumulative distribution function of the standard normal distribution.

Remark 9.3 *From (9.8), it is seen that the censoring probability $\bar{\gamma}_k^m$ is dependent on the state x_k. By employing the local approximation in (9.8), this probability can be adequately estimated via one-step prediction $\hat{x}_{k|k-1}$.*

For the notation simplicity, we define

$$\mathcal{I} = \begin{bmatrix} \mathcal{I}^1 & \mathcal{I}^2 & \cdots & \mathcal{I}^{n_y} \end{bmatrix}^T, \quad \Upsilon_k = \text{diag}\{\gamma_k^1, \gamma_k^2, \ldots, \gamma_k^{n_y}\}.$$

Then, the Tobit measurement model (9.6) can be rewritten in the following compact form

$$y_k = \Upsilon_k y_k^* + (I - \Upsilon_k)\mathcal{I}. \tag{9.9}$$

It is easy to know from (9.7) that the expectation of Υ_k is $\bar{\Upsilon}_k = \mathbb{E}\{\Upsilon_k\} = \text{diag}\{\bar{\gamma}_k^1, \bar{\gamma}_k^2, \ldots, \bar{\gamma}_k^{n_y}\}$.

In order to reduce the measurement transmission frequency, the dynamic ET mechanism proposed in [86] is adopted. Denote the transmission instants by $0 = t_0 \leq t_1 \leq t_2 \leq \cdots \leq t_l \leq \cdots$. Then, the transmission instants $\{t_k\}_{k=0}^{\infty}$ are computed by

$$t_{l+1} = \min\{k \in \mathbb{N} | k > t_l, \ \frac{1}{\theta}\eta_k + \sigma - \|\varepsilon_k\| \leq 0\} \tag{9.10}$$

where $\varepsilon_k \triangleq y_k^t - y_k$, y_k^t is the transmitted measurement at the latest event time, θ and σ are positive scalars, and η_k is an internal dynamical variable which evolves according to the following rule

$$\eta_{k+1} = \lambda\eta_k + \sigma - \|\varepsilon_k\|. \tag{9.11}$$

In (9.11), λ is a given scalar and the initial value is assumed to satisfy $\eta_0 \geq 0$. Generally, the parameters λ and θ are required to satisfy $\lambda\theta > 1$, and thus ensure $\eta_k \geq 0$. As such, the measurement transmission frequency can be further reduced comparing the static ET mechanism [86].

Remark 9.4 *Compared with the static ET mechanism, the dynamic ET mechanism provides with the superiority of producing less transmission instants. In other words, the trigger frequency in the dynamic ET mechanism is less than that in the static ET mechanism, which is capable of further saving the limited communication resource.*

Based on the received measurement y_k^t, we construct a recursive filter of the following form

$$\hat{x}_{k+1|k} = A_k\hat{x}_{k|k}, \tag{9.12}$$

$$\hat{x}_{k+1|k+1} = \hat{x}_{k+1|k} + K_{k+1}(y_{k+1}^t - \hat{y}_{k+1|k}) \tag{9.13}$$

with initial value $\hat{x}_{0|0} = \bar{x}_0$, where $\hat{x}_{k+1|k}$ and $\hat{x}_{k|k}$ represent, respectively, the prediction and the filtering of the x_k, $\hat{y}_{k+1|k}$ is the measurement estimate (will be given in the following section), and K_{k+1} is the filter gain to be designed.

Let $\tilde{x}_{k+1|k+1} = x_{k+1} - \hat{x}_{k+1|k+1}$ be the filtering error and $P_{k+1|k+1} = \mathbb{E}\{\tilde{x}_{k+1|k+1}\tilde{x}_{k+1|k+1}^T\}$ be the FEC. The main objective of this section is to design a filter of the form (9.12) and (9.13) under the dynamic ET mechanism such that, for all PUs and CMs, the FEC $P_{k+1|k+1}$ is bounded by a minimized upper bound matrix $\bar{P}_{k+1|k+1}$. Moreover, we will provide a sufficient condition to verify the boundedness of estimation error in mean square sense.

9.1.2 Main Results

In this subsection, the bound of the FEC is expressed by a recursion. Subsequently, the desired explicit form of the estimator gain is given based on the solutions to two difference equations. At last, a sufficient condition is established to guarantee the boundedness of the error covariance.

In order to derive our main results, the following lemmas are helpful.

Lemma 9.1 *[6] On the condition that x_k and R_k are known, the mathematical expectation and variance of y_k^m $(m = 1, 2, \ldots, n_y)$ are obtained* *as follows*

$$\begin{aligned}\mathbb{E}\{y_k^m|x_k, R_k^{m,m}\} =&\Phi\left(\frac{\Sigma_{l=1}^{n_x} C_k^{m,l} x_k^l - \mathcal{I}^m}{\sqrt{R_k^{m,m}}}\right)\left[\Sigma_{l=1}^{n_x} C_k^{m,l} x_k^l\right.\\ &\left.+\sqrt{R_k^{m,m}}\lambda\left(\frac{\mathcal{I}^m - \Sigma_{l=1}^{n_x} C_k^{m,l} x_k^l}{\sqrt{R_k^{m,m}}}\right)\right]\\ &+\Phi\left(\frac{\mathcal{I}^m - \Sigma_{l=1}^{n_x} C_k^{m,l} x_k^l}{\sqrt{R_k^{m,m}}}\right)\mathcal{I}^m,\end{aligned} \tag{9.14}$$

$$\mathrm{Var}(y_k^m|x_k, R_k^{m,m}) = R_k^{m,m}\left[1-\psi\left(\frac{\mathcal{I}^m - \Sigma_{l=1}^{n_x} C_k^{m,l} x_k^l}{\sqrt{R_k^{m,m}}}\right)\right] \tag{9.15}$$

where

$$\psi(\alpha) = \lambda(\alpha)(\lambda(\alpha)-\alpha), \quad \lambda(\alpha) = \frac{\phi(\alpha)}{1-\Phi(\alpha)} \tag{9.16}$$

and $\phi(\cdot)$ is the probability density function of the standard normal distribution.

Lemma 9.2 *[58] Suppose that $X = X^T > 0$, $Y = Y^T > 0$ and $\mathscr{H}_k(\cdot)$: $\mathbb{R}^n \to \mathbb{R}^n$. If*

$$\mathscr{H}_k(X) \le \mathscr{H}_k(Y), \quad \forall X \le Y \tag{9.17}$$

then, the solutions M_k and N_k to the following difference equations

$$N_{k+1} = \mathscr{H}_k(N_k), \quad M_{k+1} \le \mathscr{H}_k(M_k), \quad M_0 = N_0 \tag{9.18}$$

satisfy

$$M_{k+1} \le N_{k+1}. \tag{9.19}$$

Lemma 9.3 *Suppose that the positive scalars d_k, e_k and f_k are given. An upper bound of $\mathcal{Y}_k = \mathbb{E}\{\eta_k^2\}$ is obtained by the following recursion*

$$\begin{aligned}\bar{\mathcal{Y}}_{k+1} \triangleq&\Xi_k(\bar{\mathcal{Y}}_k)\\ =&\left((1+d_k)(1+e_k)\lambda^2 + \frac{(1+f_k)(1+d_k^{-1})}{\theta^2}\right)\bar{\mathcal{Y}}_k\\ &+\left((1+d_k)(1+e_k^{-1}) + (1+d_k^{-1})(1+f_k^{-1})\right)\sigma^2\end{aligned}$$

with the initial value $\bar{\mathcal{Y}}_0 = \eta_0^2$, i.e., $\mathcal{Y}_k \le \bar{\mathcal{Y}}_k$.

Proof *By Lemma 7.1, it follows from (9.10) that*

$$\begin{aligned}\varepsilon_k^T\varepsilon_k \le&\left(\frac{1}{\theta}\eta_k + \sigma\right)^2\\ \le&\frac{1+f_k}{\theta^2}\eta_k^2 + (1+f_k^{-1})\sigma^2.\end{aligned} \tag{9.20}$$

From (9.11) and (9.20), we obtain

$$\begin{aligned}\mathcal{Y}_{k+1} =&\mathbb{E}\{\eta_{k+1}^2\}\\ =&\mathbb{E}\{(\lambda\eta_k+\sigma- \parallel \varepsilon_k \parallel)^2\}\\ \leq&\mathbb{E}\{(1+d_k)(\lambda\eta_k+\sigma)^2+(1+d_k^{-1})\varepsilon_k^T\varepsilon_k\}\\ \leq&(1+d_k)(1+e_k)\lambda^2\mathbb{E}\{\eta_k^2\}+(1+d_k)(1+e_k^{-1})\sigma^2\\ &+(1+d_k^{-1})\mathbb{E}\{\varepsilon_k^T\varepsilon_k\}\\ \leq&\left((1+d_k)(1+e_k)\lambda^2+\frac{(1+f_k)(1+d_k^{-1})}{\theta^2}\right)\mathcal{Y}_k\\ &+\left((1+d_k)(1+e_k^{-1})+(1+d_k^{-1})(1+f_k^{-1})\right)\sigma^2\\ =&\Xi_k(\mathcal{Y}_k).\end{aligned}$$

By Lemma 9.2, $\mathcal{Y}_{k+1}\leq\bar{\mathcal{Y}}_{k+1}$ *holds immediately. The proof is complete.*

Let $\tilde{x}_{k+1|k}=x_{k+1}-\hat{x}_{k+1|k}$ be the prediction error. From (9.1) and (9.12), the prediction error can be expressed by

$$\tilde{x}_{k+1|k}=A_k\tilde{x}_{k|k}+\Delta A_kx_k+B_kw_k. \tag{9.21}$$

In light of Lemma 9.1 and the approximation in (9.8), the estimate of the measurement y_k^m is derived by

$$\hat{y}_{k+1|k}^m=\bar{\gamma}_{k+1}^m\left(\Sigma_{l=1}^{n_x}C_{k+1}^{m,l}\hat{x}_{k+1|k}^m+\sqrt{R_{k+1}^{m,m}}\lambda_{m,k+1}\right)+\left(1-\bar{\gamma}_{k+1}^m\right)\mathcal{I}^m.$$

where $\hat{y}_{k+1|k}^m$ is the mth element of the measurement estimate $\hat{y}_{k+1|k}$ and $\lambda_{m,k}=\lambda\left(\frac{\mathcal{I}^m-\Sigma_{l=1}^{n_x}C_k^{m,l}\hat{x}_{k|k-1}^l}{\sqrt{R_k^{m,m}}}\right)$. Setting

$$\begin{aligned}\mathcal{R}_k&=\begin{bmatrix}\sqrt{R_k^{1,1}} & \sqrt{R_k^{2,2}} & \cdots & \sqrt{R_k^{n_y,n_y}}\end{bmatrix}^T,\\ \vec{\lambda}_k&=\mathrm{diag}\{\lambda_{1,k},\lambda_{2,k},\ldots,\lambda_{n_y,k}\},\end{aligned}$$

we further have

$$\hat{y}_{k+1|k}=\bar{\Upsilon}_{k+1}\left(C_{k+1}\hat{x}_{k+1|k}+\vec{\lambda}_{k+1}\mathcal{R}_{k+1}\right)+\left(I-\bar{\Upsilon}_{k+1}\right)\mathcal{I}. \tag{9.22}$$

Subsequently, from (9.1), (9.13), and (9.22), the filtering error is of the following form

$$\begin{aligned}\tilde{x}_{k+1|k+1}=&(I-K_{k+1}\bar{\Upsilon}_{k+1}C_{k+1})\tilde{x}_{k+1|k}-K_{k+1}\tilde{\Upsilon}_{k+1}C_{k+1}x_{k+1}\\ &-K_{k+1}\bar{\Upsilon}_{k+1}(v_{k+1}-\bar{v}_{k+1})-K_{k+1}\tilde{\Upsilon}_{k+1}v_{k+1}\\ &+K_{k+1}\tilde{\Upsilon}_{k+1}\mathcal{I}-K_{k+1}(y_{k+1}^t-y_{k+1})\end{aligned} \tag{9.23}$$

where $\tilde{\Upsilon}_k=\Upsilon_k-\bar{\Upsilon}_k$ and $\bar{v}_k=\vec{\lambda}_k\mathcal{R}_k$.

By noting (9.21) and (9.23), the recursions for covariances of the prediction and filtering are, respectively, given as follows

$$\begin{aligned} P_{k+1|k} =& A_k P_{k|k} A_k^T + A_k \mathbb{E}\{\tilde{x}_{k|k} x_k^T\} \Delta A_k^T \\ &+ \Delta A_k \mathbb{E}\{x_k \tilde{x}_{k|k}^T\} A_k^T + \Delta A_k \mathbb{E}\{x_k x_k^T\} \Delta A_k^T + B_k Q_k B_k^T \end{aligned} \tag{9.24}$$

and

$$\begin{aligned} P_{k+1|k+1} =& (I - K_{k+1}\bar{\Upsilon}_{k+1}C_{k+1}) P_{k+1|k} (I - K_{k+1}\bar{\Upsilon}_{k+1}C_{k+1})^T - \mathcal{F}_{k+1} \\ &- \mathcal{F}_{k+1}^T + K_{k+1}\mathbb{E}\{\tilde{\Upsilon}_{k+1}C_{k+1}x_{k+1}x_{k+1}^T C_{k+1}^T \tilde{\Upsilon}_{k+1}^T\} K_{k+1}^T \\ &- \mathcal{A}_{k+1} - \mathcal{A}_{k+1}^T + K_{k+1}\bar{\Upsilon}_{k+1}\mathbb{E}\{(v_{k+1} - \bar{v}_{k+1})(v_{k+1} - \bar{v}_{k+1})^T\} \\ &\times \bar{\Upsilon}_{k+1}^T K_{k+1}^T - \mathcal{B}_{k+1} - \mathcal{B}_{k+1}^T + K_{k+1}\mathbb{E}\{\tilde{\Upsilon}_{k+1}v_{k+1}v_{k+1}^T \tilde{\Upsilon}_{k+1}^T\} K_{k+1}^T \\ &+ K_{k+1}\mathbb{E}\{\tilde{\Upsilon}_{k+1}\mathcal{I}\mathcal{I}^T \tilde{\Upsilon}_{k+1}^T\} K_{k+1}^T + K_{k+1}\mathbb{E}\{(y_{k+1}^t - y_{k+1}) \\ &\times (y_{k+1}^t - y_{k+1})^T\} K_{k+1}^T - \mathcal{C}_{k+1} - \mathcal{C}_{k+1}^T + \mathcal{D}_{k+1} + \mathcal{D}_{k+1}^T \\ &- \mathcal{E}_{k+1} - \mathcal{E}_{k+1}^T - \mathcal{G}_{k+1} - \mathcal{G}_{k+1}^T + \mathcal{H}_{k+1} + \mathcal{H}_{k+1}^T, \\ &- \mathcal{I}_{k+1} - \mathcal{I}_{k+1}^T + \mathcal{J}_{k+1} + \mathcal{J}_{k+1}^T. \end{aligned} \tag{9.25}$$

where

$$\begin{aligned} \mathcal{A}_{k+1} =& \mathbb{E}\{(I - K_{k+1}\bar{\Upsilon}_{k+1}C_{k+1})\tilde{x}_{k+1|k}x_{k+1}^T C_{k+1}^T \tilde{\Upsilon}_{k+1}^T K_{k+1}^T\}, \\ \mathcal{B}_{k+1} =& \mathbb{E}\{(I - K_{k+1}\bar{\Upsilon}_{k+1}C_{k+1})\tilde{x}_{k+1|k}(y_{k+1}^t - y_{k+1})^T K_{k+1}^T\}, \\ \mathcal{C}_{k+1} =& \mathbb{E}\{K_{k+1}\tilde{\Upsilon}_{k+1}C_{k+1}x_{k+1}\mathcal{I}^T \tilde{\Upsilon}_{k+1}^T K_{k+1}^T\}, \\ \mathcal{D}_{k+1} =& \mathbb{E}\{K_{k+1}\tilde{\Upsilon}_{k+1}C_{k+1}x_{k+1}(y_{k+1}^t - y_{k+1})^T K_{k+1}^T\}, \\ \mathcal{E}_{k+1} =& \mathbb{E}\{K_{k+1}\tilde{\Upsilon}_{k+1}\mathcal{I}(y_{k+1}^t - y_{k+1})^T K_{k+1}^T\}, \\ \mathcal{F}_{k+1} =& \mathbb{E}\{(I - K_{k+1}\bar{\Upsilon}_{k+1}C_{k+1})\tilde{x}_{k+1|k}\mathcal{I}^T \tilde{\Upsilon}_{k+1}^T K_{k+1}^T\}, \\ \mathcal{G}_{k+1} =& \mathbb{E}\{(I - K_{k+1}\bar{\Upsilon}_{k+1}C_{k+1})\tilde{x}_{k+1|k}(v_{k+1} - \bar{v}_{k+1})^T \bar{\Upsilon}_{k+1}^T K_{k+1}^T\}, \\ \mathcal{H}_{k+1} =& \mathbb{E}\{K_{k+1}\tilde{\Upsilon}_{k+1}C_{k+1}x_{k+1}(v_{k+1} - \bar{v}_{k+1})^T \bar{\Upsilon}_{k+1}^T K_{k+1}^T\}, \\ \mathcal{I}_{k+1} =& \mathbb{E}\{K_{k+1}\tilde{\Upsilon}_{k+1}I(v_{k+1} - \bar{v}_{k+1})^T \bar{\Upsilon}_{k+1}^T K_{k+1}^T\}, \\ \mathcal{J}_{k+1} =& \mathbb{E}\{K_{k+1}(y_{k+1}^t - y_{k+1})(v_{k+1} - \bar{v}_{k+1})^T \bar{\Upsilon}_{k+1}^T K_{k+1}^T\}. \end{aligned} \tag{9.26}$$

In the following theorem, an upper bound is provided for the FEC.

Theorem 9.1 *Suppose that the positive scalars* ε_i $(i = 1, 2, \ldots, 11)$, a_k, b_k, d_k, e_k *and* f_k *are given. The PEC* $P_{k+1|k}$ *and FEC* $P_{k+1|k+1}$ *are given in (9.24) and (9.25), respectively. If the real-valued matrices* $\bar{P}_{k+1|k}$ *and* $\bar{P}_{k+1|k+1}$ *satisfy the following recursions*

$$\bar{P}_{k+1|k} = (1 + a_k) A_k \bar{P}_{k|k} A_k^T + (1 + a_k^{-1}) E_k \mathrm{tr}\{X_1\} E_k^T + B_k Q_k B_k^T \tag{9.27}$$

and

$$
\begin{aligned}
&\bar{P}_{k+1|k+1} \\
=&(1+\varepsilon_1+\varepsilon_2+\varepsilon_3+\varepsilon_7)(I-K_{k+1}\bar{\Upsilon}_{k+1}C_{k+1})\bar{P}_{k+1|k}(I-K_{k+1}\bar{\Upsilon}_{k+1}C_{k+1})^T \\
&+(1+\varepsilon_1^{-1}+\varepsilon_4+\varepsilon_5+\varepsilon_8)K_{k+1}\left\{\check{\Upsilon}_{k+1}\circ\left\{C_{k+1}\mathrm{tr}\{X_2\}C_{k+1}^T\right\}\right\}K_{k+1}^T \\
&+(1+\varepsilon_3^{-1}+\varepsilon_5^{-1}+\varepsilon_6^{-1}+\varepsilon_{10}^{-1})K_{k+1}\bar{\Xi}_{k+1}(\bar{\mathcal{Y}}_{k+1})K_{k+1}^T \\
&+(1+\varepsilon_2^{-1}+\varepsilon_4^{-1}+\varepsilon_6+\varepsilon_9)K_{k+1}\left\{\check{\Upsilon}_{k+1}\circ\mathcal{I}\mathcal{I}^T\right\}K_{k+1}^T \\
&+K_{k+1}\left\{\check{\Upsilon}_{k+1}\circ R_{k+1}\right\}K_{k+1}^T \\
&+(1+\varepsilon_7^{-1}+\varepsilon_8^{-1}+\varepsilon_9^{-1}+\varepsilon_{10})K_{k+1}\bar{\Upsilon}_{k+1}\mathscr{R}_{k+1}\bar{\Upsilon}_{k+1}^TK_{k+1}^T
\end{aligned}
\tag{9.28}
$$

with initial $P_{0|0}=\bar{P}_{0|0}>0$, *where*

$$
\begin{aligned}
&\check{\Upsilon}_k=\mathrm{diag}\left\{\bar{\gamma}_k^1(1-\bar{\gamma}_k^1),\ldots,\bar{\gamma}_k^{n_y}(1-\bar{\gamma}_k^{n_y})\right\}, \\
&\bar{\Xi}_k(\bar{\mathcal{Y}}_k)=\left((1+f_k)\theta^{-2}\bar{\mathcal{Y}}_k+(1+f_k^{-1})\sigma^2\right)I, \\
&\mathscr{R}_k=R_k-\vec{\lambda}_k\mathcal{R}_k\mathcal{R}_k^T\vec{\lambda}_k^T, \\
&\mathrm{tr}\{X_1\}=\mathrm{tr}\left\{H_k\left((1+b_k)\bar{P}_{k|k}+(1+b_k^{-1})\hat{x}_{k|k}\hat{x}_{k|k}^T\right)H_k^T\right\}, \\
&\mathrm{tr}\{\mathrm{X}_2\}=\mathrm{tr}\left\{(1+\varepsilon_{11})\bar{P}_{k+1|k}+(1+\varepsilon_{11}^{-1})\hat{x}_{k+1|k}\hat{x}_{k+1|k}^T\right\},
\end{aligned}
$$

then, the $\bar{P}_{k+1|k+1}$ *is an upper bound of the* $P_{k+1|k+1}$, *that is,*

$$P_{k+1|k+1}\leq\bar{P}_{k+1|k+1}.$$

Proof *From (9.3) and (9.24), one obtains*

$$
\begin{aligned}
P_{k+1|k}\leq&(1+a_k)A_k\mathbb{E}\{\tilde{x}_{k|k}\tilde{x}_{k|k}^T\}A_k^T+(1+a_k^{-1})E_kF_kH_k \\
&\times\mathbb{E}\{x_kx_k^T\}H_k^TF_k^TE_k^T+B_kQ_kB_k^T.
\end{aligned}
\tag{9.29}
$$

By Lemma 7.1, one has

$$
\begin{aligned}
\mathbb{E}\{x_kx_k^T\}=&\mathbb{E}\{(\tilde{x}_{k|k}+\hat{x}_{k|k})(\tilde{x}_{k|k}+\hat{x}_{k|k})^T\} \\
\leq&\mathbb{E}\{(1+b_k)\tilde{x}_{k|k}\tilde{x}_{k|k}^T+(1+b_k^{-1})\hat{x}_{k|k}\hat{x}_{k|k}^T\} \\
=&(1+b_k)P_{k|k}+(1+b_k^{-1})\mathbb{E}\{\hat{x}_{k|k}\hat{x}_{k|k}^T\}.
\end{aligned}
\tag{9.30}
$$

By which and noting $FF^T\leq I$, *we further have*

$$
\begin{aligned}
&E_kF_kH_k\mathbb{E}\{x_kx_k^T\}H_k^TF_k^TE_k^T \\
\leq&E_kF_kH_k\left((1+b_k)P_{k|k}+(1+b_k^{-1})\mathbb{E}\{\hat{x}_{k|k}\hat{x}_{k|k}^T\}\right)H_k^TF_k^TE_k^T \\
\leq&E_k\mathrm{tr}\left\{H_k\left((1+b_k)P_{k|k}+(1+b_k^{-1})\hat{x}_{k|k}\hat{x}_{k|k}^T\right)H_k^T\right\}E_k^T \\
=&E_k\mathrm{tr}\left\{\Omega_1\right\}E_k^T
\end{aligned}
\tag{9.31}
$$

where $\Omega_1 = H_k\left((1+b_k)P_{k|k} + (1+b_k^{-1})\hat{x}_{k|k}\hat{x}_{k|k}^T\right)H_k^T$.

Thus, from (9.29) and (9.31), we obtain

$$P_{k+1|k} \leq (1+a_k)A_kP_{k|k}A_k^T + (1+a_k^{-1})E_k\text{tr}\left\{\Omega_1\right\}E_k^T + B_kQ_kB_k^T. \tag{9.32}$$

On the other hand, by using Lemma 7.1 again, it follows from (9.25) that

$$\begin{aligned} P_{k+1|k+1} \leq &(1+\varepsilon_1+\varepsilon_2+\varepsilon_3+\varepsilon_7)(I-K_{k+1}\bar{\Upsilon}_{k+1}C_{k+1})P_{k+1|k} \\ &\times(I-K_{k+1}\bar{\Upsilon}_{k+1}C_{k+1})^T + (1+\varepsilon_1^{-1}+\varepsilon_4+\varepsilon_5+\varepsilon_8)K_{k+1} \\ &\times\mathbb{E}\left\{\tilde{\Upsilon}_{k+1}C_{k+1}x_{k+1}x_{k+1}^TC_{k+1}^T\Upsilon_{k+1}^T\right\}K_{k+1}^T \\ &+(1+\varepsilon_3^{-1}+\varepsilon_5^{-1}+\varepsilon_6^{-1}+\varepsilon_{10}^{-1})K_{k+1}\mathbb{E}\{\varepsilon_{k+1}\varepsilon_{k+1}^T\}K_{k+1}^T \\ &+(1+\varepsilon_2^{-1}+\varepsilon_4^{-1}+\varepsilon_6+\varepsilon_9)K_{k+1}\mathbb{E}\{\tilde{\Upsilon}_{k+1}\mathcal{I}\mathcal{I}^T\tilde{\Upsilon}_{k+1}^T\}K_{k+1}^T \\ &+K_{k+1}\tilde{\Upsilon}_{k+1}\mathbb{E}\{v_{k+1}v_{k+1}^T\}\tilde{\Upsilon}_{k+1}^TK_{k+1}^T \\ &+(1+\varepsilon_7^{-1}+\varepsilon_8^{-1}+\varepsilon_9^{-1}+\varepsilon_{10})K_{k+1}\bar{\Upsilon}_{k+1}\mathscr{R}_{k+1}\bar{\Upsilon}_{k+1}^TK_{k+1}^T. \end{aligned} \tag{9.33}$$

Similarly, it is known that

$$\begin{aligned} \mathbb{E}\{x_{k+1}x_{k+1}^T\} =&\mathbb{E}\{(\tilde{x}_{k+1|k}+\hat{x}_{k+1|k})(\tilde{x}_{k+1|k}+\hat{x}_{k+1|k})^T\} \\ \leq&\mathbb{E}\{(1+\varepsilon_{11})\tilde{x}_{k+1|k}\tilde{x}_{k+1|k}^T+(1+\varepsilon_{11}^{-1})\hat{x}_{k+1|k}\hat{x}_{k+1|k}^T\} \\ \leq&\text{tr}\{(1+\varepsilon_{11})P_{k+1|k}+(1+\varepsilon_{11}^{-1})\hat{x}_{k+1|k}\hat{x}_{k+1|k}^T\}I \\ =&\text{tr}\left\{\Omega_2\right\}I \end{aligned} \tag{9.34}$$

where $\Omega_2 = (1+\varepsilon_{11})P_{k+1|k} + (1+\varepsilon_{11}^{-1})\hat{x}_{k+1|k}\hat{x}_{k+1|k}^T$.

Moreover, in view of the triggering condition (9.10), we have

$$\begin{aligned} \varepsilon_{k+1}\varepsilon_{k+1}^T \leq&\varepsilon_{k+1}^T\varepsilon_{k+1}I \\ =&\left(\frac{1}{\theta}\eta_{k+1}+\sigma\right)^2 I \\ \leq&\left(\frac{1+f_{k+1}}{\theta^2}\eta_{k+1}^2+(1+f_{k+1}^{-1})\sigma^2\right)I \\ =&\bar{\Xi}_{k+1}(\mathcal{Y}_{k+1}), \end{aligned}$$

which, from Lemma 9.3, implies

$$\mathbb{E}\{\varepsilon_{k+1}\varepsilon_{k+1}^T\} \leq \bar{\Xi}_{k+1}(\bar{\mathcal{Y}}_{k+1}). \tag{9.35}$$

Substituting (9.34) and (9.35) into (9.33) leads to

$$
\begin{aligned}
P_{k+1|k+1} \leq &(1+\varepsilon_1+\varepsilon_2+\varepsilon_3+\varepsilon_7)(I-K_{k+1}\bar{\Upsilon}_{k+1}C_{k+1})P_{k+1|k} \\
&\times (I-K_{k+1}\bar{\Upsilon}_{k+1}C_{k+1})^T+(1+\varepsilon_1^{-1}+\varepsilon_4+\varepsilon_5+\varepsilon_8) \\
&\times K_{k+1}\{\check{\Upsilon}_{k+1}\circ\{C_{k+1}\mathrm{tr}\{\Omega_2\}C_{k+1}^T\}\}K_{k+1}^T \\
&+(1+\varepsilon_3^{-1}+\varepsilon_5^{-1}+\varepsilon_6^{-1}+\varepsilon_{10}^{-1})K_{k+1}\bar{\Xi}_{k+1}(\bar{\mathcal{Y}}_{k+1})K_{k+1}^T \\
&+(1+\varepsilon_2^{-1}+\varepsilon_4^{-1}+\varepsilon_6+\varepsilon_9)K_{k+1}\{\check{\Upsilon}_{k+1}\circ\mathcal{I}\mathcal{I}^T\}K_{k+1}^T \\
&+K_{k+1}\{\check{\Upsilon}_{k+1}\circ R_{k+1}\}K_{k+1}^T+(1+\varepsilon_7^{-1}+\varepsilon_8^{-1}+\varepsilon_9^{-1} \\
&+\varepsilon_{10})K_{k+1}\bar{\Upsilon}_{k+1}\mathscr{R}_{k+1}\bar{\Upsilon}_{k+1}^T K_{k+1}^T.
\end{aligned} \tag{9.36}
$$

By using Lemma 9.2, we arrive at

$$
P_{k+1|k+1} \leq \bar{P}_{k+1|k+1}. \tag{9.37}
$$

The proof is now complete.

In the following theorem, the filtering parameter K_k is designed such that the upper bound of the FEC obtained in Theorem 9.1 is minimized at each time instant.

Theorem 9.2 *Suppose that the positive scalars ε_i $(i=1,2,\ldots,11)$, a_k, b_k, d_k, e_k and f_k are given. The upper bound of the FEC $\bar{P}_{k+1|k+1}$ given by (9.28) is minimized by the following filter gain*

$$
K_{k+1} = \Psi_{k+1}\Phi_{k+1}^{-1} \tag{9.38}
$$

where

$$
\begin{aligned}
\Psi_{k+1} =&(1+\varepsilon_1+\varepsilon_2+\varepsilon_3+\varepsilon_7)\bar{P}_{k+1|k}C_{k+1}^T\bar{\Upsilon}_{k+1}^T, \\
\Phi_{k+1} =&(1+\varepsilon_1+\varepsilon_2+\varepsilon_3+\varepsilon_7)\bar{\Upsilon}_{k+1}C_{k+1}\bar{P}_{k+1|k}C_{k+1}^T\bar{\Upsilon}_{k+1}^T \\
&+(1+\varepsilon_1^{-1}+\varepsilon_4+\varepsilon_5+\varepsilon_8)\{\check{\Upsilon}_{k+1}\circ\{C_{k+1}\mathrm{tr}\{\mathrm{X}_2\}C_{k+1}^T\}\} \\
&+(1+\varepsilon_3^{-1}+\varepsilon_5^{-1}+\varepsilon_6^{-1}+\varepsilon_{10}^{-1})\bar{\Xi}_{k+1}(\bar{\mathcal{Y}}_{k+1})+(1+\varepsilon_2^{-1}+\varepsilon_4^{-1} \\
&+\varepsilon_6+\varepsilon_9)\{\check{\Upsilon}_{k+1}\circ\mathcal{I}\mathcal{I}^T\}+\{\check{\Upsilon}_{k+1}\circ R_{k+1}\} \\
&+(1+\varepsilon_7^{-1}+\varepsilon_8^{-1}+\varepsilon_9^{-1}+\varepsilon_{10})\bar{\Upsilon}_{k+1}\mathscr{R}_{k+1}\bar{\Upsilon}_{k+1}^T.
\end{aligned} \tag{9.39}
$$

Proof *The trace of the upper bound $\bar{P}_{k+1|k+1}$ obtained by (9.28) is* *expressed by*

$$\begin{aligned}\mathrm{tr}\{\bar{P}_{k+1|k+1}\} =&\mathrm{tr}\left\{(1+\varepsilon_1+\varepsilon_2+\varepsilon_3+\varepsilon_7)(I-K_{k+1}\bar{\Upsilon}_{k+1}C_{k+1})\bar{P}_{k+1|k}\right.\\ &\times(I-K_{k+1}\bar{\Upsilon}_{k+1}C_{k+1})^T+(1+\varepsilon_1^{-1}+\varepsilon_4+\varepsilon_5+\varepsilon_8)\\ &\times K_{k+1}\left\{\check{\Upsilon}_{k+1}\circ\{C_{k+1}\mathrm{tr}\{X_2\}C_{k+1}^T\}\right\}K_{k+1}^T\\ &+(1+\varepsilon_3^{-1}+\varepsilon_5^{-1}+\varepsilon_6^{-1}+\varepsilon_{10}^{-1})K_{k+1}\bar{\Xi}_{k+1}(\bar{\mathcal{Y}}_{k+1})K_{k+1}^T\\ &+(1+\varepsilon_2^{-1}+\varepsilon_4^{-1}+\varepsilon_6+\varepsilon_9)K_{k+1}\{\check{\Upsilon}_{k+1}\circ\mathcal{I}\mathcal{I}^T\}K_{k+1}^T\\ &+K_{k+1}\{\check{\Upsilon}_{k+1}\circ R_{k+1}\}K_{k+1}^T+(1+\varepsilon_7^{-1}+\varepsilon_8^{-1}\\ &\left.+\varepsilon_9^{-1}+\varepsilon_{10})K_{k+1}\bar{\Upsilon}_{k+1}\mathscr{R}_{k+1}\bar{\Upsilon}_{k+1}^T K_{k+1}^T\right\}.\end{aligned}\tag{9.40}$$

The derivative of $\mathrm{tr}\{\bar{P}_{k+1|k+1}\}$ *can be computed as*

$$\begin{aligned}\frac{\partial}{\partial K_{k+1}}\mathrm{tr}\{\bar{P}_{k+1|k+1}\} =&-2(1+\varepsilon_1+\varepsilon_2+\varepsilon_3+\varepsilon_7)(I-K_{k+1}\bar{\Upsilon}_{k+1}C_{k+1})\\ &\times\bar{P}_{k+1|k}(I-K_{k+1}\bar{\Upsilon}_{k+1}C_{k+1})^T+2(1+\varepsilon_1^{-1}\\ &+\varepsilon_4+\varepsilon_5+\varepsilon_8)K_{k+1}\{\check{\Upsilon}_{k+1}\circ\{C_{k+1}\mathrm{tr}\{\mathrm{X}_2\}C_{k+1}^T\}\}\\ &+2(1+\varepsilon_3^{-1}+\varepsilon_5^{-1}+\varepsilon_6^{-1}+\varepsilon_{10}^{-1})K_{k+1}\bar{\Xi}_{k+1}(\bar{\mathcal{Y}}_{k+1})\\ &+2(1+\varepsilon_2^{-1}+\varepsilon_4^{-1}+\varepsilon_6+\varepsilon_9)K_{k+1}\left\{\check{\Upsilon}_{k+1}\circ\mathcal{I}\mathcal{I}^T\right\}\\ &+2K_{k+1}\{\check{\Upsilon}_{k+1}\circ R_{k+1}\}+2(1+\varepsilon_7^{-1}+\varepsilon_8^{-1}\\ &+\varepsilon_9^{-1}+\varepsilon_{10})K_{k+1}\bar{\Upsilon}_{k+1}\mathscr{R}_{k+1}\bar{\Upsilon}_{k+1}^T.\end{aligned}\tag{9.41}$$

Letting the above derivative be zero, we obtain the filter gain as follows

$$\begin{aligned}K_{k+1} =&(1+\varepsilon_1+\varepsilon_2+\varepsilon_3+\varepsilon_7)\bar{P}_{k+1|k}C_{k+1}^T\bar{\Upsilon}_{k+1}^T\\ &\times\left\{(1+\varepsilon_1+\varepsilon_2+\varepsilon_3+\varepsilon_7)\bar{\Upsilon}_{k+1}C_{k+1}P_{k+1|k}C_{k+1}^T\bar{\Upsilon}_{k+1}^T\right.\\ &+(1+\varepsilon_1^{-1}+\varepsilon_4+\varepsilon_5+\varepsilon_8)\left\{\check{\Upsilon}_{k+1}\circ\{C_{k+1}\mathrm{tr}\{\mathrm{X}_2\}C_{k+1}^T\}\right\}\\ &+(1+\varepsilon_3^{-1}+\varepsilon_5^{-1}+\varepsilon_6^{-1}+\varepsilon_{10}^{-1})\bar{\Xi}_{k+1}(\bar{\mathcal{Y}}_{k+1})+(1+\varepsilon_2^{-1}+\varepsilon_4^{-1}\\ &+\varepsilon_6+\varepsilon_9)\{\check{\Upsilon}_{k+1}\circ\mathcal{I}\mathcal{I}^T\}+\{\check{\Upsilon}_{k+1}\circ R_{k+1}\}\\ &+(1+\varepsilon_7^{-1}+\varepsilon_8^{-1}+\varepsilon_9^{-1}+\varepsilon_{10})\bar{\Upsilon}_{k+1}\mathscr{R}_{k+1}\bar{\Upsilon}_{k+1}^T\\ &\left.+\bar{\Upsilon}_{k+1}C_{k+1}\bar{P}_{k+1|k}C_{k+1}^T\bar{\Upsilon}_{k+1}^T\right\}^{-1},\end{aligned}\tag{9.42}$$

which completes the proof of Theorem 9.2.

So far, we have obtained an upper bound of the estimation error covariance for the discrete time-varying systems with CMs and PUs and designed a dynamic ET filter such that the upper bound is minimized. In what follows, we will provide a sufficient condition to ensure that the estimation error covariance is bounded.

Before giving the performance analysis, an assumption is introduced to place some constraints on system parameters.

Assumption 9.1 *There exist positive real numbers $\bar{a}$, $\bar{b}$, $\bar{c}$, $\underline{c}$, $\bar{e}$, $\bar{k}$, $\bar{q}$, $\bar{r}$, $\bar{\vartheta}$, $\bar{\pi}$, $\bar{\theta}$, $\underline{\varsigma}$, $\bar{\varsigma}$, $\bar{\omega}_1$, $\bar{\omega}_2$, and $\bar{R}$ such that the following parameters are bounded, i.e.,*

$$\begin{aligned}&\|A_k\| \leq \bar{a}, \quad \|B_k\| \leq \bar{b}, \quad \underline{c} \leq \|C_k\| \leq \bar{c}, \quad \|E_k\| \leq \bar{e}, \quad \|\check{\Upsilon}_k\| \leq \bar{\varsigma},\\ &\|Q_k\| \leq \bar{q}, \quad \|\mathscr{R}_k\| \leq \bar{r}, \quad \underline{\vartheta} \leq \|\bar{\Upsilon}_{k+1}\| \leq \bar{\vartheta}, \quad \underline{R} \leq \|R_k\| \leq \bar{R}, \quad \|\bar{\Xi}_k\| \leq \bar{\theta},\\ &\|\mathrm{tr}\{\Omega_1\}\| \leq \bar{\omega}_1, \quad \|a_k\| \leq \bar{\pi}, \quad \|\mathrm{tr}\{\Omega_2\}\| \leq \bar{\omega}_2.\end{aligned}$$

Theorem 9.3 *For the time-varying system (9.1) with the CMs (9.2). Under Assumption 9.1, if the following inequality holds*

$$(1+\varepsilon_1+\varepsilon_2+\varepsilon_3+\varepsilon_7)\bar{\kappa}^2(1+\bar{\pi})\bar{a}^2 < 1, \tag{9.43}$$

then, the estimation error of the filter described in (9.12) and (9.13) with gain given by (9.38) is mean-square bounded, i.e.,

$$\sup_{k\in\mathbb{N}} \mathbb{E}\{\tilde{x}_{k|k}\tilde{x}_{k|k}^T\} < \infty. \tag{9.44}$$

Proof *Under Assumption 9.1, it follows from (9.27) that*

$$\|\bar{P}_{k+1|k}\| \leq (1+\bar{\pi})\bar{a}^2\|\bar{P}_{k|k}\| + (1+\bar{\pi}^{-1})\bar{e}^2\bar{\omega}_1 + \bar{b}^2\bar{q}. \tag{9.45}$$

By noting (9.38) with (9.39), the upper bound of the filter gain K_{k+1} is estimated as

$$\|K_{k+1}\| \leq \|\Psi(k)\|\|\Phi(k)\|^{-1} \leq \frac{\bar{\vartheta}\bar{c}}{\underline{\vartheta}^2\underline{c}^2} \triangleq \bar{k} \tag{9.46}$$

from which and denoting $\bar{\kappa} \triangleq \|I - K_{k+1}\bar{\Upsilon}_{k+1}C_{k+1}\|$, it can be obtained from (9.28) that

$$\begin{aligned}\|\bar{P}_{k+1|k+1}\| \leq &(1+\varepsilon_1+\varepsilon_2+\varepsilon_3+\varepsilon_7)\|I - K_{k+1}\bar{\Upsilon}_{k+1}C_{k+1}\|^2\|\bar{P}_{k+1|k}\|\\ &+ (1+\varepsilon_1^{-1}+\varepsilon_4+\varepsilon_5+\varepsilon_8)k^2\bar{\varsigma}\bar{c}^2\bar{\omega}_2 + (1+\varepsilon_3^{-1}+\varepsilon_5^{-1}\\ &+\varepsilon_6^{-1}+\varepsilon_{10}^{-1})\bar{k}^2\bar{\theta} + (1+\varepsilon_2^{-1}+\varepsilon_4^{-1}+\varepsilon_6+\varepsilon_9)\bar{k}^2\bar{\varsigma}\\ &+\bar{k}^2\bar{\varsigma}\bar{R} + (1+\varepsilon_7^{-1}+\varepsilon_8^{-1}+\varepsilon_9^{-1}+\varepsilon_{10})\bar{k}^2\bar{\vartheta}^2\bar{r}\\ =&(1+\varepsilon_1+\varepsilon_2+\varepsilon_3+\varepsilon_7)\bar{\kappa}^2(1+\bar{\pi})\bar{a}^2\|P_{k|k}\| + (1+\varepsilon_1^{-1}\\ &+\varepsilon_4+\varepsilon_5+\varepsilon_8)k^2\bar{\varsigma}\bar{c}^2\bar{\omega}_2 + (1+\varepsilon_3^{-1}+\varepsilon_5^{-1}+\varepsilon_6^{-1}+\varepsilon_{10}^{-1})\\ &\times\bar{k}^2\bar{\theta} + (1+\varepsilon_2^{-1}+\varepsilon_4^{-1}+\varepsilon_6+\varepsilon_9)\bar{k}^2\bar{\varsigma} + (1+\varepsilon_1+\varepsilon_2\\ &+\varepsilon_3+\varepsilon_7)(1+\bar{\pi}^{-1})\bar{e}^2\bar{\omega}_1 + (1+\varepsilon_1+\varepsilon_2+\varepsilon_3+\varepsilon_7)\bar{b}^2\bar{q}\\ &+\bar{k}^2\bar{\varsigma}\bar{R} + (1+\varepsilon_7^{-1}+\varepsilon_8^{-1}+\varepsilon_9^{-1}+\varepsilon_{10})\bar{k}^2\bar{\vartheta}^2\bar{r}.\end{aligned} \tag{9.47}$$

Considering (9.45), *we finally obtain*

$$\|\bar{P}_{k+1|k+1}\| \leq (1+\varepsilon_1+\varepsilon_2+\varepsilon_3+\varepsilon_7)\bar{\kappa}^2(1+\bar{\pi})\bar{a}^2\|\bar{P}_{k|k}\| + \Pi \tag{9.48}$$

where

$$\begin{aligned}\Pi =&(1+\varepsilon_1^{-1}+\varepsilon_4+\varepsilon_5+\varepsilon_8)k^2\bar{\varsigma}\bar{c}^2\bar{\omega}_2\\&+(1+\varepsilon_3^{-1}+\varepsilon_5^{-1}+\varepsilon_6^{-1}+\varepsilon_{10}^{-1})\bar{k}^2\bar{\theta}\\&+(1+\varepsilon_2^{-1}+\varepsilon_4^{-1}+\varepsilon_6+\varepsilon_9)\bar{k}^2\bar{\varsigma}\\&+(1+\varepsilon_1+\varepsilon_2+\varepsilon_3+\varepsilon_7)(1+\bar{\pi}^{-1})\bar{e}^2\bar{\omega}_1\\&+(1+\varepsilon_1+\varepsilon_2+\varepsilon_3+\varepsilon_7)\bar{b}^2\bar{q}\\&+\bar{k}^2\bar{\varsigma}\bar{R}+(1+\varepsilon_7^{-1}+\varepsilon_8^{-1}+\varepsilon_9^{-1}+\varepsilon_{10})\bar{k}^2\bar{\vartheta}^2\bar{r}.\end{aligned} \tag{9.49}$$

Since $(1+\varepsilon_1+\varepsilon_2+\varepsilon_3+\varepsilon_7)\bar{\kappa}^2(1+\bar{\pi})\bar{a}^2 < 1$, *the upper bound* $\|\bar{P}_{k+1|k+1}\|$ *converges eventually. Note the fact that* $\bar{P}_{k+1|k+1}$ *is always the upper bound of the real estimation error covariance* $P_{k+1|k+1}$. *Therefore, we conclude that the filtering error is mean-square stable. The proof is complete now.*

Remark 9.5 *Up to now, we have addressed the recursive filtering problem for a class of uncertainty systems with CMs by means of dynamic ET mechanism. Based on the matrix analysis technique, an upper bound of the FEC is obtained at each time instant and the derived upper bound is minimized by the designed filter gain in Theorems 9.1 and 9.2. A sufficient condition has been conducted to reveal the boundedness behaviour of the estimation error in Theorem 9.3. The main challenge of designing the filtering algorithm comes from the CMs, uncertainties and the dynamic ET protocol. It's worthwhile to mention that 1) the proposed estimator structure is simple and of a recursive form, hence the proposed filtering algorithm is suitable for online application; and 2) the effects induced by PUs, dynamic ET mechanism and CMs are explicitly reflected in the algorithm.*

9.2 Dynamic Event-Triggered Distributed Filtering on GE Channels

9.2.1 Problem Formulation

Given a finite time-horizon $[0, N]$, where N is a positive integer, consider a target plant described by the following discrete time-varying nonlinear system:

$$x_{k+1} = f_k(x_k) + B_k w_k \tag{9.50}$$

where $x_k \in \mathbb{R}^{n_x}$ is the plant state, $w_k \in \mathbb{R}^{n_w}$ is the process noise, and B_k is a known time-varying matrix with appropriate dimension.

Assumption 9.2 *[122] The nonlinear function $f_k(\cdot)$ in (9.50) satisfies $f_k(0) = 0$ and the following condition:*

$$\|f_k(x) - f_k(y) - E_k(x - y)\| \leq \gamma_k \|x - y\| \tag{9.51}$$

for all $x, y \in \mathbb{R}^{n_x}$, where E_k is a known matrix and γ_k is a known time-varying nonnegative scalar.

For each sensor node i $(1 \leq i \leq n)$, the measurement output is described by

$$y_{i,k} = C_{i,k}x_k + D_{i,k}v_k \tag{9.52}$$

where $y_{i,k} \in \mathbb{R}^{n_y}$ is the measurement output of the ith node, $v_k \in \mathbb{R}^{n_v}$ denotes the measurement noise, and $C_{i,k}$ and $D_{i,k}$ are known time-varying matrices.

Assumption 9.3 *w_k and v_k are mutually uncorrelated zero-mean Gaussian white-noise sequences with respective covariances $R_k > 0$ and $Q_k > 0$.*

Assumption 9.4 *The initial value x_0 with mean $\bar{x}_0$ and covariance $\mathcal{X}_0$ is uncorrelated with both w_k and v_k.*

In this section, each sensor node in the sensor network shares its local measurement with its neighbors according to a randomly varying network topology. Let $\{\tau_{ij,k}\}_{k\geq 0}$ $(1 \leq i, j \leq n)$ be a Markov chain that takes values on 0 or 1 to describe the communication link between sensor nodes i and j, where $\tau_{ij,k} = 1$ means that sensor node i can receive the measurement from node j at time-instant k, while $\tau_{ij,k} = 0$ means that sensor node i cannot. Furthermore, $\tau_{ii,k} = 1$. The set of transition probabilities is given by

$$\begin{aligned}
&\text{Prob}\{\tau_{ij,k} = 1|\tau_{ij,k-1} = 1\} = \alpha_{ij}, \ \text{Prob}\{\tau_{ij,k} = 0|\tau_{ij,k-1} = 1\} = 1 - \alpha_{ij}, \\
&\text{Prob}\{\tau_{ij,k} = 1|\tau_{ij,k-1} = 0\} = \beta_{ij}, \ \text{Prob}\{\tau_{ij,k} = 0|\tau_{ij,k-1} = 0\} = 1 - \beta_{ij},
\end{aligned} \tag{9.53}$$

where $\alpha_{ij}, \beta_{ij} \in [0, 1]$ and the initial distribution is $\text{Prob}\{\tau_{ij,0} = 0\} = \alpha_0$ and $\text{Prob}\{\tau_{ij,0} = 1\} = 1 - \alpha_0$.

Remark 9.6 *Note that the randomly varying network topology (accounting for phenomena such as random packet dropouts or sensor failures) is taken into account in the sensor network by restoring to the Markov chain $\{\tau_{ij,k}\}$ $(1 \leq i, j \leq n)$. Such a two-state Markov chain model is known as the Gilbert-Elliott channels initiated in [49] and [31], which has been widely adopted to represent unreliable communication channels, see e.g. [17, 64, 147]. In this section, we are concerned with the distributed filtering problem in sensor network over a finite time-horizon, and it is only required that the topology structure of the communication network varies according to the Markov chain $\{\tau_{ij,k}\}$.*

For the sake of saving communication cost, a dynamic ET mechanism is employed for each sensor node to decide whether its current measurement should be broadcasted to its neighbors or not. For node i $(1 \leq i \leq n)$, denote by $0 \leq k_0^i < k_1^i < \cdots < k_l^i < \cdots$ the sequence of triggering instants that is determined by the following condition:

$$k_{l+1}^i = \min\left\{k \in \mathbb{N} | k > k_l^i, \frac{1}{\theta_i}\eta_{i,k} + \sigma_i - \|\varepsilon_{i,k}\| \leq 0\right\} \tag{9.54}$$

where σ_i and θ_i are given positive scalars, $\varepsilon_{i,k}$ is defined by $\varepsilon_{i,k} \triangleq y_{i,k} - y_{i,k_l^i}$ with y_{i,k_l^i} being the transmitted measurement at latest event time, and $\eta_{i,k}$ is an internal dynamic variable that evolves according to

$$\eta_{i,k+1} = \lambda_i \eta_{i,k} + \sigma_i - \|\varepsilon_{i,k}\|, \quad \eta_{i,0} = \eta_0^i \tag{9.55}$$

with λ_i being a given constant and $\eta_0^i \geq 0$ being the given initial condition. It is assumed that the parameters λ_i and θ_i satisfy $\lambda_i \theta_i \geq 1$. Then, the variable $\eta_{i,k}$ determined by (9.54) with (9.55) satisfies $\eta_{i,k} \geq 0$ for all $k \in [0, N]$.

Remark 9.7 *In the event triggering condition (9.54), an internal dynamic variable $\eta_{i,k}$ is introduced, which is actually a filtered version of the signal used to trigger events [50]. In practice, if the dynamic ET condition (9.54) is satisfied, then the measurement $y_{i,k}$ on node i would be transmitted to the filters on those nodes, which are connected to node i. In (9.54), the parameter θ_i determines the triggering frequency and the triggering frequency monotonically increases as θ_i increases, while the parameter $\eta_{i,k}$ can make $\sigma_i - \|\varepsilon_{i,k}\|$ not always nonnegative, and it is sufficient that it is non-negative in average. The parameters θ_i and $\eta_{i,0}$ (determining the value of $\eta_{i,k}$) can be selected according to the engineering requirement.*

Remark 9.8 *By using the ET transmission mechanism (9.54), the redundant executions could be effectively reduced and hence the communication burden is alleviated. Indeed, the system performance may deteriorate to a certain extent because of the introduction of ET strategy.*

Under the ET strategy, the broadcasted measurement from sensor node i $(1 \leq i \leq n)$ can be expressed by $y_{i,k_{s_i(k)}^i}$ with $s_i(k) = \arg\min_{\{l \in \mathbb{N} | k \geq k_l^i\}} \{k - k_l^i\}$. Then, in view of the given network topology, for node i $(1 \leq i \leq n)$, the actual input to the corresponding filter is denoted by

$$\vec{y}_{i,k} = \bar{\tau}_{i,k}\bar{y}_k, \tag{9.56}$$

where $\bar{\tau}_{i,k} = \text{diag}\{\tau_{i1,k}I, \tau_{i2,k}I, \ldots, \tau_{in,k}I\}$ and $\bar{y}_k = \text{col}_n\{y_{i,k_{s_i(k)}^i}\}$.

For node i $(1 \leq i \leq n)$, we construct the following distributed filter:

$$\hat{x}_{i,k+1} = K_{i,k}\hat{x}_{i,k} + G_{i,k}\vec{y}_{i,k} \tag{9.57}$$

where $\hat{x}_{i,k} \in \mathbb{R}^{n_x}$ is the estimate of state x_k on the node i, $G_{i,k} = \text{vec}_n^i\{G_{ij,k}\}$, and $K_{i,k}$ and $G_{ij,k}$ $(1 \leq j \leq n)$ are the filter parameters to be determined.

Remark 9.9 *The filter (9.57) reflects a scenario that, at a certain time instant, the filter on node i only receives the measurements from the sensors on node j when nodes i and j are connected. Such a scenario is characterized by the Markov chain $\{\tau_{ij,k}\}_{k\geq 0}$ $(1 \leq i, j \leq n)$. It is seen from (9.56) and (9.57) that, when $\tau_{ij,k} = 0$ (which means there are no communications between nodes j and i), the filter on node i cannot receive the measurements from node j. In this case, for the filter on node i, the measurements from the nodes j $(j \neq i)$ are replaced by zero measurements. As such, the phenomena of asequent and delayed arrivals [123, 126] are avoided.*

Letting the filtering error be $e_{i,k} = x_k - \hat{x}_{i,k}$, for each i $(1 \leq i \leq n)$, we obtain the following filtering error dynamics:

$$e_{i,k+1} = f_k(x_k) + B_k w_k - K_{i,k}\hat{x}_{i,k} - G_{i,k}\vec{y}_{i,k}. \tag{9.58}$$

In this section, the filtering problem to be solved is stated as follows. Under the dynamic ET mechanism (9.54)–(9.55), we are interested in designing distributed filter parameters $K_{i,k}$ and $G_{i,k}$ $(1 \leq i \leq n)$ such that 1) there exists an upper bound on the FEC $\mathcal{P}_k^i = \mathbb{E}\{e_{i,k}e_{i,k}^T\}$; and 2) such an upper bound is minimized at each time instant.

Remark 9.10 *Comparing with the state estimation algorithms under intermittent communication proposed/used in [77, 146, 147], three differences that our filtering algorithm possesses can be identified: 1) our filtering algorithm is effective for the nonlinear system satisfying condition (9.51), 2) a dynamic ET mechanism is introduced in our filtering algorithm, and 3) our filtering algorithm is achieved by minimizing an upper bound of the FEC at each time instant.*

9.2.2 Main Results

In order to derive the main results, the following lemmas are established.

Lemma 9.4 *For all $1 \leq i \leq n$ and $0 \leq k \leq N$, suppose that $\lambda_i\theta_i \geq 1$ and let the positive scalars $b_{i,k}$, $c_{i,k}$, and d_k be given. The covariance matrices $\mathcal{Y}_k^i = \mathbb{E}\{\eta_{i,k}^2\}$ and $\mathcal{X}_k = \mathbb{E}\{x_k x_k^T\}$, respectively, satisfy $\mathcal{Y}_k^i \leq \bar{\mathcal{Y}}_k^i$ and $\mathcal{X}_k^i \leq \bar{\mathcal{X}}_k^i$, where*

$$\begin{aligned}\bar{\mathcal{Y}}_{k+1}^i \triangleq & \Xi_k^i(\bar{\mathcal{Y}}_k^i) \\ = & \Big((1+b_{i,k})(1+c_{i,k})\lambda_i^2 + (1+\theta_i)(1+b_{i,k}^{-1})\theta_i^{-2}\Big)\bar{\mathcal{Y}}_k^i \\ & + \Big((1+b_{i,k})(1+c_{i,k}^{-1}) + (1+b_{i,k}^{-1})(1+\theta_i^{-1})\Big)\sigma_i^2,\end{aligned} \tag{9.59}$$

$$\begin{aligned}\bar{\mathcal{X}}_{k+1} \triangleq & \Psi_k(\bar{\mathcal{X}}_k) \\ = & (1+d_k)E_k\bar{\mathcal{X}}_k E_k^T + (1+d_k^{-1})\gamma_k^2 tr[\bar{\mathcal{X}}_k]I + B_k R_k B_k^T\end{aligned} \tag{9.60}$$

with the initial conditions $\bar{\mathcal{Y}}_0^i = (\eta_0^i)^2$ *and* $\bar{\mathcal{X}}_0 = \mathcal{X}_0$.

Proof: By using the following elementary inequality:

$$UV^T + VU^T \leq \alpha UU^T + \alpha^{-1}VV^T \tag{9.61}$$

for any positive scalar α, where U and V are matrices with compatible dimensions, it follows from (9.54) that

$$\varepsilon_{i,k}^T\varepsilon_{i,k} \leq \left(\frac{1}{\theta_i}\eta_{i,k} + \sigma_i\right)^2 \leq \frac{(1+\theta_i)}{\theta_i^2}\eta_{i,k}^2 + (1+\theta_i{}^{-1})\sigma_i^2. \tag{9.62}$$

Furthermore, it is obtained from (9.50), (9.55), and (9.62) that

$$\begin{aligned}
\mathcal{Y}_{k+1}^i &= \mathbb{E}\left\{\eta_{i,k+1}^2\right\} \\
&= \mathbb{E}\left\{\left(\lambda_i\eta_{i,k} + \sigma_i - \|\varepsilon_{i,k}\|\right)^2\right\} \\
&\leq \mathbb{E}\left\{(1+b_{i,k})\left(\lambda_i\eta_{i,k}+\sigma_i\right)^2 + (1+b_{i,k}^{-1})\varepsilon_{i,k}^T\varepsilon_{i,k}\right\} \\
&\leq (1+b_{i,k})(1+c_{i,k})\lambda_i^2\mathbb{E}\left\{\eta_{i,k}^2\right\} + (1+b_{i,k}) \\
&\quad \times(1+c_{i,k}^{-1})\sigma_i^2 + (1+b_{i,k}^{-1})\mathbb{E}\left\{\varepsilon_{i,k}^T\varepsilon_{i,k}\right\} \\
&\leq \Xi_k^i(\mathcal{Y}_k^i),
\end{aligned} \tag{9.63}$$

$$\begin{aligned}
\mathcal{X}_{k+1} &= \mathbb{E}\left\{x_{k+1}x_{k+1}^T\right\} \\
&= \mathbb{E}\Big\{E_kx_kx_k^TE_k^T + \left(f_k(x_k) - E_kx_k\right) \\
&\quad \times\left(f_k(x_k) - E_kx_k\right)^T + E_kx_k\left(f_k(x_k)\right. \\
&\quad \left.-E_kx_k\right)^T + \left(f_k(x_k) - E_kx_k\right)x_k^TE_k^T\Big\} + B_kR_kB_k^T \\
&\leq \mathbb{E}\Big\{(1+d_k^{-1})\left(f_k(x_k) - E_kx_k\right)\left(f_k(x_k)\right. \\
&\quad \left.-E_kx_k\right)^T + (1+d_k)E_kx_kx_k^TE_k^T\Big\} + B_kR_kB_k^T \\
&\leq \mathbb{E}\Big\{(1+d_k^{-1})\|f_k(x_k) - E_kx_k\|^2I + B_kR_kB_k^T \\
&\quad +(1+d_k)E_kx_kx_k^TE_k^T\Big\} \\
&\leq \Psi_k(\mathcal{X}_k).
\end{aligned} \tag{9.64}$$

It then follows from Lemma 9.2 that $\mathcal{Y}_{k+1}^i \leq \bar{\mathcal{Y}}_{k+1}^i$ and $\mathcal{X}_{k+1} \leq \bar{\mathcal{X}}_{k+1}$, which ends the proof. □

Theorem 9.4 *For all* $1 \leq i \leq n$ *and* $0 \leq k \leq N$, *suppose that* $\lambda_i\theta_i \geq 1$ *and let the positive scalars* $f_{i,k}$, $g_{i,k}$, $h_{i,k}$ *and* $m_{i,k}$ *be given. The FEC* $\mathcal{P}_k^i$ *satisfies* $\mathcal{P}_k^i \leq \bar{\mathcal{P}}_k^i$, *where*

$$\bar{\mathcal{P}}_{k+1}^i = \tilde{\Phi}(\bar{\mathcal{P}}_k^i)$$

$$
\begin{aligned}
&= \sum_{l=1}^{n} \Big((1 + f_{i,k}^{-1})(1 + h_{i,k}^{-1}) + m_{i,k}^{-1} \Big) \Big(G_{i,k}\bar{\tau}_{i,k} \Big) \bar{\Xi}_k^l(\bar{\mathcal{Y}}_k^l) \Big(G_{i,k}\bar{\tau}_{i,k} \Big)^T \\
&\quad + (1 + f_{i,k}^{-1})(1 + h_{i,k}) K_{i,k} \bar{\mathcal{P}}_k^i K_{i,k}^T + (1 + m_{i,k}) \Big(G_{i,k}\bar{\tau}_{i,k}\bar{D}_k \Big) \\
&\quad \times Q_k \Big(G_{i,k}\bar{\tau}_{i,k}\bar{D}_k \Big)^T + \bar{\Psi}_k(\bar{\mathcal{X}}_k) + B_k R_k B_k^T, \qquad (9.65)
\end{aligned}
$$

$$
\bar{\Xi}_k^i(\bar{\mathcal{Y}}_k^i) = \Big((1 + \theta_i)\theta_i^{-2} \bar{\mathcal{Y}}_k^i + (1 + \theta_i^{-1})\sigma_i^2 \Big) I, \qquad (9.66)
$$

$$
\begin{aligned}
\bar{\Psi}_k(\bar{\mathcal{X}}_k) &= (1 + f_{i,k})(1 + g_{i,k}) \Big(E_k - K_{i,k} - G_{i,k}\bar{\tau}_{i,k}\bar{C}_k \Big) \bar{\mathcal{X}}_k \Big(E_k - K_{i,k} \\
&\quad - G_{i,k}\bar{\tau}_{i,k}\bar{C}_k \Big)^T + (1 + f_{i,k})(1 + g_{i,k}^{-1}) \gamma_k^2 tr[\bar{\mathcal{X}}_k] I, \qquad (9.67)
\end{aligned}
$$

with the initial condition $\bar{\mathcal{P}}_0^i = \mathcal{P}_0^i$.

Proof: By noting (9.58), the FEC at time-instant $k+1$ is computed as follows:

$$
\begin{aligned}
\mathcal{P}_{k+1}^i =& \mathbb{E}\left\{ e_{i,k+1} e_{i,k+1}^T \right\} \\
=& \mathbb{E}\Big\{ \Big(f_k(x_k) - E_k x_k + K_{i,k} e_{i,k} + G_{i,k}\bar{\tau}_{i,k}\varepsilon_k + (E_k - K_{i,k} \\
& - G_{i,k}\bar{\tau}_{i,k}\bar{C}_k) x_k \Big) \Big(f_k(x_k) - E_k x_k + K_{i,k} e_{i,k} + G_{i,k}\bar{\tau}_{i,k}\varepsilon_k \\
& + (E_k - K_{i,k} - G_{i,k}\bar{\tau}_{i,k}\bar{C}_k) x_k \Big)^T \Big\} + B_k R_k B_k^T + \Big(G_{i,k}\bar{\tau}_{i,k}\bar{D}_k \Big) \\
& \times Q_k \Big(G_{i,k}\bar{\tau}_{i,k}\bar{D}_k \Big)^T - G_{i,k}\bar{\tau}_{i,k} \mathbb{E}\{\varepsilon_k v_k^T\} \Big(G_{i,k}\bar{\tau}_{i,k}\bar{D}_k \Big)^T \\
& - \Big(G_{i,k}\bar{\tau}_{i,k}\bar{D}_k \Big) \mathbb{E}\{v_k \varepsilon_k^T\} \bar{\tau}_{i,k} G_{i,k}^T
\end{aligned}
$$

where $\varepsilon_k = \text{col}_n\{\varepsilon_{i,k}\}$, $\bar{C}_k = \text{col}_n\{C_{i,k}\}$ and $\bar{D}_k = \text{col}_n\{D_{i,k}\}$. Using (9.61) again, we further obtain

$$
\begin{aligned}
& \mathcal{P}_{k+1}^i \\
\leq & \mathbb{E}\Big\{ (1 + f_{i,k}) \Big(f_k(x_k) - E_k x_k + (E_k - K_{i,k} - G_{i,k}\bar{\tau}_{i,k}\bar{C}_k) x_k \Big) \\
& \times \Big(f_k(x_k) - E_k x_k + (E_k - K_{i,k} - G_{i,k}\bar{\tau}_{i,k}\bar{C}_k) x_k \Big)^T + (1 + f_{i,k}^{-1}) \\
& \times \Big(K_{i,k} e_{i,k} + G_{i,k}\bar{\tau}_{i,k}\varepsilon_k \Big) \Big(K_{i,k} e_{i,k} + G_{i,k}\bar{\tau}_{i,k}\varepsilon_k \Big)^T \Big\} \\
& + B_k R_k B_k^T + m_{i,k}^{-1} \Big(G_{i,k}\bar{\tau}_{i,k} \Big) \mathbb{E}\{\varepsilon_k \varepsilon_k^T\} \Big(G_{i,k}\bar{\tau}_{i,k} \Big)^T \\
& + (1 + m_{i,k}) \Big(G_{i,k}\bar{\tau}_{i,k}\bar{D}_k \Big) Q_k \Big(G_{i,k}\bar{\tau}_{i,k}\bar{D}_k \Big)^T \\
\leq & \bar{\Psi}_k(\mathcal{X}_k) + (1 + f_{i,k}^{-1})(1 + h_{i,k}) K_{i,k} \mathbb{E}\{e_{i,k} e_{i,k}^T\} K_{i,k}^T + (1 + f_{i,k}^{-1}) \\
& \times (1 + h_{i,k}^{-1}) \Big(G_{i,k}\bar{\tau}_{i,k} \Big) \mathbb{E}\{\varepsilon_k \varepsilon_k^T\} \Big(G_{i,k}\bar{\tau}_{i,k} \Big)^T + B_k R_k B_k^T
\end{aligned}
$$

$$+ m_{i,k}^{-1}\Big(G_{i,k}\bar{\tau}_{i,k}\Big)\mathbb{E}\{\varepsilon_k\varepsilon_k^T\}\Big(G_{i,k}\bar{\tau}_{i,k}\Big)^T + (1+m_{i,k})$$
$$\times \Big(G_{i,k}\bar{\tau}_{i,k}\bar{D}_k\Big)Q_k\Big(G_{i,k}\bar{\tau}_{i,k}\bar{D}_k\Big)^T. \tag{9.68}$$

On the other hand, it is easy to verify from (9.62) that

$$\begin{aligned}\varepsilon_k\varepsilon_k^T &\le \varepsilon_k^T\varepsilon_k I = \sum_{i=1}^{n}\varepsilon_{i,k}^T\varepsilon_{i,k}I\\ &\le \sum_{i=1}^{n}\Big(\frac{1+\theta_i}{\theta_i^2}\eta_{i,k}^2 + (1+\theta_i{}^{-1})\sigma_i^2\Big)I,\end{aligned} \tag{9.69}$$

which implies $\mathbb{E}\{\varepsilon_k\varepsilon_k^T\} \le \sum_{l=1}^{n}\Xi_k^l(\mathcal{Y}_k^l)$.

From Lemma 9.2, it is easily obtained that $\mathbb{E}\{\varepsilon_k\varepsilon_k^T\} \le \sum_{l=1}^{n}\bar{\Xi}_k^l(\bar{\mathcal{Y}}_k^l)$ and $\bar{\Psi}_k(\mathcal{X}_k) \le \bar{\Psi}_k(\bar{\mathcal{X}}_k)$. Then, we have

$$\begin{aligned}\mathcal{P}_{k+1}^i \le &\sum_{l=1}^{n}\Big((1+f_{i,k}^{-1})(1+h_{i,k}^{-1}) + m_{i,k}^{-1}\Big)\Big(G_{i,k}\bar{\tau}_{i,k}\Big)\bar{\Xi}_k^l(\bar{\mathcal{Y}}_k^l)\Big(G_{i,k}\bar{\tau}_{i,k}\Big)^T\\ &+ \bar{\Psi}_k(\bar{\mathcal{X}}_k) + B_kR_kB_k^T + (1+m_{i,k})\Big(G_{i,k}\bar{\tau}_{i,k}\bar{D}_k\Big)Q_k\Big(G_{i,k}\bar{\tau}_{i,k}\bar{D}_k\Big)^T\\ &+ (1+f_{i,k}^{-1})(1+h_{i,k})K_{i,k}\mathcal{P}_k^iK_{i,k}^T\\ \le&\tilde{\Phi}(\mathcal{P}_k^i).\end{aligned}$$

It can now be derived from Lemma 9.2 that $\mathcal{P}_{k+1}^i \le \bar{\mathcal{P}}_{k+1}^i$, and the proof is now complete. □

Remark 9.11 *In Theorem 9.4, an upper bound of the FEC is given in terms of the recursion (9.66). Such an upper bound might be a bit conservative and the main conservatism results from the ET mechanism (9.54) and (9.55) and the nonlinearity characterized by (9.51). In order to reduce the conservatism as much as possible, we have introduced the adjustable parameter α in the elementary inequality (9.61) and employed Lemma 9.2 to derive the upper bound of the FEC as tightly as possible.*

Theorem 9.5 *For all $1 \le i \le n$ and $0 \le k \le N$, suppose that $\lambda_i\theta_i \ge 1$ and define the following real-valued matrices $\Pi_{1i,k}$, $\Pi_{2i,k}$, $\Pi_{3i,k}$, $\Pi_{4i,k}$, and $\Pi_{5i,k}$:*

$$\begin{aligned}\Pi_{1i,k} &= (1+f_{i,k})(1+g_{i,k})\bar{\mathcal{X}}_k + (1+f_{i,k}^{-1})(1+h_{i,k})\bar{\mathcal{P}}_k^i,\\ \Pi_{2i,k} &= (1+f_{i,k})(1+g_{i,k})E_k\bar{\mathcal{X}}_k,\\ \Pi_{3i,k} &= (1+f_{i,k})(1+g_{i,k})\Big(\bar{\tau}_{i,k}\bar{C}_k\Big)\bar{\mathcal{X}}_k\Big(\bar{\tau}_{i,k}\bar{C}_k\Big)^T + (1+m_{i,k})\\ &\quad\times\Big(\bar{\tau}_{i,k}\bar{D}_k\Big)Q_k\Big(\bar{\tau}_{i,k}\bar{D}_k\Big)^T + \sum_{l=1}^{n}\Big((1+f_{i,k}^{-1})(1+h_{i,k}^{-1}) + m_{i,k}^{-1}\Big)\end{aligned}$$

$$\begin{aligned}
&\times\bar{\tau}_{i,k}\bar{\Xi}_k^l(\bar{\mathcal{Y}}_k^l)\bar{\tau}_{i,k} - \Pi_{5i,k}\Pi_{1i,k}^{-1}\Pi_{5i,k}^T,\\
\Pi_{4i,k} &= (1+f_{i,k})(1+g_{i,k})E_k\left(\bar{\tau}_{i,k}\bar{C}_k\bar{\mathcal{X}}_k\right)^T - \Pi_{2i,k}\Pi_{1i,k}^{-1}\Pi_{5i,k}^T,\\
\Pi_{5i,k} &= (1+f_{i,k})(1+g_{i,k})\left(\bar{\tau}_{i,k}\bar{C}_k\bar{\mathcal{X}}_k\right).
\end{aligned} \tag{9.70}$$

Then, the upper bound $\bar{\mathcal{P}}_k^i$ for the FEC $\mathcal{P}_k^i$ can be minimized with the parameters designed as

$$K_{i,k} = \Pi_{2i,k}\Pi_{1i,k}^{-1} - G_{i,k}\Pi_{5i,k}\Pi_{1i,k}^{-1}, \tag{9.71}$$

$$G_{ij,k} = \begin{cases} 0, & \text{if } \tau_{ij,k}=0 \\ (\tilde{\Pi}_{4i,k}\tilde{\Pi}_{3i,k}^{-1})_j^* & \text{if } \tau_{ij,k}=1 \end{cases} \tag{9.72}$$

where $\tilde{\Pi}_{4i,k}$ and $\tilde{\Pi}_{3i,k}$ are, respectively, the simplified matrices by removing the zero-columns of $\Pi_{4i,k}$ and by removing the zero-columns and zero-rows of $\Pi_{3i,k}$. Here, $(A)_{j_l}^*$ means the sub-matrix A_l extracted from matrix $A = \begin{bmatrix} A_1 & A_2 & \cdots & A_m \end{bmatrix}$ for all j_l satisfying $\tau_{ij_l,k}=1$.

Proof: It follows from (9.67) that

$$\begin{aligned}
&\mathrm{tr}(\bar{\mathcal{P}}_{k+1}^i)\\
=&(1+f_{i,k})(1+g_{i,k})\mathrm{tr}\Big\{(E_k - K_{i,k} - G_{i,k}\bar{\tau}_{i,k}\bar{C}_k)\bar{\mathcal{X}}_k(E_k - K_{i,k} - G_{i,k}\bar{\tau}_{i,k}\bar{C}_k)^T\Big\}\\
&+(1+f_{i,k})(1+g_{i,k}^{-1})\gamma_k^2\mathrm{tr}[\bar{\mathcal{X}}_k]n_x + (1+f_{i,k}^{-1})(1+h_{i,k})\mathrm{tr}\Big\{K_{i,k}\bar{\mathcal{P}}_k^iK_{i,k}^T\Big\}\\
&+\sum_{l=1}^n\Big((1+f_{i,k}^{-1})(1+h_{i,k}^{-1})+m_{i,k}^{-1}\Big)\mathrm{tr}\Big\{G_{i,k}\bar{\tau}_{i,k}\bar{\Xi}_k^l(\bar{\mathcal{Y}}_k^l)\bar{\tau}_{i,k}G_{i,k}^T\Big\}\\
&+(1+m_{i,k})\mathrm{tr}\Big\{\left(G_{i,k}\bar{\tau}_{i,k}\bar{D}_k\right)Q_k\left(G_{i,k}\bar{\tau}_{i,k}\bar{D}_k\right)^T\Big\} + \mathrm{tr}\Big\{B_kR_kB_k^T\Big\}.
\end{aligned}$$

To determine the filter parameters $K_{i,k}$ and $G_{i,k}$, we take the partial derivation of the trace of $\bar{\mathcal{P}}_{k+1}^i$ with respect to $K_{i,k}$ and $G_{i,k}$, respectively, and then have

$$\begin{aligned}
\frac{\partial}{\partial K_{i,k}}\mathrm{tr}(\bar{\mathcal{P}}_{k+1}^i) =&2(1+f_{i,k})(1+g_{i,k})\big(-E_k\bar{\mathcal{X}}_k + K_{i,k}\bar{\mathcal{X}}_k\\
&+G_{i,k}\bar{\tau}_{i,k}\bar{C}_k\bar{\mathcal{X}}_k\big) + 2(1+f_{i,k}^{-1})(1+h_{i,k})K_{i,k}\bar{\mathcal{P}}_k^i = 0,
\end{aligned} \tag{9.73}$$

and

$$\begin{aligned}
\frac{\partial}{\partial G_{i,k}}\mathrm{tr}(\bar{\mathcal{P}}_{k+1}^i) =\ & 2(1+f_{i,k})(1+g_{i,k})\Big(-E_k\bar{\mathcal{X}}_k\bar{C}_k^T\bar{\tau}_{i,k} + K_{i,k}\bar{\mathcal{X}}_k\\
&\times\bar{C}_k^T\bar{\tau}_{i,k} + G_{i,k}\bar{\tau}_{i,k}\bar{C}_k\bar{\mathcal{X}}_k\bar{C}_k^T\bar{\tau}_{i,k}\Big)\\
&+2(1+m_{i,k})\Big(G_{i,k}\bar{\tau}_{i,k}\bar{D}_k\Big)Q_k\Big(\bar{\tau}_{i,k}\bar{D}_k\Big)^T
\end{aligned}$$

$$+\sum_{l=1}^{n}2\Big(m_{i,k}^{-1}(1+f_{i,k}^{-1})(1+h_{i,k}^{-1})\Big)G_{i,k}\bar{\tau}_{i,k}$$
$$\times\bar{\Xi}_k^l(\bar{\mathcal{Y}}_k^l)\bar{\tau}_{i,k}=0. \tag{9.74}$$

By noting (9.70), the optimal filter parameter $K_{i,k}$ is determined as follows:

$$K_{i,k}=\Pi_{2i,k}\Pi_{1i,k}^{-1}-G_{i,k}\Pi_{5i,k}\Pi_{1i,k}^{-1}. \tag{9.75}$$

Next, our task is to obtain another filter gain $G_{i,k}$. It is derived from (9.74) that

$$G_{i,k}\bar{\Pi}_{3i,k}=(1+f_{i,k})(1+g_{i,k})\Big(E_k\bar{\mathcal{X}}_k\bar{C}_k^T\bar{\tau}_{i,k}-K_{i,k}\bar{\mathcal{X}}_k\bar{C}_k^T\bar{\tau}_{i,k}\Big) \tag{9.76}$$

where

$$\begin{aligned}\bar{\Pi}_{3i,k}=&(1+f_{i,k})(1+g_{i,k})\Big(\bar{\tau}_{i,k}\bar{C}_k\Big)\bar{\mathcal{X}}_k\Big(\bar{\tau}_{i,k}\bar{C}_k\Big)^T\\&+(1+m_{i,k})\Big(\bar{\tau}_{i,k}\bar{D}_k\Big)Q_k\Big(\bar{\tau}_{i,k}\bar{D}_k\Big)^T\\&+\sum_{l=1}^{n}\Big((1+f_{i,k}^{-1})(1+h_{i,k}^{-1})+m_{i,k}^{-1}\Big)\bar{\tau}_{i,k}\bar{\Xi}_k^l(\bar{\mathcal{Y}}_k^l)\bar{\tau}_{i,k}.\end{aligned}$$

Substituting (9.75) to (9.76), the optimal filter gain $G_{i,k}$ is given by

$$G_{i,k}\Pi_{3i,k}=\Pi_{4i,k}. \tag{9.77}$$

Furthermore, recalling the definition of $\bar{\tau}_{i,k}$, it is easily seen that there are some difficulties to obtain the gain $G_{i,k}$ directly, which is due to the fact that $\Pi_{3i,k}$ may be non-invertible at time-instant k. In this case, the technique developed in [101] can be used to deal with this issue. Then, we have

$$G_{ij,k}=\begin{cases}0, & \text{if } \tau_{ij,k}=0\\(\tilde{\Pi}_{4i,k}\tilde{\Pi}_{3i,k}^{-1})_j^* & \text{if } \tau_{ij,k}=1\end{cases} \tag{9.78}$$

which completes the proof. □

Remark 9.12 *In order to reduce the difference between the bound and the exact error covariance, in Theorem 9.5, the filter gain matrices are designed such that the upper bound of FEC derived in Theorem 9.4 is minimized. Furthermore, it is worthwhile to note that the minimized bound obtained in Theorem 9.5 is closely related to the parameters $b_{i,k}$, $c_{i,k}$, d_k, $f_{i,k}$, $g_{i,k}$, $h_{i,k}$ and $m_{i,k}$. Hence the possible conservatism of the upper bound can be reduced further by appropriately selecting these parameters.*

9.3 Dynamic Event-Triggered Resilient H_∞ State Estimation

9.3.1 Problem Formulation

Consider the following discrete-time delayed neural networks

$$\begin{cases} x(k+1) = Ax(k) + Df(x(k)) + Wg(x(k-\tau(k))) + L\mu(k) \\ y(k) = Mx(k) + E\mu(k) \\ z(k) = Cx(k) \end{cases} \tag{9.79}$$

where $x(k) = \begin{bmatrix} x_1(k) & \cdots & x_{n_x}(k) \end{bmatrix}^T \in \mathbb{R}^{n_x}$ is the neural state vector; $y(k) = \begin{bmatrix} y_1(k) & \cdots & y_{n_y}(k) \end{bmatrix}^T \in \mathbb{R}^{n_y}$ is the measurement output; $z(k) = \begin{bmatrix} z_1(k) & \cdots & z_{n_z}(k) \end{bmatrix}^T \in \mathbb{R}^{n_z}$ is the output to be estimated; $A = \text{diag}\{a_1, a_2, \cdots, a_{n_x}\}$ is the coefficient matrix with entries $|a_i| < 1$; $D = [d_{ij}]_{n_x \times n_x}$ and $W = [w_{ij}]_{n_x \times n_x}$ stand for the connection weight matrix and the delayed connection weight matrix, respectively; $f(x(k)) = \begin{bmatrix} f_1(x_1(k)) & \cdots & f_{n_x}(x_{n_x}(k)) \end{bmatrix}^T$ and $g(x(k - \tau(k))) = \begin{bmatrix} g_1(x_1(k-\tau(k))) & \cdots & g_{n_x}(x_{n_x}(k-\tau(k))) \end{bmatrix}^T$ are the neuron activation functions; $\tau(k)$ denotes the TVD satisfying $0 < \tau_m \leq \tau(k) \leq \tau_M$; L, M, E and C are known constant matrices with appropriate dimensions; $\mu(k) \in \mathbb{R}^{n_\mu}$ is the exogenous disturbance input which satisfies $\sum_{k=0}^{N} \mu^T(k)\mu(k) \leq \hbar$.

Assumption 9.5 *The neuron activation functions $f(\cdot)$ and $g(\cdot)$ are continuous and satisfy $f(0) = 0$, $g(0) = 0$ and the following sector-bounded conditions*

$$[f(a) - f(b) - \mathcal{L}_1(a-b)]^T [f(a) - f(b) - \mathcal{L}_2(a-b)] \leq 0, \quad a, b \in \mathbb{R}^{n_x},$$
$$[g(a_\tau) - g(b_\tau) - \mathcal{T}_1(a_\tau - b_\tau)]^T [g(a_\tau) - g(b_\tau) - \mathcal{T}_2(a_\tau - b_\tau)] \leq 0,$$
$$\tag{9.80}$$

for $a_\tau, b_\tau \in \mathbb{R}^{n_x}$, where $\mathcal{L}_1$, $\mathcal{L}_2$, $\mathcal{T}_1$, and $\mathcal{T}_2$ are known constant matrices of appropriate dimensions.

In order to save computation and communication resources, a dynamic ET mechanism is employed before the measured signal enters into the estimator. Denote the transmission time instants as $0 = r_0 < r_1 < r_2 < \cdots < r_p < \cdots$, which are determined based on the following ET condition

$$r_{p+1} = \min\{k \in \mathbb{N} \mid k > r_p, \frac{1}{\theta}\delta(k) + \sigma y^T(k)y(k) - \xi^T(k)\xi(k) \leq 0\} \tag{9.81}$$

where θ and σ are known positive scalars, $\xi(k)$ is defined by $\xi(k) = y(k) - y(r_p)$, and $\delta(k)$ is an internal dynamic variable satisfying

$$\delta(k+1) = \varrho\delta(k) + \sigma y^T(k)y(k) - \xi^T(k)\xi(k) \tag{9.82}$$

with $\varrho \in (0,1)$ being a known constant and $\delta(0) = \delta_0 \geq 0$ being the initial condition.

Remark 9.13 *Note that the dynamic ET mechanism has been first proposed in [50] for solving the stability problem of nonlinear systems. As shown in [50], by employing the dynamic ET mechanism, the energy consumption can be further reduced as the triggering times under the dynamic ET mechanism are much less than the one for the static ET mechanism. As such, under the dynamic ET mechanism, the control problem has been investigated in [174] for nonlinear stochastic systems and the distributed set-membership estimation problem has been studied in [46] for linear system. Nevertheless, the state estimation problem of neural networks under dynamic ET mechanism has not obtained adequate research attention despite its practical insight.*

In this section, we are interested in constructing the following state estimator

$$\begin{aligned} \hat{x}(k+1) =& (A_f + \Delta A_f(k))\hat{x}(k) + (B_f + \Delta B_f(k))(y(r_p) - M\hat{x}(k)) \\ \hat{z}(k) =& (C_f + \Delta C_f(k))\hat{x}(k) \end{aligned} \tag{9.83}$$

for $k \in [r_p, r_{p+1})$, where $\hat{x}(k)$ is the state estimate, the matrices A_f, B_f and C_f are the state estimator gains to be designed, and $\Delta A_f(k)$, $\Delta B_f(k)$, and $\Delta C_f(k)$ stand for the gain variations. The uncertain gains are assumed to be of the following form

$$\Delta A_f(k) = P_1\Gamma_1(k)Q_1, \quad \Delta B_f(k) = P_2\Gamma_2(k)Q_2, \quad \Delta C_f(k) = P_3\Gamma_3(k)Q_3 \tag{9.84}$$

where P_1, P_2, P_3, Q_1, Q_2, and Q_3 are known real constant matrices, and $\Gamma_1(k) \in \mathbb{R}^{n_s \times n_t}$, $\Gamma_2(k) \in \mathbb{R}^{n_s \times n_t}$ and $\Gamma_3(k) \in \mathbb{R}^{n_s \times n_t}$ denote unknown time-varying matrices function satisfying

$$\Gamma_1^T(k)\Gamma_1(k) \leq I, \quad \Gamma_2^T(k)\Gamma_2(k) \leq I, \quad \Gamma_3^T(k)\Gamma_3(k) \leq I. \tag{9.85}$$

Defining $\vec{x}(k) = \begin{bmatrix} x^T(k) & \hat{x}^T(k) \end{bmatrix}^T$ and $\vec{z}(k) = z(k) - \hat{z}(k)$, the state estimation error system can be obtained from (9.79) and (9.83) as follows

$$\begin{aligned} \vec{x}(k+1) =& (\Xi_1 + \mathscr{P}_1\Gamma_1(k)\mathscr{Q}_1 + \mathscr{P}_2\Gamma_2(k)\mathscr{Q}_2)\vec{x}(k) + \mathcal{W}g(\vec{x}(k-\tau(k))) + \mathcal{D} \\ & \times f(\vec{x}(k)) + (\Xi_2 + \mathscr{P}_2\Gamma_2(k)Q_2E)\mu(k) + (\Xi_3 - \mathscr{P}_2\Gamma_2(k)Q_2)\xi(k) \\ \vec{z}(k) =& (\Xi_4 - P_3\Gamma_3(k)\mathscr{Q}_3)\vec{x}(k) \end{aligned} \tag{9.86}$$

where

$$\Xi_1 = \begin{bmatrix} A & 0 \\ B_fM & A_f - B_fM \end{bmatrix}, \quad \Xi_2 = \begin{bmatrix} L \\ B_fE \end{bmatrix},$$

$$\Xi_3 = \begin{bmatrix} 0 \\ -B_f \end{bmatrix}, \quad \Xi_4 = \begin{bmatrix} C & -C_f \end{bmatrix}, \quad \mathscr{P}_1 = \begin{bmatrix} 0 \\ P_1 \end{bmatrix},$$

$$\mathscr{P}_2 = \begin{bmatrix} 0 \\ P_2 \end{bmatrix}, \quad \mathscr{Q}_1 = \begin{bmatrix} 0 & Q_1 \end{bmatrix}, \quad \mathscr{Q}_2 = \begin{bmatrix} Q_2 M & -Q_2 M \end{bmatrix},$$

$$\mathscr{Q}_3 = \begin{bmatrix} 0 & Q_3 \end{bmatrix}, \quad \mathcal{D} = \begin{bmatrix} D & 0 \\ 0 & 0 \end{bmatrix}, \quad \mathcal{W} = \begin{bmatrix} W & 0 \\ 0 & 0 \end{bmatrix},$$

$$f(\vec{x}(k)) = \begin{bmatrix} f(x(k)) \\ f(\hat{x}(k)) \end{bmatrix}, \quad g(\vec{x}(k-\tau(k))) = \begin{bmatrix} g(x(k-\tau(k))) \\ g(\hat{x}(k-\tau(k))) \end{bmatrix}.$$

In addition, it is assumed that the initial conditions of the system (9.86) satisfy $\vec{x}^T(\epsilon_1)\mathcal{R}\vec{x}(\epsilon_1) \le \alpha_1$ ($\epsilon_1 \in [-\tau_M, 0]$) and $(\vec{x}(\epsilon_2+1)-\vec{x}(\epsilon_2))^T(\vec{x}(\epsilon_2+1)-\vec{x}(\epsilon_2)) \le \beta$ ($\epsilon_2 \in [-\tau_M, -1]$), where $\mathcal{R}$ is a known positive definite matrix, and α_1 and β are known positive constant scalars.

Definition 9.1 *[160] The system (9.86) is said to be finite-time bounded with respect to $(\alpha_1, \alpha_2, N, \mathcal{R}, \hbar)$, where $\alpha_2 > \alpha_1 \ge 0$, $\hbar > 0$, $N \in \mathbb{N}^+$ and $\mathcal{R}$ is a positive definite matrix, if*

$$\begin{cases} \vec{x}^T(k_1)\mathcal{R}\vec{x}(k_1) \le \alpha_1 \\ \sum_{k=0}^{N} \mu^T(k)\mu(k) \le \hbar \end{cases} \forall k_1 \in [-\tau_M, 0] \Longrightarrow \vec{x}^T(k_2)\mathcal{R}\vec{x}(k_2) \le \alpha_2, \forall k_2 \in [1, N]. \tag{9.87}$$

The aim of this section is to design a estimator with the form (9.83) for the discrete-time delayed neural networks (9.79), such that the dynamics of the system (9.86) is finite-time bounded and satisfies a prescribed H_∞ performance level. Specifically, the following two requirements need to be achieved simultaneously.

1. The system (9.86) is finite-time bounded with respect to $(\alpha_1, \alpha_2, N, \mathcal{R}, \hbar)$.
2. Under the zero-initial condition, for a fixed integer $N > 0$, a given disturbance attenuation level γ, the output estimation error $\vec{z}(k)$ satisfies
$$\sum_{k=0}^{N} \vec{z}^T(k)\vec{z}(k) < \gamma^2 \sum_{k=0}^{N} \mu^T(k)\mu(k). \tag{9.88}$$

Remark 9.14 *Note that the existing ET state estimation methods are only applicable to neural networks with the exactly known estimator gain parameters and the adopted ET mechanism is static. With respect to the case that the estimator gains are subject to unexpected fluctuation, the corresponding results have not yet been reported in the existing literature, not to mention that the dynamic ET mechanism is considered simultaneously. Moreover, in the existing results on the ET state estimation problem for neural networks, the* *estimation performance is expected to be achieved in a infinite time. Therefore, to the best of our knowledge, this section represents one of the first few attempts to design a finite-time resilient state estimator for neural networks under the dynamic ET mechanism.*

9.3.2 Main Results

In this subsection, the finite-time boundedness and H_∞ performance analysis problems are firstly conducted for the estimation error system (9.86). Then, according to the conducted analysis results, the desired resilient H_∞ state estimator is designed for the system (9.79). First of all, we introduce the following lemmas that will be useful in establishing our main results.

Lemma 9.5 *[86] For the dynamic ET mechanism given by (9.81) and (9.82) with the initial value $\delta_0 \geq 0$, if the parameters $\varrho \in (0,1)$ and $\theta > 0$ are chosen to satisfy the constraint $\varrho\theta \geq 1$, then we have $\delta(k) \geq 0$ for all $k \geq 0$.*

Lemma 9.6 *[121] Denote*

$$\aleph_{\vec{x}}(k, s_1, s_2) = \begin{cases} \frac{1}{s_2 - s_1}[(2\sum_{j=k-s_2}^{k-s_1-1} \vec{x}(j)) + \vec{x}(k-s_1) - \vec{x}(k-s_2)], & s_1 < s_2, \\ 2\vec{x}(k-s_1), \quad s_1 = s_2 \end{cases} \tag{9.89}$$

where $\vec{x}(k) \in \mathbb{R}^{2n_x}$ and s_1, s_2, k are given nonnegative integers satisfying $s_1 \leq s_2 \leq k$. Then, for a given positive definite matrix $\mathcal{M} \in \mathbb{R}^{2n_x \times 2n_x}$, we have

$$\begin{aligned} &-(s_2 - s_1) \sum_{j=k-s_2}^{k-s_1-1} (\vec{x}(j+1) - \vec{x}(j))^T \mathcal{M}(\vec{x}(j+1) - \vec{x}(j)) \\ &\leq - \begin{bmatrix} \mathfrak{T}_1 \\ \mathfrak{T}_2 \end{bmatrix}^T \begin{bmatrix} \mathcal{M} & 0 \\ 0 & 3\mathcal{M} \end{bmatrix} \begin{bmatrix} \mathfrak{T}_1 \\ \mathfrak{T}_2 \end{bmatrix} \end{aligned} \tag{9.90}$$

where

$$\begin{aligned} \mathfrak{T}_1 &= \vec{x}(k-s_1) - \vec{x}(k-s_2), \\ \mathfrak{T}_2 &= \vec{x}(k-s_1) + \vec{x}(k-s_2) - \aleph_{\vec{x}}(k, s_1, s_2). \end{aligned}$$

Lemma 9.7 *[132] For given positive integers n, m, a scalar $\varpi \in (0,1)$, a positive definite matrix $\Re \in \mathbb{R}^{n\times n}$, two matrices $\mathfrak{M}_1 \in \mathbb{R}^{n\times m}$ and $\mathfrak{M}_2 \in \mathbb{R}^{n\times m}$, define*

$$\Im(\varpi, \Re) = \frac{1}{\varpi}\psi^T \mathfrak{M}_1^T \Re \mathfrak{M}_1 \psi + \frac{1}{1-\varpi}\psi^T \mathfrak{M}_2^T \Re \mathfrak{M}_2 \psi \tag{9.91}$$

for all vector $\psi \in \mathbb{R}^m$. *Then, if there exists a matrix* $\mathcal{X} \in \mathbb{R}^{n\times n}$ *such that* $\begin{bmatrix} \Re & \mathcal{X} \\ * & \Re \end{bmatrix} > 0$, *the following inequality holds*

$$\min_{\varpi\in(0,1)} \Im(\varpi,\Re) \geq \begin{bmatrix} \mathfrak{M}_1\psi \\ \mathfrak{M}_2\psi \end{bmatrix}^T \begin{bmatrix} \Re & \mathcal{X} \\ * & \Re \end{bmatrix} \begin{bmatrix} \mathfrak{M}_1\psi \\ \mathfrak{M}_2\psi \end{bmatrix}. \tag{9.92}$$

Lemma 9.8 *[25] Suppose that* $\mathcal{E} = \mathcal{E}^T$, $\mathcal{K}$ *and* $\mathcal{N}$ *are known matrices of appropriate dimensions and* $\mathcal{O}$ *is a matrix satisfying* $\mathcal{O}^T\mathcal{O} \leq I$. *Then,* $\mathcal{E} + \mathcal{K}\mathcal{O}\mathcal{N} + (\mathcal{K}\mathcal{O}\mathcal{N})^T < 0$ *if and only if there exist a positive scalar* $\varsigma > 0$ *such that*

$$\begin{bmatrix} \mathcal{E} & \mathcal{K} & \varsigma\mathcal{N}^T \\ * & -\varsigma I & 0 \\ * & * & -\varsigma I \end{bmatrix} < 0. \tag{9.93}$$

The following theorem provides a sufficient condition under which the estimation error system (9.86) is finite-time bounded with respect to $(\alpha_1, \alpha_2, N, \mathcal{R}, \hbar)$.

Theorem 9.6 *Assume that* ϱ $(0 < \varrho < 1)$ *and* θ $(\theta > 0)$ *satisfy* $\varrho\theta \geq 1$ *and let the state estimator gains* A_f, B_f, *and* C_f *be given. For a given scalar* λ $(\lambda > 1)$, *the estimation error system (9.86) is finite-time bounded with respect to* $(\alpha_1, \alpha_2, N, \mathcal{R}, \hbar)$ *if there exist positive definite matrices* $\mathcal{P}$, $\mathcal{Q}_1$, $\mathcal{Q}_2$, $\mathcal{Q}_3$, $\mathcal{S}_1$, $\mathcal{S}_2$, *and* $\mathcal{J}$, *positive constant scalars* c_i $(i = 0,1,2,3,4,5,6)$, ω_i $(i = 1,2,3)$, *and* e_i $(i = 1,2)$, *and matrix* $\mathcal{Y}$ *satisfying*

$$\Omega = \begin{bmatrix} \tilde{\mathcal{S}}_1 & \mathcal{Y} \\ * & \tilde{\mathcal{S}}_1 \end{bmatrix} > 0, \tag{9.94}$$

$$\begin{bmatrix} \Pi & \Theta_1 \\ * & \Theta_2 \end{bmatrix} < 0, \tag{9.95}$$

$$c_0\mathcal{R} < \mathcal{P} < c_1\mathcal{R}, \tag{9.96}$$

$$0 < \mathcal{Q}_1 < c_2\mathcal{R}, \quad 0 < \mathcal{Q}_2 < c_3\mathcal{R}, \quad 0 < \mathcal{Q}_3 < c_4\mathcal{R}, \tag{9.97}$$

$$0 < \mathcal{S}_1 < c_5 I, \quad 0 < \mathcal{S}_2 < c_6 I, \tag{9.98}$$

$$\lambda^N(\kappa_1\alpha_1 + \kappa_2\beta + \frac{\delta_0}{\theta} + \lambda_{\max}\{\mathcal{J}\}\hbar) \leq c_0\alpha_2 \tag{9.99}$$

where

$$\Pi = \begin{bmatrix} \Sigma & \bar{\mathscr{X}}_1^T & \bar{\mathscr{X}}_2^T \\ * & -\mathcal{P}^{-1} & 0 \\ * & * & -\mathscr{S}^{-1} \end{bmatrix}, \; \Sigma = \begin{bmatrix} \Sigma_{11} & \Sigma_{12} & 0 & 0 \\ * & \Sigma_{22} & 0 & 0 \\ * & * & \Sigma_{33} & 0 \\ * & * & * & \Sigma_{44} \end{bmatrix},$$

$$\Theta_1 = \begin{bmatrix} \mathcal{U}_1 & e_1\mathcal{V}_1^T & \mathcal{U}_2 & e_2\mathcal{V}_2^T \end{bmatrix}, \ \Theta_2 = \text{diag}\{-e_1 I, -e_2 I\},$$
$$\Sigma_{11} = \Phi_1 + \Phi_2 + \Phi_3 + \Phi_4 + \Phi_5, \ \Sigma_{12} = (\frac{\sigma}{\theta} + \omega_1\sigma)\Lambda_1 H^T M^T E,$$
$$\Sigma_{22} = (\frac{\sigma}{\theta} + \omega_1\sigma)E^T E - \mathcal{J}, \ \Sigma_{33} = -(\frac{1}{\theta} + \omega_1)I, \ \Sigma_{44} = \frac{\varrho - \lambda + \omega_1}{\theta},$$
$$\Phi_1 = -\lambda\Lambda_1\mathcal{P}\Lambda_1^T, \ \Phi_3 = -\lambda\pi_1\tilde{\mathcal{S}}_2\pi_1^T - \lambda^{\tau_m+1}\tilde{\pi}\Omega\tilde{\pi}^T, \ H = \begin{bmatrix} I & 0 \end{bmatrix},$$
$$\Phi_2 = \Lambda_1(\mathcal{Q}_1 + \mathcal{Q}_2 + (\tau_M - \tau_m + 1)\mathcal{Q}_3)\Lambda_1^T - \lambda^{\tau_m}\Lambda_2\mathcal{Q}_1\Lambda_2^T - \lambda^{\tau_m}\Lambda_3\mathcal{Q}_3\Lambda_3^T - \lambda^{\tau_M}\Lambda_4\mathcal{Q}_2\Lambda_4^T, \ \mathscr{S} = (\tau_M - \tau_m)^2\mathcal{S}_1 + \tau_m^2\mathcal{S}_2,$$
$$\Phi_4 = -\omega_2\pi_4\mathscr{L}\pi_4^T - \omega_3\pi_5\mathscr{T}\pi_5^T, \ \Phi_5 = (\frac{\sigma}{\theta} + \omega_1\sigma)\Lambda_1 H^T M^T M H \Lambda_1^T,$$
$$\pi_1 = \begin{bmatrix} \Lambda_1 - \Lambda_2 & \Lambda_1 + \Lambda_2 - \Lambda_5 \end{bmatrix}, \ \pi_2 = \begin{bmatrix} \Lambda_3 - \Lambda_4 & \Lambda_3 + \Lambda_4 - \Lambda_7 \end{bmatrix},$$
$$\pi_3 = \begin{bmatrix} \Lambda_2 - \Lambda_3 & \Lambda_2 + \Lambda_3 - \Lambda_6 \end{bmatrix}, \ \pi_4 = \begin{bmatrix} \Lambda_1 & \Lambda_8 \end{bmatrix}, \ \pi_5 = \begin{bmatrix} \Lambda_3 & \Lambda_9 \end{bmatrix},$$
$$\Lambda_i = \begin{bmatrix} 0_{2n_x \times (i-1)2n_x} & I_{2n_x} & 0_{2n_x \times (9-i)2n_x} \end{bmatrix}^T (i = 1, 2, \cdots, 9),$$
$$\mathscr{L} = \begin{bmatrix} \mathscr{L}_{11} & \mathscr{L}_{12} \\ * & I \end{bmatrix}, \ \mathscr{L}_{11} = \text{diag}\{\tilde{\mathcal{L}}_1, \tilde{\mathcal{L}}_1\}, \ \mathscr{L}_{12} = \text{diag}\{\tilde{\mathcal{L}}_2, \tilde{\mathcal{L}}_2\},$$
$$\mathscr{T} = \begin{bmatrix} \mathscr{T}_{11} & \mathscr{T}_{12} \\ * & I \end{bmatrix}, \ \mathscr{T}_{11} = \text{diag}\{\tilde{\mathcal{T}}_1, \tilde{\mathcal{T}}_1\}, \ \mathscr{T}_{12} = \text{diag}\{\tilde{\mathcal{T}}_2, \tilde{\mathcal{T}}_2\},$$
$$\tilde{\mathcal{L}}_1 = \frac{\mathcal{L}_1^T\mathcal{L}_2 + \mathcal{L}_2^T\mathcal{L}_1}{2}, \ \tilde{\mathcal{L}}_2 = -\frac{\mathcal{L}_1^T + \mathcal{L}_2^T}{2}, \ \tilde{\mathcal{T}}_1 = \frac{\mathcal{T}_1^T\mathcal{T}_2 + \mathcal{T}_2^T\mathcal{T}_1}{2},$$
$$\bar{\mathscr{X}}_1 = \Big[\Xi_1, \underbrace{0, 0, \cdots, 0}_{6}, \mathcal{D}, \mathcal{W}, \Xi_2, \Xi_3, 0\Big], \ \tilde{\mathcal{S}}_1 = \text{diag}\{\mathcal{S}_1, 3\mathcal{S}_1\},$$
$$\bar{\mathscr{X}}_2 = \Big[\Xi_1 - I, \underbrace{0, 0, \cdots, 0}_{6}, \mathcal{D}, \mathcal{W}, \Xi_2, \Xi_3, 0\Big], \ \tilde{\mathcal{T}}_2 = -\frac{\mathcal{T}_1^T + \mathcal{T}_2^T}{2},$$
$$\mathcal{U}_1 = \Big[\underbrace{0, 0, \cdots, 0}_{12}, \mathscr{P}_1^T, \mathscr{P}_1^T\Big]^T, \ \mathcal{V}_1 = \Big[\mathscr{Q}_1, \underbrace{0, 0, \cdots, 0}_{13}\Big], \ \tilde{\pi} = \begin{bmatrix} \pi_2 & \pi_3 \end{bmatrix},$$
$$\mathcal{U}_2 = \Big[\underbrace{0, 0, \cdots, 0}_{12}, \mathscr{P}_2^T, \mathscr{P}_2^T\Big]^T, \ \mathcal{V}_2 = \Big[\mathscr{Q}_2, \underbrace{0, 0, \cdots, 0}_{8}, Q_2 E, -Q_2, 0, 0, 0\Big],$$
$$\kappa_1 = c_1 + \tau_m\lambda^{\tau_m - 1}c_2 + \tau_M\lambda^{\tau_M - 1}c_3 + \tau_M\lambda^{\tau_M - 1}c_4 + \frac{(\tau_M - \tau_m)(\tau_M + \tau_m - 1)}{2}\lambda^{\tau_M - 2}c_4, \ \tilde{\mathcal{S}}_2 = \text{diag}\{\mathcal{S}_2, 3\mathcal{S}_2\},$$
$$\kappa_2 = \frac{(\tau_M - \tau_m)^2(\tau_M + \tau_m + 1)}{2}\lambda^{\tau_M - 1}c_5 + \frac{\tau_m^2(\tau_m + 1)}{2}\lambda^{\tau_m - 1}c_6. \quad (9.100)$$

Proof *Consider the following Lyapunov-Krasovskii functional*

$$V(k) = V_1(k) + V_2(k) + V_3(k) \quad (9.101)$$

where

$$
\begin{aligned}
V_1(k) &= \vec{x}^T(k)\mathcal{P}\vec{x}(k) + \frac{1}{\theta}\delta(k), \\
V_2(k) &= \sum_{l=k-\tau_m}^{k-1} \lambda^{k-l-1}\vec{x}^T(l)\mathcal{Q}_1\vec{x}(l) + \sum_{l=k-\tau_M}^{k-1} \lambda^{k-l-1}\vec{x}^T(l)\mathcal{Q}_2\vec{x}(l) \\
&\quad + \sum_{l=k-\tau(k)}^{k-1} \lambda^{k-l-1}\vec{x}^T(l)\mathcal{Q}_3\vec{x}(l) + \sum_{j=k-\tau_M+1}^{k-\tau_m} \sum_{l=j}^{k-1} \lambda^{k-l-1}\vec{x}^T(l)\mathcal{Q}_3\vec{x}(l), \\
V_3(k) &= (\tau_M - \tau_m) \sum_{j=k-\tau_M}^{k-\tau_m-1} \sum_{l=j}^{k-1} \lambda^{k-l-1}\eta^T(l)\mathcal{S}_1\eta(l) \\
&\quad + \tau_m \sum_{j=k-\tau_m}^{k-1} \sum_{l=j}^{k-1} \lambda^{k-l-1}\eta^T(l)\mathcal{S}_2\eta(l)
\end{aligned}
$$

with $\eta(k) = \vec{x}(k+1) - \vec{x}(k)$.

For notational simplicity, set

$$
\begin{aligned}
\phi(k) &= \begin{bmatrix} \rho^T(k) & \mu^T(k) & \xi^T(k) & \delta^{\frac{1}{2}}(k) \end{bmatrix}^T, \\
\rho(k) &= \big[\vec{x}^T(k) \quad \vec{x}^T(k-\tau_m) \quad \vec{x}^T(k-\tau(k)) \quad \vec{x}^T(k-\tau_M) \quad \aleph_{\vec{x}}^T(k,0,\tau_m) \\
&\qquad \aleph_{\vec{x}}^T(k,\tau_m,\tau(k)) \quad \aleph_{\vec{x}}^T(k,\tau(k),\tau_M) \quad f^T(\vec{x}(k)) \quad g^T(\vec{x}(k-\tau(k)))\big]^T
\end{aligned}
\tag{9.102}
$$

where $\aleph_{\vec{x}}(k,0,\tau_m)$, $\aleph_{\vec{x}}(k,\tau_m,\tau(k))$ *and* $\aleph_{\vec{x}}(k,\tau(k),\tau_M)$ *are defined in Lemma 9.6.*

Then, we can obtain

$$
\begin{aligned}
&V_1(k+1) - \lambda V_1(k) \\
=&\vec{x}^T(k+1)\mathcal{P}\vec{x}(k+1) - \lambda\vec{x}^T(k)\mathcal{P}\vec{x}(k) \\
&+ \frac{\varrho-\lambda}{\theta}\delta(k) + \frac{\sigma}{\theta}y^T(k)y(k) - \frac{1}{\theta}\xi^T(k)\xi(k),
\end{aligned}
\tag{9.103}
$$

$$
\begin{aligned}
&V_2(k+1) - \lambda V_2(k) \\
=& \sum_{l=k+1-\tau_m}^{k} \lambda^{k-l}\vec{x}^T(l)\mathcal{Q}_1\vec{x}(l) - \lambda \sum_{l=k-\tau_m}^{k-1} \lambda^{k-l-1}\vec{x}^T(l)\mathcal{Q}_1\vec{x}(l) \\
&+ \sum_{l=k+1-\tau_M}^{k} \lambda^{k-l}\vec{x}^T(l)\mathcal{Q}_2\vec{x}(l) - \lambda \sum_{l=k-\tau_M}^{k-1} \lambda^{k-l-1}\vec{x}^T(l)\mathcal{Q}_2\vec{x}(l) \\
&+ \sum_{l=k+1-\tau(k+1)}^{k} \lambda^{k-l}\vec{x}^T(l)\mathcal{Q}_3\vec{x}(l) - \lambda \sum_{l=k-\tau(k)}^{k-1} \lambda^{k-l-1}\vec{x}^T(l)\mathcal{Q}_3\vec{x}(l)
\end{aligned}
$$

$$
\begin{aligned}
&+\sum_{j=k+2-\tau_M}^{k+1-\tau_m}\sum_{l=j}^{k}\lambda^{k-l}\vec{x}^T(l)\mathcal{Q}_3\vec{x}(l)-\lambda\sum_{j=k-\tau_M+1}^{k-\tau_m}\sum_{l=j}^{k-1}\lambda^{k-l-1}\vec{x}^T(l)\mathcal{Q}_3\vec{x}(l)\\
\leq&\vec{x}^T(k)(\mathcal{Q}_1+\mathcal{Q}_2+(\tau_M-\tau_m+1)\mathcal{Q}_3)\vec{x}(k)-\lambda^{\tau_m}\vec{x}^T(k-\tau_m)\mathcal{Q}_1\vec{x}(k-\tau_m)\\
&-\lambda^{\tau_M}\vec{x}^T(k-\tau_M)\mathcal{Q}_2\vec{x}(k-\tau_M)-\lambda^{\tau_m}\vec{x}^T(k-\tau(k))\mathcal{Q}_3\vec{x}(k-\tau(k)) \quad (9.104)
\end{aligned}
$$

and

$$
\begin{aligned}
&V_3(k+1)-\lambda V_3(k)\\
=&(\tau_M-\tau_m)\sum_{j=k+1-\tau_M}^{k-\tau_m}\sum_{l=j}^{k}\lambda^{k-l}\eta^T(l)\mathcal{S}_1\eta(l)\\
&-\lambda(\tau_M-\tau_m)\sum_{j=k-\tau_M}^{k-\tau_m-1}\sum_{l=j}^{k-1}\lambda^{k-l-1}\eta^T(l)\mathcal{S}_1\eta(l)\\
&+\tau_m\sum_{j=k+1-\tau_m}^{k}\sum_{l=j}^{k}\lambda^{k-l}\eta^T(l)\mathcal{S}_2\eta(l)-\lambda\tau_m\sum_{j=k-\tau_m}^{k-1}\sum_{l=j}^{k-1}\lambda^{k-l-1}\eta^T(l)\mathcal{S}_2\eta(l)\\
\leq&\eta^T(k)((\tau_M-\tau_m)^2\mathcal{S}_1+\tau_m^2\mathcal{S}_2)\eta(k)\\
&-\lambda\tau_m\sum_{l=k-\tau_m}^{k-1}\eta^T(l)\mathcal{S}_2\eta(l)-\lambda^{\tau_m+1}(\tau_M-\tau_m)\sum_{l=k-\tau_M}^{k-\tau_m-1}\eta^T(l)\mathcal{S}_1\eta(l). \quad (9.105)
\end{aligned}
$$

From Lemma 9.6, we have

$$
\begin{aligned}
&-\lambda\tau_m\sum_{l=k-\tau_m}^{k-1}\eta^T(l)\mathcal{S}_2\eta(l)\\
\leq&-\lambda\begin{bmatrix}\Upsilon_1\\ \Upsilon_2\end{bmatrix}^T\begin{bmatrix}\mathcal{S}_2 & 0\\ 0 & 3\mathcal{S}_2\end{bmatrix}\begin{bmatrix}\Upsilon_1\\ \Upsilon_2\end{bmatrix} \quad (9.106)\\
=&-\lambda\rho^T(k)\pi_1\tilde{\mathcal{S}}_2\pi_1^T\rho(k)
\end{aligned}
$$

and

$$
\begin{aligned}
&-\lambda^{\tau_m+1}(\tau_M-\tau_m)\sum_{l=k-\tau_M}^{k-\tau_m-1}\eta^T(l)\mathcal{S}_1\eta(l)\\
=&-\lambda^{\tau_m+1}(\tau_M-\tau_m)\sum_{l=k-\tau_M}^{k-\tau(k)-1}\eta^T(l)\mathcal{S}_1\eta(l)\\
&-\lambda^{\tau_m+1}(\tau_M-\tau_m)\sum_{l=k-\tau(k)}^{k-\tau_m-1}\eta^T(l)\mathcal{S}_1\eta(l)\\
\leq&-\lambda^{\tau_m+1}\frac{\tau_M-\tau_m}{\tau_M-\tau(k)}\begin{bmatrix}\Upsilon_3\\ \Upsilon_4\end{bmatrix}^T\begin{bmatrix}\mathcal{S}_1 & 0\\ 0 & 3\mathcal{S}_1\end{bmatrix}\begin{bmatrix}\Upsilon_3\\ \Upsilon_4\end{bmatrix} \quad (9.107)
\end{aligned}
$$

$$
\begin{aligned}
& -\lambda^{\tau_m+1}\frac{\tau_M-\tau_m}{\tau(k)-\tau_m}\begin{bmatrix}\Upsilon_5\\ \Upsilon_6\end{bmatrix}^T\begin{bmatrix}\mathcal{S}_1 & 0\\ 0 & 3\mathcal{S}_1\end{bmatrix}\begin{bmatrix}\Upsilon_5\\ \Upsilon_6\end{bmatrix}\\
= & -\lambda^{\tau_m+1}\frac{\tau_M-\tau_m}{\tau_M-\tau(k)}\rho^T(k)\pi_2\tilde{\mathcal{S}}_1\pi_2^T\rho(k)\\
& -\lambda^{\tau_m+1}\frac{\tau_M-\tau_m}{\tau(k)-\tau_m}\rho^T(k)\pi_3\tilde{\mathcal{S}}_1\pi_3^T\rho(k)
\end{aligned}
$$

where π_1, π_2, π_3, $\tilde{\mathcal{S}}_1$, and $\tilde{\mathcal{S}}_2$ are defined in (9.100), and

$$
\begin{aligned}
&\Upsilon_1=\vec{x}(k)-\vec{x}(k-\tau_m),\ \Upsilon_2=\vec{x}(k)+\vec{x}(k-\tau_m)-\aleph_{\vec{x}}(k,0,\tau_m),\\
&\Upsilon_3=\vec{x}(k-\tau(k))-\vec{x}(k-\tau_M),\ \Upsilon_5=\vec{x}(k-\tau_m)-\vec{x}(k-\tau(k)),\\
&\Upsilon_4=\vec{x}(k-\tau(k))+\vec{x}(k-\tau_M)-\aleph_{\vec{x}}(k,\tau(k),\tau_M),\\
&\Upsilon_6=\vec{x}(k-\tau_m)+\vec{x}(k-\tau(k))-\aleph_{\vec{x}}(k,\tau_m,\tau(k)).
\end{aligned}
$$

Noting (9.94), by using Lemma 9.7, it can be further obtained that

$$
\begin{aligned}
& -\lambda^{\tau_m+1}\frac{\tau_M-\tau_m}{\tau_M-\tau(k)}\rho^T(k)\pi_2\tilde{\mathcal{S}}_1\pi_2^T\rho(k)\\
& -\lambda^{\tau_m+1}\frac{\tau_M-\tau_m}{\tau(k)-\tau_m}\rho^T(k)\pi_3\tilde{\mathcal{S}}_1\pi_3^T\rho(k)\\
\leq & -\lambda^{\tau_m+1}\rho^T(k)\tilde{\pi}\Omega\tilde{\pi}^T\rho(k)
\end{aligned}
\tag{9.108}
$$

where $\tilde{\pi}$ is defined in (9.100).

On the other hand, from the triggering condition (9.81), we have

$$
\frac{1}{\theta}\delta(k)+\sigma y^T(k)y(k)-\xi^T(k)\xi(k)\geq 0. \tag{9.109}
$$

Moreover, from Assumption 9.5, it is easy to obtain the following inequalities

$$
\begin{bmatrix}\vec{x}(k)\\ f(\vec{x}(k))\end{bmatrix}^T\begin{bmatrix}\mathscr{L}_{11} & \mathscr{L}_{12}\\ * & I\end{bmatrix}\begin{bmatrix}\vec{x}(k)\\ f(\vec{x}(k))\end{bmatrix}\leq 0, \tag{9.110}
$$

$$
\begin{bmatrix}\vec{x}(k-\tau(k))\\ g(\vec{x}(k-\tau(k)))\end{bmatrix}^T\begin{bmatrix}\mathscr{T}_{11} & \mathscr{T}_{12}\\ * & I\end{bmatrix}\begin{bmatrix}\vec{x}(k-\tau(k))\\ g(\vec{x}(k-\tau(k)))\end{bmatrix}\leq 0 \tag{9.111}
$$

where $\mathscr{L}_{11}$, $\mathscr{L}_{12}$, $\mathscr{T}_{11}$, and $\mathscr{T}_{12}$ are defined in (9.100).

By taking (9.109)–(9.111) into account, it follows from (9.101)–(9.108) that

$$
\begin{aligned}
& V(k+1)-\lambda V(k)-\mu^T(k)\mathcal{J}\mu(k)\\
\leq & \vec{x}^T(k+1)\mathcal{P}\vec{x}(k+1)-\lambda\vec{x}^T(k)\mathcal{P}\vec{x}(k)+\frac{\varrho-\lambda+\omega_1}{\theta}\delta(k)+(\frac{\sigma}{\theta}+\omega_1\sigma)\\
& \times y^T(k)y(k)-(\frac{1}{\theta}+\omega_1)\xi^T(k)\xi(k)-\mu^T(k)\mathcal{J}\mu(k)+\vec{x}^T(k)(\mathcal{Q}_1+\mathcal{Q}_2
\end{aligned}
$$

$$
\begin{aligned}
&+(\tau_M-\tau_m+1)\mathcal{Q}_3)\vec{x}(k)-\lambda^{\tau_m}\vec{x}^T(k-\tau_m)\mathcal{Q}_1\vec{x}(k-\tau_m)\\
&-\lambda^{\tau_M}\vec{x}^T(k-\tau_M)\mathcal{Q}_2\vec{x}(k-\tau_M)-\lambda^{\tau_m}\vec{x}^T(k-\tau(k))\mathcal{Q}_3\vec{x}(k-\tau(k))\\
&+\eta^T(k)((\tau_M-\tau_m)^2\mathcal{S}_1+\tau_m^2\mathcal{S}_2)\eta(k)-\rho^T(k)(\lambda\pi_1\tilde{\mathcal{S}}_2\pi_1^T\\
&+\lambda^{\tau_m+1}\tilde{\pi}\Omega\tilde{\pi}^T)\rho(k)-\omega_2\begin{bmatrix}\vec{x}(k)\\ f(\vec{x}(k))\end{bmatrix}^T\begin{bmatrix}\mathscr{L}_{11} & \mathscr{L}_{12}\\ * & I\end{bmatrix}\begin{bmatrix}\vec{x}(k)\\ f(\vec{x}(k))\end{bmatrix}\\
&-\omega_3\begin{bmatrix}\vec{x}(k-\tau(k))\\ g(\vec{x}(k-\tau(k)))\end{bmatrix}^T\begin{bmatrix}\mathscr{T}_{11} & \mathscr{T}_{12}\\ * & I\end{bmatrix}\begin{bmatrix}\vec{x}(k-\tau(k))\\ g(\vec{x}(k-\tau(k)))\end{bmatrix}\\
=&\phi^T(k)(\Sigma+\mathscr{X}_1^T(k)\mathcal{P}\mathscr{X}_1(k)+\mathscr{X}_2^T(k)\mathscr{S}\mathscr{X}_2(k))\phi(k)
\end{aligned}
\tag{9.112}
$$

where Σ *and* $\mathscr{S}$ *are defined in (9.100), and*

$$
\begin{aligned}
\mathscr{X}_1(k)=&\Big[\Xi_1+\mathscr{P}_1\Gamma_1(k)\mathscr{Q}_1+\mathscr{P}_2\Gamma_2(k)\mathscr{Q}_2,\underbrace{0,0,\cdots,0}_{6},\mathcal{D},\mathcal{W},\\
&\qquad\Xi_2+\mathscr{P}_2\Gamma_2(k)Q_2E,\Xi_3-\mathscr{P}_2\Gamma_2(k)Q_2,0\Big],\\
\mathscr{X}_2(k)=&\Big[\Xi_1+\mathscr{P}_1\Gamma_1(k)\mathscr{Q}_1+\mathscr{P}_2\Gamma_2(k)\mathscr{Q}_2-I,\underbrace{0,0,\cdots,0}_{6},\mathcal{D},\mathcal{W},\\
&\qquad\Xi_2+\mathscr{P}_2\Gamma_2(k)Q_2E,\Xi_3-\mathscr{P}_2\Gamma_2(k)Q_2,0\Big].
\end{aligned}
$$

Denote

$$
\mathfrak{F}(k)=\begin{bmatrix}\Sigma & \mathscr{X}_1^T(k) & \mathscr{X}_2^T(k)\\ * & -\mathcal{P}^{-1} & 0\\ * & * & -\mathscr{S}^{-1}\end{bmatrix}
\tag{9.113}
$$

and rewrite it as

$$
\mathfrak{F}(k)=\Pi+\mathcal{U}_1\Gamma_1(k)\mathcal{V}_1+(\mathcal{U}_1\Gamma_1(k)\mathcal{V}_1)^T+\mathcal{U}_2\Gamma_2(k)\mathcal{V}_2+(\mathcal{U}_2\Gamma_2(k)\mathcal{V}_2)^T
\tag{9.114}
$$

where Π, $\mathcal{U}_1$, $\mathcal{U}_2$, $\mathcal{V}_1$, *and* $\mathcal{V}_2$ *are defined in (9.100).*

With help of Lemma 9.8, it can be obtained from (9.95) that $\mathfrak{F}(k)<0$ *which, by using the Schur complement, is equivalent to* $(\Sigma+\mathscr{X}_1^T(k)\mathcal{P}\mathscr{X}_1(k)+\mathscr{X}_2^T(k)\mathscr{S}\mathscr{X}_2(k))<0$. *Therefore, from (9.112), we have*

$$
V(k+1)<\lambda V(k)+\mu^T(k)\mathcal{J}\mu(k).
\tag{9.115}
$$

Based on the above inequality, we further have

$$
\begin{aligned}
V(k)&<\lambda V(k-1)+\lambda_{\max}\{\mathcal{J}\}\mu^T(k-1)\mu(k-1)\\
&<\lambda^N V(0)+\lambda_{\max}\{\mathcal{J}\}\sum_{l=0}^{N-1}\lambda^{N-l-1}\mu^T(l)\mu(l)\\
&<\lambda^N V(0)+\lambda^N\lambda_{\max}\{\mathcal{J}\}\hbar.
\end{aligned}
\tag{9.116}
$$

According to (9.96)–(9.98) and the definition of $V(k)$, we can obtain

$$V(0) < \kappa_1\alpha_1 + \kappa_2\beta + \frac{\delta_0}{\theta} \tag{9.117}$$

where κ_1 and κ_2 are defined in (9.100). Furthermore, it can be obtained from (9.96) and (9.101) that

$$V(k) \geq c_0\vec{x}^T(k)\mathcal{R}\vec{x}(k). \tag{9.118}$$

Substituting (9.117)–(9.118) into (9.116) and noting (9.99), we finally have

$$\vec{x}^T(k)\mathcal{R}\vec{x}(k) \leq \frac{\lambda^N(\kappa_1\alpha_1 + \kappa_2\beta + \frac{\delta_0}{\theta} + \lambda_{\max}\{\mathcal{J}\}\hbar)}{c_0} \leq \alpha_2. \tag{9.119}$$

Therefore, according to Definition 9.1, the estimation error system (9.86) is finite-time bounded with respect to $(\alpha_1, \alpha_2, N, \mathcal{R}, \hbar)$. This accomplishes the proof of Theorem 9.6.

Now, we are ready to deal with the H_∞ performance analysis issue for the estimation error system (9.86). In the following theorem, a sufficient condition is proposed to ensure both the finite-time boundedness and the H_∞ performance of the estimation error system (9.86).

Theorem 9.7 *Suppose that ϱ $(0 < \varrho < 1)$ and θ $(\theta > 0)$ satisfy $\varrho\theta \geq 1$. Given the state estimator gains A_f, B_f and C_f, a scalar λ $(\lambda > 1)$ and the H_∞ performance index γ, the estimation error system (9.86) is finite-time bounded with respect to $(\alpha_1, \alpha_2, N, \mathcal{R}, \hbar)$ and the H_∞ performance constraint is satisfied if there exist positive definite matrices $\mathcal{P}$, $\mathcal{Q}_1$, $\mathcal{Q}_2$, $\mathcal{Q}_3$, $\mathcal{S}_1$, $\mathcal{S}_2$, and $\mathcal{J}$, positive constant scalars c_i $(i = 0, 1, 2, 3, 4, 5, 6)$, ω_i $(i = 1, 2, 3)$, and $\tilde{e}_i$ $(i = 1, 2, 3)$, and matrix $\mathcal{Y}$ satisfying (9.94), (9.96)–(9.99) and*

$$\lambda^{-N}\gamma^2 I \leq \mathcal{J}, \tag{9.120}$$

$$\begin{bmatrix} \tilde{\Pi} & \tilde{\Theta}_1 \\ * & \tilde{\Theta}_2 \end{bmatrix} < 0 \tag{9.121}$$

where

$$\tilde{\Pi} = \begin{bmatrix} \tilde{\Sigma} & \bar{\mathscr{X}}_1^T & \bar{\mathscr{X}}_2^T & \bar{\mathscr{X}}_3^T \\ * & -\mathcal{P}^{-1} & 0 & 0 \\ * & * & -\mathscr{S}^{-1} & 0 \\ * & * & * & -I \end{bmatrix}, \quad \tilde{\Sigma} = \begin{bmatrix} \Sigma_{11} & \Sigma_{12} & 0 & 0 \\ * & \tilde{\Sigma}_{22} & 0 & 0 \\ * & * & \Sigma_{33} & 0 \\ * & * & * & \Sigma_{44} \end{bmatrix},$$

$$\tilde{\Theta}_1 = \begin{bmatrix} \tilde{\mathcal{U}}_1 & \tilde{e}_1\tilde{\mathcal{V}}_1^T & \tilde{\mathcal{U}}_2 & \tilde{e}_2\tilde{\mathcal{V}}_2^T & \tilde{\mathcal{U}}_3 & \tilde{e}_3\tilde{\mathcal{V}}_3^T \end{bmatrix}, \ \tilde{\Theta}_2 = \mathrm{diag}\{-\tilde{e}_1 I, -\tilde{e}_2 I, -\tilde{e}_3 I\},$$

$$\tilde{\Sigma}_{22} = (\frac{\sigma}{\theta} + \omega_1\sigma)E^T E - \lambda^{-N}\gamma^2 I, \ \tilde{\mathcal{U}}_3 = \Big[\underbrace{0, 0, \cdots, 0}_{14}, P_3^T\Big]^T,$$

$$\tilde{\mathscr{X}}_3 = \Big[\Xi_4, \underbrace{0, 0, \cdots, 0}_{11}\Big], \quad \tilde{\mathcal{V}}_1 = \Big[\mathscr{Q}_1, \underbrace{0, 0, \cdots, 0}_{14}\Big],$$
$$\tilde{\mathcal{U}}_1 = \Big[\underbrace{0, 0, \cdots, 0}_{12}, \mathscr{P}_1^T, \mathscr{P}_1^T, 0\Big]^T, \tilde{\mathcal{U}}_2 = \Big[\underbrace{0, 0, \cdots, 0}_{12}, \mathscr{P}_2^T, \mathscr{P}_2^T, 0\Big]^T,$$
$$\tilde{\mathcal{V}}_2 = \Big[\mathscr{Q}_2, \underbrace{0, 0, \cdots, 0}_{8}, Q_2E, -Q_2, 0, 0, 0, 0\Big], \tilde{\mathcal{V}}_3 = \Big[-\mathscr{Q}_3, \underbrace{0, 0, \cdots, 0}_{14}\Big]. \tag{9.122}$$

and other parameters are defined in (9.100).

Proof *Consider the same Lyapunov-Krasovskii functional as (9.101). By the similar methods used in Theorem 9.6, we can have that*

$$\begin{aligned}
&\vec{z}^T(k)\vec{z}(k) - \lambda^{-N}\gamma^2\mu^T(k)\mu(k) + V(k+1) - \lambda V(k)\\
=&\vec{x}^T(k)(\Xi_4 - P_3\Gamma_3(k)\mathscr{Q}_3)^T(\Xi_4 - P_3\Gamma_3(k)\mathscr{Q}_3)\vec{x}(k)\\
&- \lambda^{-N}\gamma^2\mu^T(k)\mu(k) + V(k+1) - \lambda V(k)\\
\leq&\phi^T(k)(\tilde{\Sigma} + \mathscr{X}_1^T(k)\mathcal{P}\mathscr{X}_1(k) + \mathscr{X}_2^T(k)\mathscr{S}\mathscr{X}_2(k) + \mathscr{X}_3^T(k)I\mathscr{X}_3(k))\phi(k)
\end{aligned} \tag{9.123}$$

where

$$\mathscr{X}_3(k) = \Big[\Xi_4 - P_3\Gamma_3(k)\mathscr{Q}_3, \underbrace{0, 0, \cdots, 0}_{11}\Big]. \tag{9.124}$$

Denote

$$\tilde{\mathfrak{F}}(k) = \begin{bmatrix} \tilde{\Sigma} & \mathscr{X}_1^T(k) & \mathscr{X}_2^T(k) & \mathscr{X}_3^T(k) \\ * & -\mathcal{P}^{-1} & 0 & 0 \\ * & * & -\mathscr{S}^{-1} & 0 \\ * & * & * & -I \end{bmatrix}. \tag{9.125}$$

It is not difficult to write (9.125) as

$$\begin{aligned}
\tilde{\mathfrak{F}}(k) =&\tilde{\Pi} + \tilde{\mathcal{U}}_1\Gamma_1(k)\tilde{\mathcal{V}}_1 + (\tilde{\mathcal{U}}_1\Gamma_1(k)\tilde{\mathcal{V}}_1)^T + \tilde{\mathcal{U}}_2\Gamma_2(k)\tilde{\mathcal{V}}_2\\
&+ (\tilde{\mathcal{U}}_2\Gamma_2(k)\tilde{\mathcal{V}}_2)^T + \tilde{\mathcal{U}}_3\Gamma_3(k)\tilde{\mathcal{V}}_3 + (\tilde{\mathcal{U}}_3\Gamma_3(k)\tilde{\mathcal{V}}_3)^T
\end{aligned} \tag{9.126}$$

where $\tilde{\Pi}$, $\tilde{\mathcal{U}}_1$, $\tilde{\mathcal{U}}_2$, $\tilde{\mathcal{U}}_3$, $\tilde{\mathcal{V}}_1$, $\tilde{\mathcal{V}}_2$, and $\tilde{\mathcal{V}}_3$ are defined in (9.122).

Then, by Lemma 9.8, it follows from (9.121) that $\tilde{\mathfrak{F}}(k) < 0$. By using the Schur complement, it is easy to obtain that $(\tilde{\Sigma} + \mathscr{X}_1^T(k)\mathcal{P}\mathscr{X}_1(k) + \mathscr{X}_2^T(k)\mathscr{S}\mathscr{X}_2(k) + \mathscr{X}_3^T(k)I\mathscr{X}_3(k)) < 0$, which means

$$\vec{z}^T(k)\vec{z}(k) - \lambda^{-N}\gamma^2\mu^T(k)\mu(k) + V(k+1) - \lambda V(k) < 0. \tag{9.127}$$

Noticing (9.120), it is obvious to see from the above inequality that

$$\begin{aligned}
&V(k+1) - \lambda V(k) - \mu^T(k)\mathcal{J}\mu(k)\\
<&-\vec{z}^T(k)\vec{z}(k) + \lambda^{-N}\gamma^2\mu^T(k)\mu(k) - \mu^T(k)\mathcal{J}\mu(k)\\
<&0.
\end{aligned} \tag{9.128}$$

Therefore, it follows directly from Theorem 9.6 that the estimation error system (9.86) is finite-time bounded with respect to $(\alpha_1, \alpha_2, N, \mathcal{R}, \hbar)$.

Moreover, under the zero initial condition, it follows from (9.127) that

$$\begin{aligned} V(k+1) <& \lambda V(k) + \lambda^{-N}\gamma^2\mu^T(k)\mu(k) - \vec{z}^T(k)\vec{z}(k) \\ <& \lambda^{k+1}V(0) + \lambda^{-N}\gamma^2\sum_{l=0}^{k}\lambda^{k-l}\mu^T(l)\mu(l) - \sum_{l=0}^{k}\lambda^{k-l}\vec{z}^T(l)\vec{z}(l) \\ =& \lambda^{-N}\gamma^2\sum_{l=0}^{k}\lambda^{k-l}\mu^T(l)\mu(l) - \sum_{l=0}^{k}\lambda^{k-l}\vec{z}^T(l)\vec{z}(l). \end{aligned} \tag{9.129}$$

Considering $V(N+1) \geq 0$ *and* $\lambda > 1$, *we obtain*

$$\sum_{l=0}^{N}\vec{z}^T(l)\vec{z}(l) < \gamma^2\sum_{l=0}^{N}\mu^T(l)\mu(l). \tag{9.130}$$

The proof is accomplished.

Finally, the design method of the desired resilient H_∞ state estimator is proposed in the following theorem.

Theorem 9.8 *Assume that* ϱ $(0 < \varrho < 1)$ *and* θ $(\theta > 0)$ *satisfy* $\varrho\theta \geq 1$. *Given a scalar* λ $(\lambda > 1)$ *and the* H_∞ *performance index* γ, *the estimation error system (9.86) is finite-time bounded with respect to* $(\alpha_1, \alpha_2, N, \mathcal{R}, \hbar)$ *and the* H_∞ *performance constraint is met if there exist positive definite matrices* $\mathcal{P}$, $\mathcal{Q}_1$, $\mathcal{Q}_2$, $\mathcal{Q}_3$, $\mathcal{S}_1$, $\mathcal{S}_2$, *and* $\mathcal{J}$, *positive constant scalars* c_i $(i = 0, 1, 2, 3, 4, 5, 6)$, ω_i $(i = 1, 2, 3)$, *and* $\tilde{e}_i$ $(i = 1, 2, 3)$, *and matrices* $\mathcal{Y}$, $\mathcal{Z}_1$, $\mathcal{Z}_2$, $\mathcal{Z}_3$, *and* $\mathcal{G} = \begin{bmatrix} \mathcal{G}_1 & \mathcal{G}_3 \\ \mathcal{G}_2 & \mathcal{G}_3 \end{bmatrix}$ *satisfying (9.94), (9.96)–(9.99), (9.120) and*

$$\begin{bmatrix} \bar{\Pi} & \tilde{\Theta}_1^{\mathcal{G}} \\ * & \tilde{\Theta}_2 \end{bmatrix} < 0 \tag{9.131}$$

where

$$\bar{\Pi} = \begin{bmatrix} \tilde{\Sigma} & X_1^{\mathcal{G}T} & X_2^{\mathcal{G}T} & X_3^T \\ * & \mathcal{P} - \mathcal{G} - \mathcal{G}^T & 0 & 0 \\ * & * & \mathscr{S} - \mathcal{G} - \mathcal{G}^T & 0 \\ * & * & * & -I \end{bmatrix},$$

$$\tilde{\Theta}_1^{\mathcal{G}} = \begin{bmatrix} \tilde{\mathcal{U}}_1^{\mathcal{G}} & \tilde{e}_1\tilde{\mathcal{V}}_1^T & \tilde{\mathcal{U}}_2^{\mathcal{G}} & \tilde{e}_2\tilde{\mathcal{V}}_2^T & \tilde{\mathcal{U}}_3 & \tilde{e}_3\tilde{\mathcal{V}}_3^T \end{bmatrix},$$

$$X_1^{\mathcal{G}} = \Big[\Xi_1^{\mathcal{G}}, \underbrace{0, 0, \cdots, 0}_{6}, \mathcal{D}^{\mathcal{G}}, \mathcal{W}^{\mathcal{G}}, \Xi_2^{\mathcal{G}}, \Xi_3^{\mathcal{G}}, 0\Big], \quad \mathcal{D}^{\mathcal{G}} = \begin{bmatrix} \mathcal{G}_1 D & 0 \\ \mathcal{G}_2 D & 0 \end{bmatrix},$$

$$X_2^{\mathcal{G}} = \Big[\Xi_1^{\mathcal{G}} - \mathcal{G}, \underbrace{0, 0, \cdots, 0}_{6}, \mathcal{D}^{\mathcal{G}}, \mathcal{W}^{\mathcal{G}}, \Xi_2^{\mathcal{G}}, \Xi_3^{\mathcal{G}}, 0\Big], \quad \mathcal{W}^{\mathcal{G}} = \begin{bmatrix} \mathcal{G}_1 W & 0 \\ \mathcal{G}_2 W & 0 \end{bmatrix},$$

$$X_3 = \begin{bmatrix} \Xi_4, \underbrace{0, 0, \cdots, 0}_{11} \end{bmatrix}, \quad \Xi_1^{\mathcal{G}} = \begin{bmatrix} \mathcal{G}_1 A + \mathcal{Z}_2 M & \mathcal{Z}_1 - \mathcal{Z}_2 M \\ \mathcal{G}_2 A + \mathcal{Z}_2 M & \mathcal{Z}_1 - \mathcal{Z}_2 M \end{bmatrix},$$

$$\Xi_2^{\mathcal{G}} = \begin{bmatrix} \mathcal{G}_1 L + \mathcal{Z}_2 E \\ \mathcal{G}_2 L + \mathcal{Z}_2 E \end{bmatrix}, \quad \Xi_3^{\mathcal{G}} = \begin{bmatrix} -\mathcal{Z}_2 \\ -\mathcal{Z}_2 \end{bmatrix}, \quad \Xi_4 = \begin{bmatrix} C & -\mathcal{Z}_3 \end{bmatrix},$$

$$\tilde{\mathcal{U}}_1^{\mathcal{G}} = \begin{bmatrix} \underbrace{0, 0, \cdots, 0}_{12}, \mathscr{P}_1^{\mathcal{G}T}, \mathscr{P}_1^{\mathcal{G}T}, 0 \end{bmatrix}^T, \quad \mathscr{P}_1^{\mathcal{G}} = \begin{bmatrix} \mathcal{G}_3 P_1 \\ \mathcal{G}_3 P_1 \end{bmatrix},$$

$$\tilde{\mathcal{U}}_2^{\mathcal{G}} = \begin{bmatrix} \underbrace{0, 0, \cdots, 0}_{12}, \mathscr{P}_2^{\mathcal{G}T}, \mathscr{P}_2^{\mathcal{G}T}, 0 \end{bmatrix}^T, \quad \mathscr{P}_2^{\mathcal{G}} = \begin{bmatrix} \mathcal{G}_3 P_2 \\ \mathcal{G}_3 P_2 \end{bmatrix}. \tag{9.132}$$

and other parameters are defined in (9.122). Moreover, if the inequality (9.131) is solvable, the desired state estimator gains are determined by

$$A_f = \mathcal{G}_3^{-1} \mathcal{Z}_1, \quad B_f = \mathcal{G}_3^{-1} \mathcal{Z}_2 \quad \text{and} \quad C_f = \mathcal{Z}_3. \tag{9.133}$$

Proof *First of all, it is easy to verify that the following inequalities*

$$-\mathcal{G}\mathcal{P}^{-1}\mathcal{G}^T \leq \mathcal{P} - \mathcal{G} - \mathcal{G}^T, \quad -\mathcal{G}\mathscr{S}^{-1}\mathcal{G}^T \leq \mathscr{S} - \mathcal{G} - \mathcal{G}^T \tag{9.134}$$

hold.

According to the above inequalities, it follows from (9.131) that the following inequality holds

$$\begin{bmatrix} \vec{\Pi} & \tilde{\Theta}_1^{\mathcal{G}} \\ * & \tilde{\Theta}_2 \end{bmatrix} < 0 \tag{9.135}$$

where

$$\vec{\Pi} = \begin{bmatrix} \tilde{\Sigma} & X_1^{\mathcal{G}T} & X_2^{\mathcal{G}T} & X_3^T \\ * & -\mathcal{G}\mathcal{P}^{-1}\mathcal{G}^T & 0 & 0 \\ * & * & -\mathcal{G}\mathscr{S}^{-1}\mathcal{G}^T & 0 \\ * & * & * & -I \end{bmatrix}. \tag{9.136}$$

Performing a congruence transformation to (9.135) with $\mathrm{diag}\{I, \mathcal{G}^{-1}, \mathcal{G}^{-1}, I, I\}$*, we can easily obtain (9.121). Therefore, the rest of the proof of Theorem 9.8 can be accomplished by following that of Theorem 9.7.*

Remark 9.15 *In this section, it is the estimator parameter variations, dynamic ET mechanism as well as the finite-time H_∞ performance constraints that bring essential distinctions of our proposed algorithm from the existing ones. The advantages of the estimation algorithm proposed can be summarized as follows: i) the estimation algorithm proposed is capable of handling the case of gain uncertainties which caters for more requirements of practical engineering; ii) the estimation algorithm proposed involves the dynamic ET mechanism, which can effectively save the computation resources; and iii)* *the estimation algorithm proposed guarantees that the estimation performance is achieved within finite time, which is more significant in the practical application.*

To this end, the finite-time resilient H_∞ state estimation problem has been solved for a class of discrete-time delayed neural networks under the dynamic ET mechanism and the desired state estimator gains have been obtained in terms of the solutions to a set of LMIs.

9.4 Illustrative Examples

We use some simulation examples in this section to validate the effectiveness of the filtering/estimation algorithms proposed in this chapter.

9.4.1 Example 1

Consider a time-varying stochastic system described by (9.1) and (9.2) with the following parameters:

$$
\begin{aligned}
A_k &= \begin{bmatrix} 0.12 & -0.01 + 0.01\cos(k/30) \\ 0.01 & 0.12 \end{bmatrix}, \\
B_k &= \begin{bmatrix} 0.15 \\ 0.18 \end{bmatrix}, \; E_k = \begin{bmatrix} 1 \\ 1 \end{bmatrix}, \; F_k = \sin(k/30), \\
C_k &= \begin{bmatrix} 1.2 & 0.3 + 0.05\cos(k/30) \end{bmatrix}, \\
H_k &= \begin{bmatrix} 0.01\cos(k/30) & 0.02 \end{bmatrix}.
\end{aligned}
$$

The parameters of the dynamically triggering conditions (9.10) and (9.11) are chosen as $\sigma = 0.03$, $\lambda = 0.2$, and $\theta = 6$. The covariances of the process noise w_k and the measurement noise v_k are set as $R_k = 0.1$ and $Q_k = 0.3$, respectively. The threshold of measurements is taken as $\mathcal{I} = \begin{bmatrix} -0.75 & -0.5 & -0.25 \end{bmatrix}^T$. In this example, we set $a_k = 0.9$, $b_k = 0.5$, $d_k = e_k = f_k = 1$, $\varepsilon_1 = \varepsilon_{10} = 1$, $\varepsilon_2 = \varepsilon_3 = \varepsilon_4 = \varepsilon_5 = 0.01$, $\varepsilon_6 = 0.05$, and $\varepsilon_7 = \varepsilon_8 = \varepsilon_9 = \varepsilon_{11} = 0.7$. Based on the above parameters, the probability $\bar{\gamma}_k^m$ and the filter gain K_k can be computed according to (9.8) and (9.38), respectively.

The simulation results are shown in Figs. 9.1–9.6. Figs. 9.1 and 9.2 plot the states and their estimates. Figs. 9.3–9.5 depict the trace of MSE_k and its minimum upper bound $\bar{P}_{k|k}$ in the sense of logarithm (lg), where the mean square error (MSE) is defined by $\text{MSE}_k \triangleq \frac{1}{300}\sum_{t=1}^{300}\sum_{s=1}^{2}(x_k^s - \hat{x}_{k|k}^s)(x_k^s - \hat{x}_{k|k}^s)^T$. The event-based release instants for the measurements are displayed in Fig. 9.6. The simulation results illustrate the feasibility of the filtering algorithm proposed in Section 9.1.

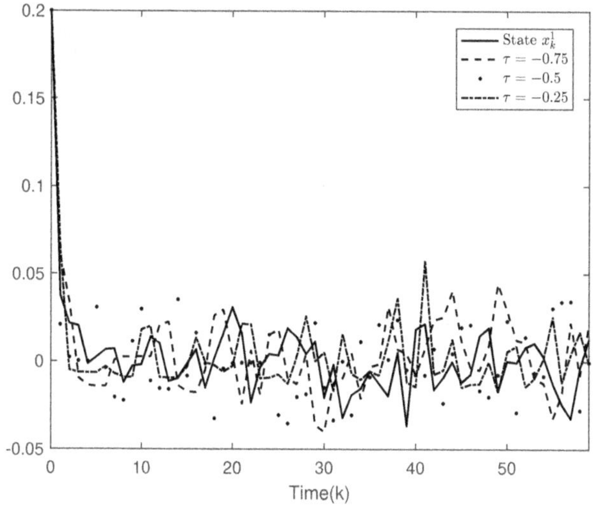

FIGURE 9.1
State x_k^1 and its estimate.

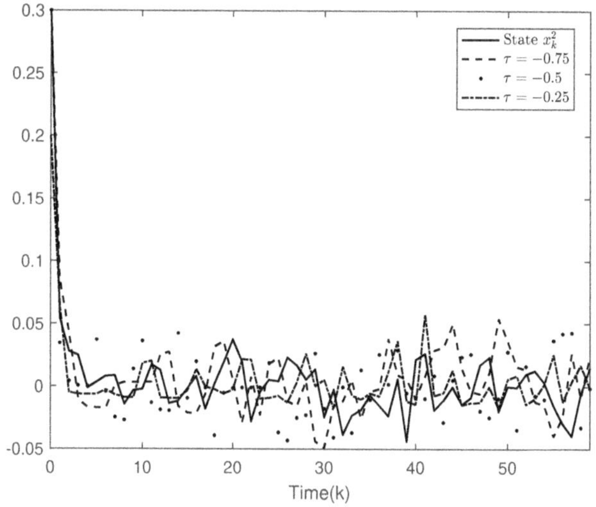

FIGURE 9.2
State x_k^2 and its estimate.

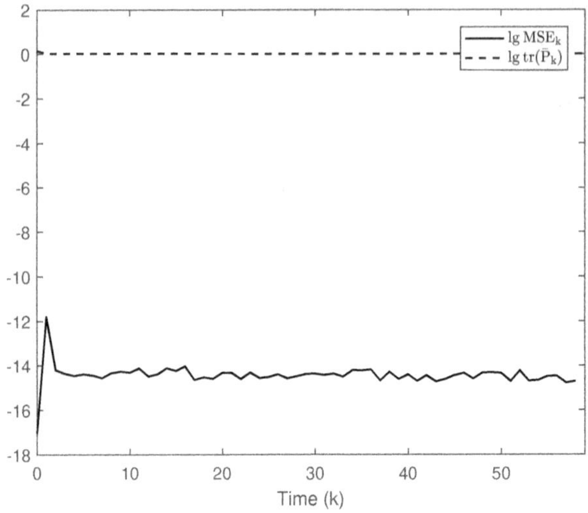

FIGURE 9.3
The logarithm of MSE_k and $\mathrm{tr}(\bar{P}_k)$.

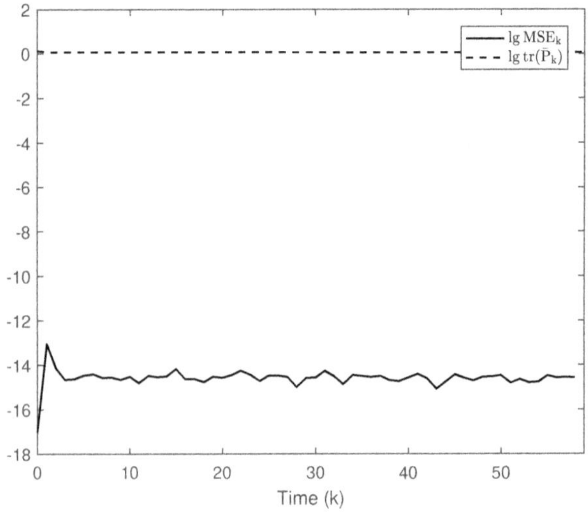

FIGURE 9.4
The logarithm of MSE_k and $\mathrm{tr}(\bar{P}_k)$.

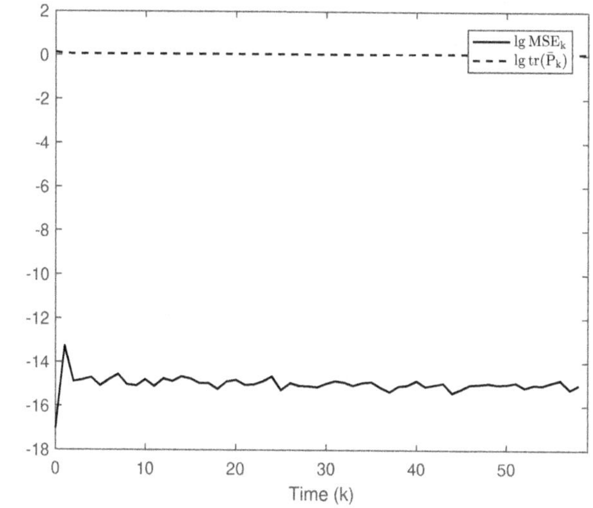

FIGURE 9.5
The logarithm of MSE_k and $\text{tr}(\bar{P}_k)$.

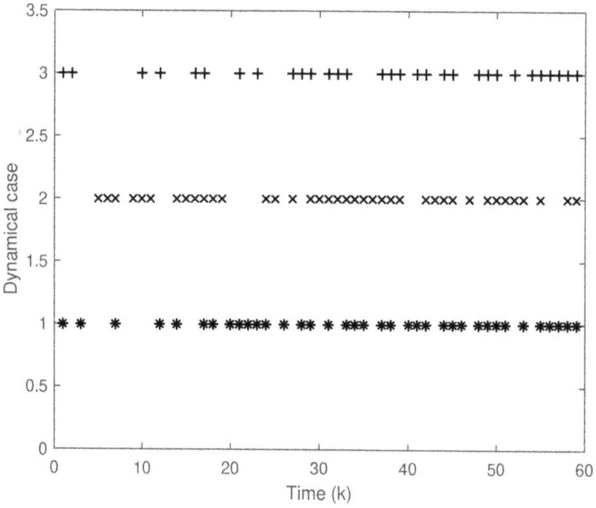

FIGURE 9.6
Event-based release instants.

9.4.2 Example 2

A. Demonstrations of Results

Given the finite time-horizon $[0, N]$ with $N = 60$, the parameters of the target plant (9.50) with the sensor measurement model (9.52) are set as follows:

$$\begin{aligned} &B_k = \begin{bmatrix}0.15 & 0.18\end{bmatrix}, \quad C_{1,k} = \begin{bmatrix}1 & 0.3+0.05\cos(k)\end{bmatrix}, \\ &D_{1,k} = 0.6, \quad C_{2,k} = \begin{bmatrix}0.8+0.05\sin(k) & 0.7\end{bmatrix}, \\ &D_{2,k} = 0.8, \quad D_{3,k} = 0.7, \quad D_{4,k} = 0.8, \\ &C_{3,k} = \begin{bmatrix}1+0.05\sin(k) & 0.9+0.05\cos(k)\end{bmatrix}, \, C_{4,k} = \begin{bmatrix}1.2 & 0.8\end{bmatrix}. \end{aligned}$$

The nonlinear function in the plant (9.50) is selected as $f_k(x_k) = E_k x_k + F_k(x_k)$, where

$$E_k = \begin{bmatrix}0.5 & -0.01+0.01\sin(k) \\ 0.15 & 0.4\end{bmatrix}, \quad F_k(x_k) = 0.1\sin(x_k).$$

Clearly, (9.51) is satisfied with $\gamma_k = 0.1$.

For the dynamic triggering condition (9.54) with (9.55), the thresholds are chosen as $\sigma_1 = \sigma_2 = \sigma_3 = \sigma_4 = 0.2$, and the other parameters are taken as $\lambda_1 = \lambda_2 = \lambda_3 = \lambda_4 = 0.2$ and $\theta_1 = \theta_2 = \theta_3 = \theta_4 = 5$. The covariances of the process noise w_k and the measurement noise v_k are set as $R_k = 0.6$ and $Q_k = 0.6$, respectively. Let the probabilities be $\alpha_{ij} = \alpha_0 = 0.6$ and $\beta_{ij} = \beta_0 = 0.6$ for $i, j = 1, 2, 3, 4$. The initial value of the state x_0 is uniformly distributed over $[-0.1, 0.1]$, and the initial value of the internal dynamic variable is taken as $\eta_0^1 = \eta_0^2 = \eta_0^3 = \eta_0^4 = 0$. Moreover, the adjusting parameters are chosen as $h_{i,k} = b_{i,k} = c_{i,k} = d_k = f_{i,k} = g_{i,k} = m_{i,k} = 1$. Based on these parameters, the filter parameters $K_{i,k}$ and $G_{i,k}$, for $i = 1, 2, 3, 4$, can be calculated at each iteration according to (9.71) and (9.72).

Note that the trace of the minimal upper bound $\bar{\mathcal{P}}_k^i$ is an upper bound of the mean square error for the estimation at time instant k on node i, which is defined by $\text{MSE}_{i,k} = \frac{1}{n}\sum_{t=1}^{n}\sum_{s=1}^{2}(x_{s,k} - \hat{x}_{is,k})^2$, where n is the implementation number of the independent experiments. In the simulation, the experiments are independently implemented $n = 300$ times and each experiment is run from time instant $k = 0$ to $k = N = 60$. Simulation results are shown in Figs. 9.7–9.9. Fig. 9.7 depicts the state trajectories and their estimates for state $x_{1,k}$ and $x_{2,k}$. For node i $(i = 1, 2, 3, 4)$, Fig. 9.8 plots the trace of the minimal upper bound $\bar{\mathcal{P}}_k^i$ and the mean square error (MSE) for the estimation of the state. The triggering instants of each sensor node under the dynamic ET mechanism can be seen in Fig. 9.9. The simulation results have illustrated the feasibility of the dynamic ET filtering algorithm proposed in this section.

B. Comparisons on the different triggering parameters

For the purpose of comparison, we consider three cases including the dynamic event-triggering, static event-triggering and periodic sampling on

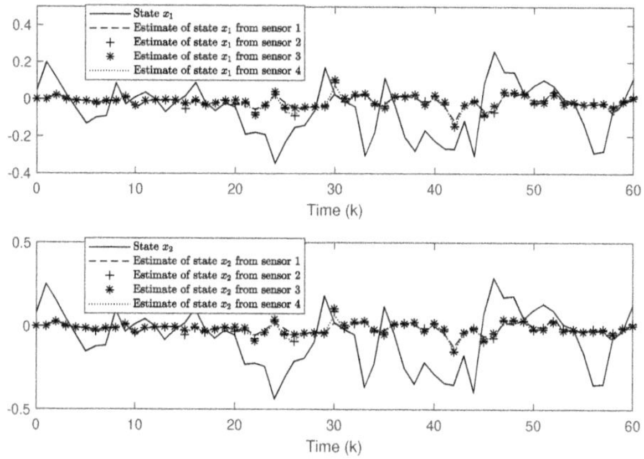

FIGURE 9.7
States x_j $(j = 1, 2)$ and their estimates.

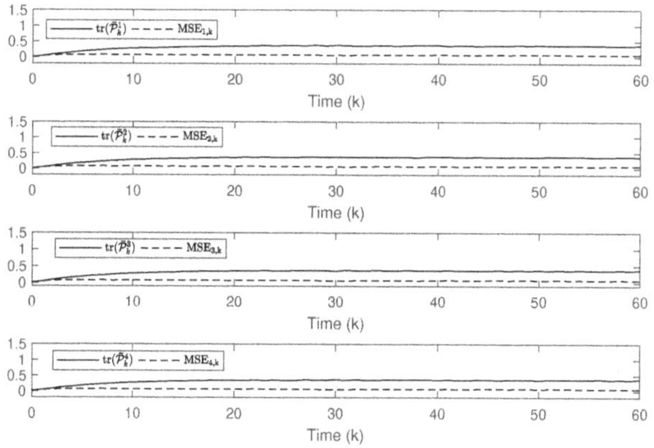

FIGURE 9.8
Trace of error variance and its upper bound for nodes 1, 2, 3, and 4.

the distributed filtering problem over sensor network described in the above subsection. Actually, the dynamic triggering case considered in this chapter reduces to the static triggering one when $\theta_i \to +\infty$ and the periodic sampling case is a special case of the static triggering case when $\sigma_i = 0$. As such, our experiments are implemented when the parameters are set as i) $\theta_i = 5$, $\sigma_i = 0.2$; ii) $\theta_i = 10$, $\sigma_i = 0.2$; iii) $\theta_i = +\infty$, $\sigma_i = 0.2$; and iv) $\theta_i = +\infty$,

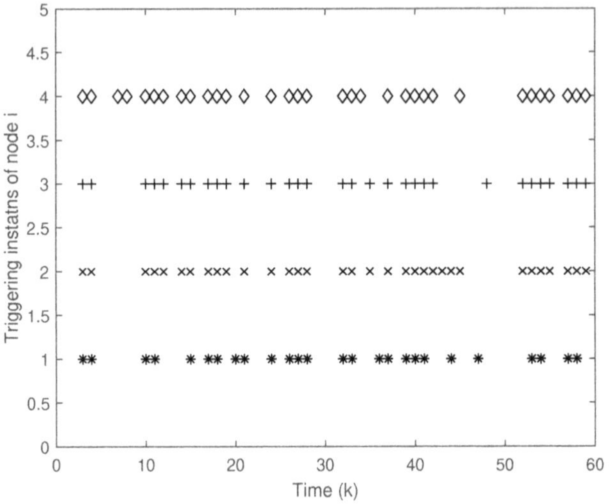

FIGURE 9.9
Triggering instants of sensor node i $(i = 1, 2, 3, 4)$.

$\sigma_i = 0$. In the comparison, the communication saving level is characterized by the triggering rate and the trace of the minimized upper bound of the FEC is employed to reflect the estimation performance.

Table 9.1 shows the triggering rates under different triggering cases over sensor nodes i $(i = 1, 2, 3, 4)$. From Table 9.1, it can be seen that the triggering rate monotonically increases as θ_i increases and the triggering rate in the dynamic event-triggering case is less than the static event-triggering and periodic sampling cases, which confirms that the dynamic event-triggering mechanism indeed improves the communication saving level.

The traces of the minimized upper bounds of the FEC under different triggering cases over sensor nodes i $(i = 1, 2, 3, 4)$ are shown in Fig. 9.10. It can be seen from Fig. 9.10 that the trace of the minimized upper bound monotonically decreases as θ_i increases and the trace of the minimized upper bound in the dynamic ET case is bigger than the ones in other two cases. That is, the estimation performance in the dynamic ET case deteriorates a bit comparing with the static ET and periodic sampling cases, which conforms to our intuition since less measurement data is used in the dynamic ET case.

9.4.3 Example 3

In this subsection, a numerical simulation example is given to illustrate the effectiveness of the proposed finite-time resilient H_∞ state estimation

	Node 1	Node 2	Node 3	Node 4
Dynamic ET case ($\theta_i = 5, \sigma_i = 0.2$)	43.3%	55%	51.7%	55%
Dynamic ET case ($\theta_i = 10, \sigma_i = 0.2$)	43.3%	56.7%	53.3%	56.7%
Static ET case ($\theta_i = +\infty, \sigma_i = 0.2$)	45%	61.7%	58.3%	56.7%
Periodic sampling case ($\theta_i = +\infty, \sigma_i = 0$)	100%	100%	100%	100%

TABLE 9.1
Triggering rates for node i ($i = 1, 2, 3, 4$).

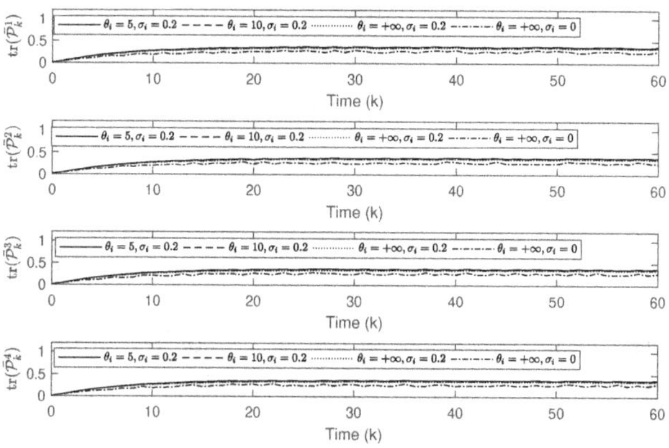

FIGURE 9.10
Traces of the minimal upper bounds with different triggering parameters.

scheme for the discrete-time delayed neural networks under the dynamic ET mechanism.

Consider a class of discrete-time delayed neural networks described by (9.79) with the following parameters

$$A = \begin{bmatrix} 0.42 & 0 & 0 \\ 0 & 0.31 & 0 \\ 0 & 0 & -0.2 \end{bmatrix}, \quad D = \begin{bmatrix} -0.3 & 0.2 & 0.2 \\ 0.3 & -0.2 & 0.4 \\ 0.2 & 0.3 & -0.3 \end{bmatrix},$$

$$W = \begin{bmatrix} -0.4 & 0.2 & 0.3 \\ 0.4 & -0.4 & 0.2 \\ 0.2 & 0.3 & -0.4 \end{bmatrix}, \quad L = \begin{bmatrix} 0.0068 \\ 0.0068 \\ 0.0068 \end{bmatrix},$$

$$M = \begin{bmatrix} 0.32 & 0.21 & 0.21 \end{bmatrix}, \quad E = 0.021, \quad C = \begin{bmatrix} 0.12 & -0.22 & 0.08 \end{bmatrix},$$

$$P_1 = \begin{bmatrix} 0.4 \\ 0.3 \\ -0.2 \end{bmatrix}, \quad P_2 = \begin{bmatrix} -0.3 \\ 0.5 \\ 0.2 \end{bmatrix}, \quad P_3 = -0.2,$$

$$Q_1 = \begin{bmatrix} 0.35 & -0.3 & 0.4 \end{bmatrix}, \quad Q_2 = 0.3, \quad Q_3 = \begin{bmatrix} 0.4 & -0.43 & 0.5 \end{bmatrix}.$$

The activation functions $f(x(k))$ and $g(x(k-\tau(k)))$ are chosen as follows

$$f(x(k)) = \begin{bmatrix} 0.2x_1(k) - \tanh(0.1x_1(k)) \\ -0.2x_2(k) - \tanh(0.2x_2(k)) \\ 0.3x_3(k) - \tanh(0.2x_3(k)) \end{bmatrix},$$

$$g(x(k-\tau(k))) = \begin{bmatrix} 0.3x_1(k-\tau(k)) - \tanh(0.2x_1(k-\tau(k))) \\ -0.2x_2(k-\tau(k)) - \tanh(0.1x_2(k-\tau(k))) \\ 0.1x_3(k-\tau(k)) - \tanh(0.3x_3(k-\tau(k))) \end{bmatrix}$$

from which it can be seen that the above nonlinear function satisfy (9.80) with

$$\mathcal{L}_1 = \begin{bmatrix} 0.2 & 0 & 0 \\ 0 & -0.2 & 0 \\ 0 & 0 & 0.3 \end{bmatrix}, \quad \mathcal{T}_1 = \begin{bmatrix} 0.3 & 0 & 0 \\ 0 & -0.2 & 0 \\ 0 & 0 & 0.1 \end{bmatrix},$$

$$\mathcal{L}_2 = \begin{bmatrix} 0.1 & 0 & 0 \\ 0 & -0.4 & 0 \\ 0 & 0 & 0.1 \end{bmatrix}, \quad \mathcal{T}_2 = \begin{bmatrix} 0.1 & 0 & 0 \\ 0 & -0.3 & 0 \\ 0 & 0 & -0.2 \end{bmatrix}.$$

Furthermore, the TVD is taken as $\tau(k) = 1+\cos^2(\frac{k\pi}{2})$ from which we have $\tau_m = 1$ and $\tau_M = 2$. For the dynamic triggering conditions (9.81) and (9.82), the initial value of the internal dynamic variable is set as $\delta_0 = 0$, the threshold is chosen as $\sigma = 0.6$ and the other parameters are taken as $\varrho = 0.8$ and $\theta = 6$. The H_∞ performance attenuation level is set as $\gamma = 0.9$. Several parameters in Theorem 9.8 are chosen as $\lambda = 1.1$, $\alpha_1 = 0.05$, $\beta = 0.05$, $\alpha_2 = 1.45$, $N = 18$, $\mathcal{R} = I$, and $\hbar = 0.25$. Then, according to (9.133), we can obtain the following parameters of the desired state estimator

$$A_f = \begin{bmatrix} 0.4685 & -0.0543 & 0.0637 \\ 0.0074 & 0.3233 & 0.0422 \\ 0.1929 & -0.0073 & 0.0084 \end{bmatrix}, \quad B_f = \begin{bmatrix} 0.0159 \\ 0.0013 \\ 0.0207 \end{bmatrix},$$

$$C_f = \begin{bmatrix} -0.1451 & 0.1832 & -0.0912 \end{bmatrix}.$$

In the simulation, we set $\mu(k) = 0.08\cos(k)e^{-0.1k}$, $\Gamma_1(k) = \Gamma_2(k) = \Gamma_3(k) = 0.5\cos(k)$. The initial state values are selected as $x_1(l) = 0.056$, $x_2(l) = 0.078$, and $x_3(l) = 0.12$ for $l \in [-2, 0]$. Simulation results are shown in Figs. 9.11–9.12. Fig. 9.11 shows the output $z(k)$ and its estimate. Fig. 9.12 shows the estimation error $\vec{z}(k)$. The simulation results have verified the usefulness of the estimation scheme.

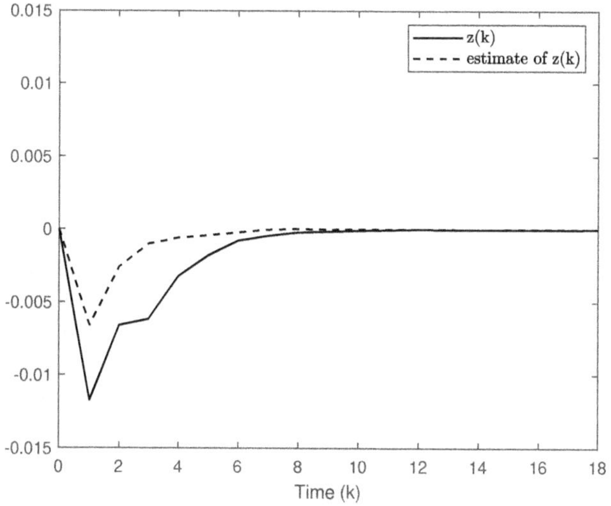

FIGURE 9.11
Output $z(k)$ and its estimate.

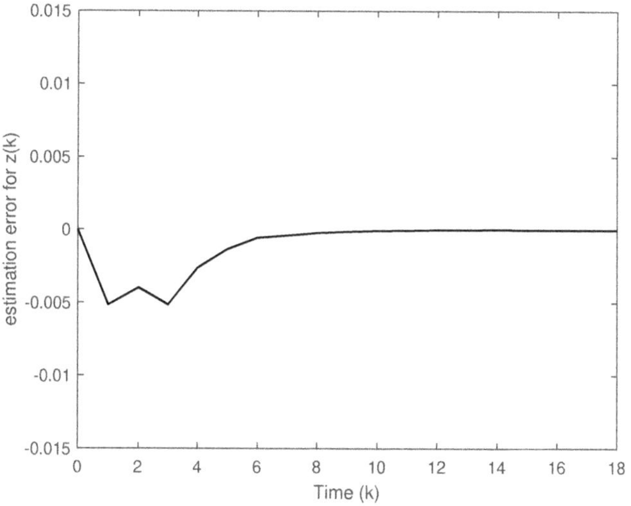

FIGURE 9.12
Estimation error for $z(k)$.

9.5 Summary

In this chapter, we have firstly investigated the dynamic ET filtering problem for a class of discrete time-varying systems with CMs and PUs. The CMs have been described by the Tobit measurement model and the PUs have been assumed to be norm bounded. A recursive filtering scheme has been proposed, where a recursion of the upper bound for the FEC has been obtained and minimized at each time instant by designing the filter gain. A sufficient condition has also been derived to guarantee the boundedness of estimation errors. Moreover, by recurring to the similar methods, some parallel results have been obtained for sensor networks subject to a time-varying topology through the Gilbert-Elliott channels characterized by a set of Markov chains. Furthermore, we also have studied the finite-time resilient H_∞ state estimation problem for a class of delayed neural networks under the dynamic ET mechanism. By using the Lyapunov functional approach, a sufficient condition has been derived to guarantee that the estimation error system is finite-time bounded and a prescribed H_∞ performance level is satisfied. Finally, via some simulation examples, we have shown the effectiveness of the dynamic ET filtering and state estimation approaches proposed in this chapter.

10

Conclusions and Future Work

This chapter draws conclusions on the book, and points out some possible research directions related to the work done in this paper.

10.1 Conclusions

The focus of this book has been placed on the control and state estimation problems for dynamical network systems with complex samplings. Specifically, several research problems have been investigated in detail.

- In Chapter 2, the stabilization problem has been studied for a class of sampled-data systems under noisy sampling intervals. We have converted the stochastic sampled-data control system into a discrete-time system. With the help of the matrix theories including matrix exponential computation, Vandermonde matrix, and Kronecker product operation, the sampled-data stabilization controller has been designed such that the stochastic sampled-data control system is stochastically stable. In addition, the stabilization problem has been discussed for a general class of stochastic sampled-data systems with multiple inputs under quantization and/or saturation effects, and a set of parallel results has been derived by using similar analysis techniques. Finally, some numerical simulation examples have been provided to demonstrate the effectiveness of the proposed design approach.
- In Chapter 3, the distributed sampled-data H_∞ state estimation problem has been discussed for a class of continuous-time systems with infinite-distributed delays and estimator parameter variations. A set of sensors has been deployed to measure the output of the plant and collaboratively share the measurement with their neighbors according to the given network topology. Then, the distributed state estimators have been designed based on the sampled measurement received by each sensor node. By taking advantage of the input delay approach, the effect of sampling intervals has been converted into an equivalent bounded TVD, and then a sufficient condition has been established to ensure both the asymptotical stability and the H_∞ performance requirement of the estimation error dynamics. By resorting

DOI: 10.1201/9781003307648-10

to the LMI approach, the parameters of the desired distributed estimators have been obtained. Finally, we have used a numerical simulation example to illustrate the validity of the design approach proposed.

- In Chapter 4, we have dealt with the ET control problem for two classes of switched systems with exogenous disturbances. In order to save the communication resource, the ET control strategy has been proposed by which the control signal is updated only when a certain condition is violated. The Lyapunov functional approach has been conducted to derive sufficient conditions for the existence of the desired controllers whose gain matrices have been obtained by solving two sets of LMIs. In the end, the results of this chapter have been demonstrated by some simulation examples.

- In Chapter 5, we have studied the ET H_∞ state estimation problem for state-saturated systems. Firstly, the distributed ET H_∞ state estimation problem has been considered for a class of state-saturated systems subject to ROMDs over sensor networks. The mixed delays comprising both discrete and distributed delays have been allowed to occur in a random way governed by two Bernoulli distributed random variables. Sufficient conditions for the exponential mean-square stability of the estimation error dynamics have been established and, at the same time, the prescribed H_∞ performance level has been guaranteed. Then, some parallel results have also been derived for a class of state-saturated complex networks in the simultaneous presence of quantization effects and randomly occurring distributed delays by using similar analysis techniques. Finally, we have used some illustrative examples to show the validity of the ET estimator design algorithm proposed in this chapter.

- In Chapter 6, the ET state estimation problem has been solved for discrete-time neural networks. For the purpose of energy saving, the ET mechanism has been adopted and the measurement outputs are only transmitted to the estimator when a certain triggered condition is met. Firstly, a new ET estimation technique has been developed for the multi-delayed neural networks with SPs and IMs. By using the Lyapunov functional approach, a sufficient condition has been given under which the estimation error dynamics is exponentially ultimately bounded in the mean square and the ultimate boundedness of the error dynamics has been given. Also, in this chapter, we have investigated the ET H_∞ state estimation problem for a class of discrete-time stochastic GRNs with MJPs and TVDs. By recurring to the similar methods, sufficient conditions have been derived to guarantee that the estimation error dynamics is exponentially stable and the H_∞ performance requirement is satisfied. At last, some numerical simulation examples are presented to illustrate the effectiveness of the proposed ET state estimation scheme.

- In Chapter 7, we have firstly investigated the ET robust fusion estimation problem for a class of uncertain multi-rate sampled-data systems. Both the stochastic nonlinearities and the coloured measurement noises have been taken into account. A new augmentation approach has been proposed to transform the multi-rate sampled-data system into a single-rate system. The local filters have been constructed, where the recursions of the upper bounds on the FECs have been guaranteed. By properly designing the filter parameters, such upper bounds have been then minimized at each iteration. Subsequently, by virtue of the modified CI method, the received local estimates have been fused together in the fusion centre. Furthermore, the ET fusion estimation has also been considered for multi-rate systems with sensor degradations. The phenomenon of sensor degradations has been characterized by a set of random variables with known probability distributions. The ET filter design problem has also been discussed. Finally, some simulation examples have been provided to verify the effectiveness of the fusion estimation scheme proposed.

- In Chapter 8, the dynamic ET synchronization control problem has been studied for a class of discrete time-delay complex dynamical networks. For the sake of energy saving, a new discrete-time version dynamic ET mechanism has been proposed and utilized in the design of synchronization controller. By constructing an appropriate Lyapunov functional, the dynamics of network nodes and the ET mechanism has been analyzed first in a unified way, and then a sufficient condition has been given to ensure exponentially ultimate boundedness of the synchronization error. With the obtained sufficient condition, the controller gains have been parameterised by means of the feasibility of a matrix inequality. A simulation example has been given to illustrate the effectiveness of the proposed dynamic ET synchronization control approach.

- In Chapter 9, we have firstly investigated the dynamic ET filtering problem for a class of discrete time-varying systems with CMs and PUs. The CMs have been described by the Tobit measurement model and the PUs have been assumed to be norm bounded. A recursive filtering scheme has been proposed, where a recursion of the upper bound for the FEC has been obtained and minimized at each time instant by designing the filter gain. A sufficient condition has also been derived to guarantee the boundedness of estimation errors. Moreover, by recurring to the similar methods, some parallel results have been obtained for sensor networks subject to a time-varying topology through the Gilbert-Elliott channels characterized by a set of Markov chains. Furthermore, we also have studied the finite-time resilient H_∞ state estimation problem for a class of delayed neural networks under the dynamic ET mechanism. By using the Lyapunov functional approach, a sufficient condition has been derived to guarantee that the estimation error system is finite-time bounded and a prescribed H_∞ performance level is satisfied. Finally, via some simulation examples, we have shown the effectiveness of the dynamic ET filtering and state estimation approaches proposed in this chapter.

10.2 Future Work

Related topics for the future research work are listed below:

- Further investigate the other new network-induced phenomena such as time-correlated fading channels and attempt to establish a unified measurement model to account for these phenomena simultaneously.
- Investigate the sampled-data control problem under the general noisy sampling interval by taking into account the network-induced phenomena.
- Investigate the problems of fault detection and fault tolerant control under ET or dynamic ET mechanisms.
- Investigate the security control and state estimation problems for dynamical network systems in the presence network attacks.
- Apply the theories and methodologies established in this research to study some practical engineering problems such as the mobile robot navigation.

Bibliography

[1] M. Abdelrahim, R. Postoyan, J. Daafouz, and D. Nešić. Covariance control problems over martingales with fixed terminal distribution arising from game theory. *SIAM Journal on Control and Optimization*, 51(2):1152–1185, 2013.

[2] M. Adès, P. E. Caines, and R. P. Malhamé. Stochastic optimal control under Poisson-distributed observations. *IEEE Transactions Automatic Control*, 45(1):3–13, 2000.

[3] T. Ahmed-Ali, E. Fridman, F. Giri, L. Burlion, and F. Lamnabhi-Lagarrigue. Using exponential time-varying gains for sampled-data stabilization and estimation. *Automatica*, 64:244–251, 2016.

[4] M. S. Ali, N. Gunasekaran, and Q. Zhu. State estimation of TCS fuzzy delayed neural networks with Markovian jumping parameters using sampled-data control. *Fuzzy Sets and Systems*, 361:87–104, 2017.

[5] M. S. Ali, M. Usha, Z. Orman, and S. Arik. Improved result on state estimation for complex dynamical networks with time varying delays and stochastic sampling via sampled-data control. *Neural Networks*, 114:28–37, 2019.

[6] B. Allik, C. Miller, M. J. Piovoso, and R. Zurakowski. The Tobit Kalman filter: An estimator for censored measurements. *IEEE Transactions on Control Systems Technology*, 24(1):365–371, 2016.

[7] A. Amir, A. Amir, and M. Arash. A unified optimization for resilient dynamic event-triggering consensus under denial of service. *IEEE Transactions on Cybernetics*, 2020. DOI: 10.1109/TCYB.2020.3022568.

[8] R. Anbuvithya, K. Mathiyalagan, and R. Sakthivel. Sampled-data state estimation for genetic regulatory networks with time-varying delays. *Neurocomputing*, 151:737–744, 2015.

[9] K.-E. Årzén. A simpled event based pid controller. In *Proceedings of the 14th IFAC World Congress, Bejing, China*, pages 423–428, 1999.

[10] N. W. Bauer, P. J. H. Maas, and W. P. M. H. Heemels. Stability analysis of networked control systems: A sum of squares approach. *Automatica*, 48(8):1514–1524, 2012.

[11] D. P. Borgers, V. S. Dolk, and W. P. M. H. Heemels. Riccati-based design of event-triggered controllers for linear systems with delays. *IEEE Transactions on Automatic Control*, 63(1):74–188, 2018.

[12] C. Briat. Convex conditions for robust stability analysis and stabilization of linear aperiodic impulsive and sampled-data systems under dwelltime constraints. *Automatica*, 49:3449–3457, 2013.

[13] Z. Cao, Y. Niu, H.-K. Lam, and J. Zhao. Sliding mode control of Markovian jump fuzzy systems: A dynamic event-triggered method. *IEEE Transactions on Fuzzy Systems*, 29(10):2902–2915, 2021.

[14] D. Chatterjee and D. Liberzon. Stabilizing randomly switched systems. *SIAM Journal on Control and Optimization*, 49(5):2008–2031, 2001.

[15] D. Chatterjee and D. Liberzon. Stability analysis of deterministic and stochastic switched systems via a comparison principle and multiple Lyapunov functions. *SIAM Journal on Control and Optimization*, 45(1):174–206, 2006.

[16] T. Chen and B. A. Francis. *Optimal Sampled-Data Control Systems*. Springer, 1999.

[17] W. Chen, J. Wang, D. Shi, and L. Shi. Event-based state estimation of hidden Markov models through a Gilbert-Elliott channel. *IEEE Transactions on Automatic Control*, 62(7):3626–3633, 2017.

[18] W. H. Chen and W. Zheng. Exponential stability of nonlinear time-delay systems with delayed impulse effects. *Automatica*, 47(5):1075–1083, 2011.

[19] W. H. Chen and W. Zheng. An improved stabilization method for sampled-data control systems with control packet loss. *IEEE Transactions on Automatic Control*, 57(9):2378–2384, 2012.

[20] C. Deng, M. Er, G.-H. Yang, and N. Wang. Event-triggered consensus of linear multiagent systems with time-varying communication delays. *IEEE Transactions on Cybernetics*, 50(7):2916–2925, 2020.

[21] Z. Deng, P. Zhang, Y. Gao, W. Qi, and J. Liu. The accuracy comparison of multisensor covariance intersection fuser and three weighting fusers. *Information Fusion*, 14(2):177–185, 2013.

[22] D. V. Dimarogonas, E. Frazzol, and K. H. Johansson. Distributed event-triggered control for multi-agent system. *IEEE Transactions on Automatic Control*, 57(5):1291–1297, 2012.

[23] D. Ding, Z. Wang, and Q.-L. Han. Neural-network-based consensus control for multiagent systems with input constraints: The event-triggered case. *IEEE Transactions on Cybernetics*, 50(8):3719–3730, 2020.

[24] H. Dong, X. Bu, N. Hou, Y. Liu, F. E. Alsaadi, and T. Hayat. Event-triggered distributed state estimation for a class of time-varying systems over sensor networks with redundant channels. *Information Fusion*, 36:243–250, 2017.

[25] H. Dong, Z. Wang, and H. Gao. Robust H_∞ filtering for a class of nonlinear networked systems with multiple stochastic communication delays and packet dropouts. *IEEE Transactions on Signal Processing*, 58(4):1957–1966, 2010.

[26] H. Dong, Z. Wang, and H. Gao. Distributed H_∞ filtering for a class of Markovian jump nonlinear time-delay systems over lossy sensor networks. *IEEE Transactions on Industrial Electronics*, 60(10):4665–4672, 2013.

[27] S. Dong, H. Zhu, S. Zhong, K. Shi, J. Cheng, and W. Kang. New result on reliable H_∞ performance state estimation for memory static neural networks with stochastic sampled-data communication. *Applied Mathematics and Computation*, 364:Art. No. 124619, 2020.

[28] M. C. F. Donkers and W. P. M. H. Heemels. Output-based event-triggered control with guaranteed $\mathcal{L}_\infty$-gain and improved and decentralized event-triggering. *IEEE Transactions on Automatic Control*, 57(6):1362–1376, 2012.

[29] S. Du, W. Xia, W. Ren, X.-M. Sun, and W. Wang. Observer-based consensus for multiagent systems under stochastic sampling mechanism. *IEEE Transactions on Systems, Man, and Cybernetics-Systems*, 48(12):2328–2338, 2018.

[30] S.-L. Du, T. Liu, and D. W. C. Ho. Dynamic event-triggered control for leader-following consensus of multiagent systems. *IEEE Transactions on Systems, Man, and Cybernetics-Systems*, 50(9):3243–3251, 2018.

[31] E. O. Elliott. Estimates of error rates for codes on burst-noise channels. *Bell System Technical Journal*, 42(5):1977–1997, 1963.

[32] L. Etienne, L. Hetel, D. Efimov, and M. Petreczky. Observer synthesis under time-varying sampling for Lipschitz nonlinear systems. *Automatica*, 85:433–440, 2017.

[33] S. Fan, H. Yan, X. Zhang, Z. Ge, and K. Shi. Distributed set-membership estimation for state-saturated systems with mixed time-delays via a dynamic event-triggered scheme. *Journal of the Franklin Institute*, 358(18):10079–10094, 2021.

[34] M. Farza, M. M'Saad, M. L. Fall, E. Pigeon, O. Gehan, and K. Busawon. Continuous-discrete time observers for a class of MIMO nonlinear systems. *IEEE Transactions on Automatic Control*, 59(4):1060–1065, 2014.

[35] Z. Fei, S. Shi, C. K Ahn, and M. V. Basin. Finite-time control for switched TšCS fuzzy systems via a dynamic event-triggered mechanism. *IEEE Transactions on Fuzzy Systems*, 298(12):3899–3909, 2021.

[36] E. Fridmana. A refined input delay approach to sampled-data control. *Automatica*, 46(2):421–427, 2010.

[37] E. Fridmana, A. Seuretb, and J.-P. Richardb. Robust sampled-data stabilization of linear systems: An input delay approach. *Automatica*, 40:1441–1446, 2004.

[38] M. Fu and L. Xie. The sector bound approach to quantized feedback control. *IEEE Transactions on Automatic Control*, 50(11):1689–1711, 2005.

[39] H. Fujioka. A discrete-time approach to stability analysis of systems with aperiodic sample-and-hold devices. *IEEE Transactions on Automatic Control*, 54(10):2440–2445, 2009.

[40] H. Fujioka. Stability analysis of systems with aperiodic sample-and-hold devices. *Automatica*, 45(3):771–775, 2009.

[41] H. Gao, T. Chen, and J. Lam. A new delay system approach to network-based control. *Automatica*, 44(1):39–52, 2008.

[42] H. Gao, J. Wu, and P. Shi. Robust sampled-data H_∞ control with stochastic sampling. *Automatica*, 45(7):1729–1736, 2009.

[43] E. Garcia and P. Antsaklis. Model-based event-triggered control for systems with quantization and time-varying network delays. *IEEE Transactions on Automatic Control*, 58(2):422–434, 2013.

[44] X. Ge and Q.-L. Han. Distributed sampled-data asynchronous H_∞ filtering of Markovian jump linear systems over sensor networks. *Signal Processing*, 127:86–99, 2016.

[45] X. Ge, Q.-L. Han, and X. Jiang. Sampled-data H_∞ filtering of Takagi-Sugeno fuzzy systems with interval time-varying delays. *Journal of the Franklin Institute*, 351(5):2515–2542, 2014.

[46] X. Ge, Q.-L. Han, and Z. Wang. A dynamic event-triggered transmission scheme for distributed set-membership estimation over wireless sensor networks. *IEEE Transactions on Cybernetics*, 49(1):171–183, 2019.

[47] X. Ge, Q.-L. Han, and Z. Wang. A threshold-parameter-dependent approach to designing distributed event-triggered H_∞ consensus filters over sensor networks. *IEEE Transactions on Cybernetics*, 49(4):1148–1159, 2019.

[48] R. H. Gielen, S. Olaru, M. Lazar, W. P. M. H. Heemels, and N. van de Wouw. On polytopic inclusions as a modeling framework for systems with time-varying delays. *Automatica*, 46(3):615–619, 2010.

[49] E. N. Gilbert. Capacity of a burst-noise channel. *Bell System Technical Journal*, 39(5):1253–1265, 1960.

[50] A. Girard. Dynamic triggering mechanisms for event-triggered control. *IEEE Transactions on Automatic Control*, 60(9):1992–1997, 2015.

[51] D. Han, Y. Mo, J. Wu, S. Weerakkody, B. Sinopoli, and L. Shi. Stochastic event-triggered sensor schedule for remote state estimation. *IEEE Transactions on Automatic Control*, 60(10):2661–2675, 2015.

[52] W. He, B. Zhang, Q. Han, F. Qian, J. Kurths, and J. Cao. Leader-following consensus of nonlinear multiagent systems with stochastic sampling. *IEEE Transactions on Cybernetics*, 47(2):327–338, 2017.

[53] X. He, Z. Wang, and D. Zhou. Robust H_∞ filtering for networked systems with multiple state delays. *International Journal of Control*, 80(8):1217–1232, 2007.

[54] W. P. M. H. Heemels, M. C. F. Donkers, and A. R. Teel. Periodic event-triggered control for linear systems. *IEEE Transactions on Automatic Control*, 58(4):847–861, 2013.

[55] T. Henningsson, E. Johannesson, and A. Cervin. Sporadic event-based control of first-order linear stochastic systems. *Automatica*, 44(11):2890–2895, 2008.

[56] A. Hu, J. Cao, M. Hu, and L. Guo. Cluster synchronization of complex networks via event-triggered strategy under stochastic sampling. *Physica A-Statistical Mechanics and Its Applications*, 434:99–110, 2015.

[57] J. Hu, N. Li, and X. Liu. Sampled-data state estimation for delayed neural networks with Markovian jumping parameters. *Nonlinear Dynamics*, 73(1):275–284, 2013.

[58] J. Hu, Z. Wang, H. Gao, and L. K. Stergioulas. Extended Kalman filtering with stochastic nonlinearities and multiple missing measurements. *Automatica*, 48(9):2007–2015, 2012.

[59] S. Hu, D. Yue, Q.-L. Han, X. Xie, X. Chen, and C. Dou. Observer-based event-triggered control for networked linear systems subject to denial-of-service attacks. *IEEE Transactions on Cybernetics*, 50(5):1952–1964, 2020.

[60] S. Hu, D. Yue, X. Yin, and X. Xie. Adaptive event-triggered control for nonlinear discrete-time systems. *International Journal of Robust and Nonlinear Control*, 26(18):4104–4125, 2016.

[61] Z. Hu, H. Ren, and P. Shi. Synchronization of complex dynamical networks subject to noisy sampling interval and packet loss. *IEEE Transactions on Neural Networks and Learning Systems*, 2021. DOI: 10.1109/TNNLS.2021.3051052.

[62] J. Huang, D. Shi, and T. Chen. Energy-based event-triggered state estimation for hidden Markov models. *Automatica*, 79:256–264, 2017.

[63] J. Huang, D. Shi, and T. Chen. Event-triggered state estimation with an energy harvesting sensor. *IEEE Transactions on Automatic Control*, 62(9):4768–4775, 2017.

[64] M. Huang and S. Dey. Stability of Kalman filtering with Markovian packet losses. *Automatica*, 43(4):598–607, 2007.

[65] X. Ji, T. Liu, Y. Sun, and H. Su. Stability analysis and controller synthesis for discrete linear time-delay systems with state saturation nonlinearities. *International Journal of Systems Science*, 42(3):397–406, 2011.

[66] Z. Jiang and Y. Wang. Input-to-state stability for discrete-time nonlinear systems. *Automatica*, 37(6):857–869, 2001.

[67] S. J. Julier and J. K. Uhlmann. *General Decentralized Data Fusion with Covariance Intersection.* in: M.E. Liggins, D. L. Hall, J. Llinas (Eds.), Handbook of Multisensor Data Fusion, Seconded., Theory and Practice, CRC Press, 2009.

[68] A. Kanchanaharuthai and M. Wongsaisuwan. Stochastic H_2-optimal controller design for sampled-data systems with random sampled measurement. In *Proceedings of the 41st SICE Annual Conference*, pages 2042–2047, 2002.

[69] C. Y. Kao and H. Fujioka. On stability of systems with aperiodic sampling devices. *IEEE Transactions on Automatic Control*, 58(8):2085–2090, 2013.

[70] J. Y. Keller and D. D. J. Sauter. Kalman filter for discrete-time stochastic linear systems subject to intermittent unknown inputs. *IEEE Transactions on Automatic Control*, 58(7):1882–1887, 2013.

[71] H. J. Kim, J. B. Park, and Y. H. Joo. Decentralized H_∞ sampled-data fuzzy filter for nonlinear interconnected oscillating systems with uncertain interconnections. *IEEE Transactions on Fuzzy Systems*, 28(3):487–498, 2020.

[72] R. Koike, T. Endo, and F. MatsuNo. Output-based dynamic event-triggered consensus control for linear multiagent systems. *Automatica*, 59:112–119, 2015.

[73] H. K. Lam. Stabilization of nonlinear systems using sampled-data output-feedback fuzzy controller based on polynomial-fuzzy-model-based control approach. *IEEE Transactions on Systems, Man, and Cybernetics, Part B-Cybernetics*, 42(1):258–267, 2012.

[74] L. Lee, Y. Liu, J. Liang, and X. Cai. Finite time stability of nonlinear impulsive systems and its applications in sampled-data systems. *ISA Transactions*, 57:172–178, 2015.

[75] T. H. Lee, J. H. Park, O. M. Kwon, and S. M. Lee. Stochastic sampled-data control for state estimation of time-varying delayed neural networks. *Neural Networks*, 46:99–108, 2013.

[76] T. H. Lee, J. H. Park, S. M. Lee, and O. M. Kwon. Robust synchronisation of chaotic systems with randomly occurring uncertainties via stochastic sampled-data control. *International Journal of Control*, 86(1):107–119, 2013.

[77] K. Y. Leung, T. D. Barfoot, and H. Liu. Decentralized cooperative slam for sparsely-communicating robot networks: A centralized equivalent approach. *Journal of Intelligent & Robotic Systems*, 66(3):321–342, 2012.

[78] B. Li, Z. Wang, and Q-L. Han. Input-to-state stabilization of delayed differential systems with exogenous disturbances: The event-triggered case. *IEEE Transactions on Systems, Man, and Cybernetics: Systems*, 49(6):1099–1109, 2019.

[79] B. Li, Z. Wang, and L. Ma. An event-triggered pinning control approach to synchronization of discrete-time stochastic complex dynamical networks. *IEEE Transactions on Neural Networks and Learning Systems*, 29(12):5812–5822, 2018.

[80] H. Li. Sampled-data state estimation for complex dynamical networks with time-varying delay and stochastic sampling. *Neurocomputing*, 138:78–85, 2014.

[81] H. Li, X. Zhang, and G. Feng. Event-triggered output feedback control of switched nonlinear systems with input saturation. *IEEE Transactions on Cybernetics*, 51(5):2319–2326, 2021.

[82] L. Li and T. Li. Stability analysis of sample data systems with input missing: A hybrid control approach. *ISA Transactions*, 90:116–122, 2019.

[83] N. Li, J. Hu, J. Hu, and L. Li. Exponential state estimation for delayed recurrent neural networks with sampled-data. *Nonlinear Dynamics*, 60(1-2):555–564, 2012.

[84] N. Li, Q. Li, and J. Suo. Dynamic event-triggered H_∞ state estimation for delayed complex networks with randomly occurring nonlinearities. *Neurocomputing*, 421:97–104, 2021.

[85] Q. Li, B. Shen, Y. Liu, and F. E. Alsaadi. Event-triggered H_∞ state estimation for discrete-time stochastic genetic regulatory networks with Markovian jumping parameters and time-varying delays. *Neurocomputing*, 174:912–920, 2016.

[86] Q. Li, B. Shen, Z. Wang, T. Huang, and J. Luo. Synchronization control for a class of discrete time-delay complex dynamical networks: A dynamic event-triggered approach. *IEEE Transactions on Cybernetics*, 49(5):1979–1986, 2019.

[87] Q. Li, B. Shen, Z. Wang, and W. Sheng. Recursive distributed filtering over sensor networks on Gilbert-Elliott channels: A dynamic event-triggered approach. *Automatica*, 113:Art. No. 108681, 2020.

[88] Q. Li, Z. Wang, N. Li, and W. Sheng. A dynamic event-triggered approach to recursive filtering for complex networks with switching topologies subject to random sensor failures. *IEEE Transactions on Neural Networks and Learning Systems*, 31(10):4381–4388, 2020.

[89] Q. Li, Z. Wang, W. Sheng, F. E. Alsaadi, and F. E. Alsaadi. Dynamic event-triggered mechanism for H_∞ non-fragile state estimation of complex networks under randomly occurring sensor saturations. *Information Sciences*, 509:304–316, 2020.

[90] X. Li, S. Nguang, K. She, J. Cheng, and S. Zhong. Resilient controller synthesis for Markovian jump systems with probabilistic faults and gain fluctuations under stochastic sampling operational mechanism. *Applied Mathematics and Computation*, 392: Art. No.125623, 2021.

[91] Y. Li, X. Liu, H. Liu, C. Du, and P. Lu. Distributed dynamic event-triggered consensus control for multi-agent systems under fixed and switching topologies. *Journal of the Franklin Institute*, 358(8):4348–4372, 2021.

[92] Z. Li and Y. Liu. Event-triggered stabilization for continuous-time stochastic systems. *IEEE Transactions on Automatic Control*, 65(10):4031–4046, 2020.

[93] Z.-M. Li, X.-H. Chang, and J. H. Park. Quantized static output feedback fuzzy tracking control for discrete-time nonlinear networked systems with asynchronous event-triggered constraints. *IEEE Transactions on Systems, Man, and Cybernetics: Systems*, 51(6):3820–3831, 2021.

[94] D. Liberzon. *Switching in Systems and Control.* Birkhauser, Boston, 2003.

[95] D. Liu and G. Yang. A dynamic event-triggered control approach to leader-following consensus for linear multiagent systems. *IEEE Transactions on Systems, Man, and Cybernetics: Systems*, 51(10):6271–6279, 2021.

[96] H. Liu, Z. Wang, W. Fei, and J. Li. Resilient H_∞ state estimation for discrete-time stochastic delayed memristive neural networks: A dynamic event-triggered mechanism. *IEEE Transactions on Cybernetics*, 2020. DOI:10.1109/TCYB.2020.3021556.

[97] K. Liu and E. Fridman. Wirtinger's inequality and Lyapunov-based sampled-data stabilization. *Automatica*, 48(1):102–108, 2012.

[98] K. Liu, V. Suplin, and X. Ma. H_∞ filtering for sampled-data stochastic systems with limited capacity channel. *Signal Processing*, 91(8):1826–1837, 2011.

[99] L. Liu, W. Zhou, X. Li, and Y. Sun. Dynamic event-triggered approach for cluster synchronization of complex dynamical networks with switching via pinning control. *Neurocomputing*, 340:32–41, 2019.

[100] M. Liu, J. You, and E. Fridman. Stability of linear systems with general sawtooth delay. *IMA Journal of Mathematical Control and Information*, 27(4):419–436, 2010.

[101] Q. Liu, Z. Wang, X. He, and D. Zhou. Event-based recursive distributed filtering over wireless sensor networks. *IEEE Transactions on Automatic Control*, 60(9):2470–2475, 2015.

[102] S. Liu, G. Wei, Y. Song, and Y. Liu. Error-constrained reliable tracking control for discrete time-varying systems subject to quantization effects. *Neurocomputing*, 174:897–905, 2016.

[103] Y. Liu and S. M. Lee. Sampled-data synchronization of chaotic Lur'e systems with stochastic sampling. *Circuits Systems and Signal Processing*, 34(12):3725–3739, 2015.

[104] Y. Liu, B. Shen, and Q. Li. State estimation for neural networks with Markov-based nonuniform sampling: The partly unknown transition probability case. *Neurocomputing*, 357:261–270, 2019.

[105] Y. Liu, B. Shen, and H. Shu. Finite-time resilient H_∞ state estimation for discrete-time delayed neural networks under dynamic event-triggered mechanism. *Neural Networks*, 121:356–365, 2020.

[106] Y. Liu, Z. Wang, J. Liang, and X. Liu. Synchronization and state estimation for discrete-time complex networks with distributed delays. *IEEE Transactions on Systems, Man, and Cybernetics-Part B: Cybernetics*, 28(5):1314–1325, 2008.

[107] C. F. Van Loan. Computing integrals involving the matrix exponential. *IEEE Transactions on Automatic Control*, AC-23(3):395–404, 1978.

[108] J. Lunze and D. Lehmann. A state-feedback approach to event-based control. *Automatica*, 46(1):211–215, 2010.

[109] J. Luo, X. Liu, W. Tian, and S. Zhong. Nonfragile sampled-data filtering of uncertain fuzzy systems with time-varying delays. *IEEE Transactions On Systems, Man, and Cybernetics: Systems*, 51(8):4993–5004, 2021.

[110] S. Luo, F. Deng, and W. Chen. Dynamic event-triggered control for linear stochastic systems with sporadic measurements and communication delays. *Automatica*, 107:86–97, 2019.

[111] L. Ma, Y.-L. Wang, and Q.-L. Han. H_∞ cluster formation control of networked multiagent systems with stochastic sampling. *IEEE Transactions on Cybernetics*, 51(12):5761–5772, 2021.

[112] L. Ma, Z. Wang, C. Cai, and F. E. Alsaadi. Dynamic event-triggered approach to H_∞ control for discrete-time singularly perturbed systems with time-delays and sensor saturations. *IEEE Transactions On Systems, Man, and Cybernetics: Systems*, 51(11):6614–6625, 2021.

[113] W. Ma, X. Jia, F. Yang, and D. Zhang. An impulsive-switched-system approach to aperiodic sampled-data systems with time-delay control. *International Journal of Robust and Nonlinear Control*, 28(6):2484–2494, 2018.

[114] Y. Ma, Z. Li, and J. Zhao. H_∞ control for switched systems based on dynamic event-triggered strategy and quantization under state-dependent switching. *IEEE Transactions On Circuits and Systems I: Regular Papers*, 67(9):3175–3186, 2020.

[115] A. N. Michel, J. A. Farrell, and W. Porod. Qualitative analysis of neural networks. *IEEE Transactions on Circuits and Systems*, 36:229–243, 1989.

[116] L. Mirkin. Some remarks on the use of time-varying delay to model sample-and-hold circuits. *IEEE Transactions on Automatic Control*, 52(6):1009–1112, 2007.

[117] C. Moler and C. Van Loan. Nineteen dubious ways to compute the exponential of a matrix, twenty-five years later. *SIAM Review*, 45(1):3–49, 2003.

[118] G. Mustafa and T. Chen. H_∞ filtering for nonuniformly sampled systems: A Markovian jump systems approach. *Systems & Control Letters*, 60(10):871–876, 2011.

[119] P. Naghshtabrizi, J. P. Hespanha, and A. R. Teel. Exponential stability of impulsive systems with application to uncertain sampled-data systems. *Systems & Control Letters*, 57(5):378–385, 2008.

[120] P. Naghshtabrizi, J. P. Hespanha, and A. R. Teel. Stability of delay impulsive systems with application to networked control systems. *Transactions of the Institute of Measurement and Control*, 32(5):511–528, 2010.

[121] P. T. Nam, P. N. Pathirana, and H. Trinh. Discrete Wirtinger-based inequality and its application. *Journal of the Franklin Institute*, 352(5):1893–1905, 2015.

[122] W. NaNacara and E. E. Yaz. Recursive estimator for linear and nonlinear systems with uncertain observations. *Signal Processing*, 62(2):215–228, 1997.

[123] E. W. Nettleton and H. F. Durrant-Whyt. Delayed and asequent data in decentralized sensing networks. In *Proceedings of Sensor Fusion Decentralized Control Robotics Systems IV Conference*, pages 1–9, 2001.

[124] Y. Oishi and H. Fujioka. Stability and stabilization of aperiodic sampled-data control systems using robust linear matrix inequalities. *Automatica*, 46(8):1327–1333, 2010.

[125] C. Peng, Q.-L. Han, and D. Yue. To transmit or not to transmit: A discrete event-triggered communication scheme for networked Takagi-Sugeno fuzzy systems. *IEEE Transactions on Fuzzy Systems*, 21(1):164–170, 2013.

[126] W. Qian, L. Wang, and M. Z. Q. Chen. Local consensus of nonlinear multiagent systems with varying delay coupling. *IEEE Transations on Systems, Man, and Cybernetics: Systems*, 48(12):2462–2469, 2018.

[127] R. Rakkiyappan, S. Dharani, and J. Cao. Synchronization of neural networks with control packet loss and time-varying delay via stochastic sampled-data controller. *IEEE Transactions on Neural Networks and Learning Systems*, 26(12):3215–3226, 2015.

[128] R. Rakkiyappan, N. Sakthivel, and J. Cao. Stochastic sampled-data control for synchronization of complex dynamical networks with control packet loss and additive time-varying delays. *Neural Networks*, 66:46–63, 2015.

[129] V. Revathi, P. Balasubramaniam, J. H. Park, and T. H. Lee. H_∞ filtering for sample data systems with stochastic sampling and Markovian jumping parameters. *Nonlinear Dynamics*, 78(2):813–830, 2014.

[130] A. SelivaNov, F. Gouaisbaut, and E. Fridman. Stability of discrete-time systems with time-varying delays via a novel summation inequality. *IEEE Transactions on Automatic Control*, 60(10):2740–2745, 2015.

[131] A. Seuret. A novel stability analysis of linear systems under asynchronous samplings. *Automatica*, 48(1):177–182, 2012.

[132] A. Seuret and F. Gouaisbaut. Wirtinger-based integral inequality: Application to time-delay systems. *Automatica*, 49(9):2860–2866, 2013.

[133] Y. Shan, K. She, S. Zhong, J. Cheng, W. Wang, and C. Zhao. Event-triggered passive control for Markovian jump discrete-time systems with incomplete transition probability and unreliable channels. *Journal of the Franklin Institute*, 365(15):8093–8117, 2019.

[134] B. Shen, H. Tan, Z. Wang, and T. Huang. Quantized/saturated control for sampled-data systems under noisy sampling intervals: A confluent vandermonde matrix approach. *IEEE Transactions on Automatic Control*, 62(19):4753–4759, 2017.

[135] B. Shen, Z. Wang, D. Ding, and H. Shu. H_∞ state estimation for complex networks with uncertain inner coupling and incomplete measurements. *IEEE Transactions on Neural Networks and Learning Systems*, 24:2027–2037, 2013.

[136] B. Shen, Z. Wang, and T. Huang. Stabilization for sampled-data systems under noisy sampling interval. *Automatica*, 63(11):162–166, 2016.

[137] B. Shen, Z. Wang, and Y. Hung. Distributed H_∞-consensus filtering in sensor networks with multiple missing measurements: The finite-horizon case. *Automatica*, 46(10):1682–1688, 2011.

[138] B. Shen, Z. Wang, J. Liang, and X. Liu. Sampled-data H_∞ filtering for stochastic genetic regulatory networks. *International Journal of Robust and Nonlinear Control*, 21(15):1759–1777, 2011.

[139] B. Shen, Z. Wang, and X. Liu. A stochastic sampled-data approach to distributed H_∞ filtering in sensor networks. *IEEE Transactions on Circuits and Systems I-Regular Papers*, 58(9):2237–2246, 2011.

[140] B. Shen, Z. Wang, and X. Liu. Sampled-data synchronization control of dynamical networks with stochastic sampling. *IEEE Transactions on Automatic Control*, 57(10):2644–2650, 2012.

[141] B. Shen, Z. Wang, and H. Qiao. Event-triggered state estimation for discrete-time multidelayed neural networks with stochastic parameters and incomplete measurements. *IEEE Transactions on Neural Networks and Learning Systems*, 28(5):1152–1163, 2017.

[142] Y. Shen, Z. Wang, H. Dong, F. E. Alsaadi, and H. Liu. Dynamic event-based recursive filtering for multirate systems with integral measurements over sensor networks. *International Journal of Robust and Nonlinear Control*, 2021. DOI: 10.1002/rnc.5884.

[143] D. Shi, R. J. Elliott, and T. Chen. Event-based state estimation of discrete-state hidden Markov models. *Automatica*, 65:12–26, 2016.

[144] K. Shi, X. Liu, H. Zhu, and S. Zhong. On designing stochastic sampled-data controller for master-slave synchronization of chaotic Lur'e system via a novel integral inequality. *Communications in Nonlinear Science and Numerical Simulation*, 34:165–184, 2016.

[145] T. Shi and H. Su. Sampled-data MPC for LPV systems with input saturation. *IET Control Theory and Applications*, 8(17):1781–1788, 2014.

[146] B. SiNopoli, L. Schenato, M. Franceschetti, K. Poolla, M. I. Jordan, and S. S. Sastry. Kalman filtering with intermittent observations. *IEEE Transactions on Automatic Control*, 49(9):1453–1464, 2004.

[147] D. Spinello. Asymptotic agreement in a class of networked Kalman filters with intermittent stochastic communications. *IEEE Transactions on Automatic Control*, 61(4):1093–1098, 2016.

[148] H. Su, J. Zhang, and X. Chen. A stochastic sampling mechanism for time-varying formation of multiagent systems with multiple leaders and communication delays. *IEEE Transactions on Neural Networks and Learning Systems*, 30(12):3699–3707, 2019.

[149] X. Su, Y. Wen, Y.-D. Song, and T. Hayat. Dissipativity-based fuzzy control of nonlinear systems via an event-triggered mechanism. *IEEE Transactions on Systems, Man, and Cybernetics: Systems*, 49(6):1208–1217, 2019.

[150] Y. Suh. Stability and stabilization of nonuniform sampling systems. *Automatica*, 4(12):3222–3226, 2008.

[151] J. Suo, Z. Wang, and B. Shen. Pinning synchronization control for a class of discrete-time switched stochastic complex networks under event-triggered mechanism. *Nonlinear Analysis-Hybrid Systems*, 37:Art. No. 100886, 2020.

[152] P. Tabuada. Event-triggered real-time scheduling of stabilizing control tasks. *IEEE Transactions on Automatic Control*, 52(9):1680–1685, 2007.

[153] P. Tallapragada and N. Chopra. Decentralized event-triggering for control of nonlinear systems. *IEEE Transactions on Automatic Control*, 52(12):3312–3324, 2014.

[154] H. Tan, B. Shen, Q. Li, and H. Shu. Non-fragile H_∞ control for body slip angle of electric vehicles with onboard vision systems: The dynamic event-triggering approach. *Journal of the Franklin Institute*, 357:2008–2027, 2020.

[155] D. W. Tank and J. J. Hopfield. Simple "Neural" optimization networks: An A/D converter, signal decision circuit, and a linear programming circuit. *IEEE Transactions on Circuits and Systems*, CAS-33:533–541, 1986.

[156] J. Tao, R. Lu, H. Su, and Z.-G. Wu. Filtering of T-S fuzzy systems with nonuniform sampling. *IEEE Transactions on Systems Man Cybernetics: Systems*, 48(12):2442–2450, 2018.

[157] S. Tarbouriech, G. Garcia, J. M. G. da Silva Jr., and I. Queinnec. *Stability and Stabilization of Linear Systems with Saturating Actuators.* Springer-Verlag London Limited, 2011.

[158] N. van de Wouw, P. Naghshtabrizi, M. B. G. Cloosterman, and J. P. Hespanha. Tracking control for sampled-data systems with uncertain time-varying sampling intervals and delays. *International Journal of Robust and Nonlinear Control*, 20(4):387–411, 2010.

[159] N. van de Wouw, D. Nesic, and W. P. M. H. Heemels. A discrete-time framework for stability analysis of nonlinear networked control systems. *Automatica*, 48(6):1144–1153, 2012.

[160] X. Wan, Z. Wang, Q.-L. Han, and M. Wu. Finite-time H_∞ state estimation for discrete time-delayed genetic regulatory networks under stochastic communication protocols. *IEEE Transactions on Circuits and Systems I-Regular Papers*, 65(10):3481–3491, 2018.

[161] X. Wan, L. Xu, H. Fang, and G. Ling. Robust non-fragile H_∞ state estimation for discrete-time genetic regulatory networks with Markov jump delays and uncertain transition probabilities. *Neurocomputing*, 154:162–173, 2015.

[162] Y. Wan, G. Wen, J. Cao, and W. Yu. Distributed node-to-node consensus of multi-agent systems with stochastic sampling. *International Journal of Robust and Nonlinear Control*, 26(1):110–124, 2016.

[163] H. Wang, P. Shi, and R. K. Agarwal. Network-based event-triggered filtering for Markovian jump systems. *International Journal of Control*, 89(6):1096–1110, 2016.

[164] H. Wang, D. Zhang, and R. Lu. Event-triggered H_∞ filter design for Markovian jump systems with quantization. *Nonlinear Analysis-Hybrid Systems*, 28:23–41, 2018.

[165] J. Wang, X. Zhang, and Q.-L. Han. Event-triggered generalized dissipativity filtering for neural networks with time-varying delays. *IEEE Transactions on Neural Networks and Learning Systems*, 27(1):77–88, 2016.

[166] L. Wang, Z. Wang, T. Huang, and G. Wei. An event-triggered approach to state estimation for a class of complex networks with mixed time delays and nonlinearities. *IEEE Transactions on Cybernetics*, 46(11):2497–2508, 2016.

[167] S. Wang, Z. Wang, H. Dong, and Y. Chen. A dynamic event-triggered approach to recursive nonfragile filtering for complex networks with sensor saturations and switching topologies. *IEEE Transactions on Cybernetics*, 2021. DOI: 10.1109/TCYB.2021.3049461.

[168] X. Wang and M. Lemmon. Event triggering in distributed networked control systems. *IEEE Transactions on Automatic Control*, 56(3):586–601, 2011.

[169] Y. Wang, A. Arumugam, Y. Liu, and F. E. Alsaadi. Finite-time event-triggered non-fragile state estimation for discrete-time delayed neural networks with randomly occurring sensor nonlinearity and energy constraints. *Neurocomputing*, 384:115–129, 2020.

[170] Y. Wang, F. Chen, and G. Zhuang. Dynamic event-based reliable dissipative asynchronous control for stochastic Markov jump systems with general conditional probabilities. *Nonlinear Dynamics*, 101(1):465–485, 2020.

[171] Y. Wang, F. Chen, G. Zhuang, and G. Yang. Dynamic event-based mixed H_∞ and dissipative asynchronous control for Markov jump singularly perturbed systems. *Applied Mathematics and Computation*, 386:Art. No. 125443, 2020.

[172] Y. Wang, G. Song, J. Zhao, J. Sun, and G. Zhuang. Reliable mixed H_∞ and passive control for networked control systems under adaptive event-triggered scheme with actuator faults and randomly occurring nonlinear perturbations. *ISA Transactions*, 89:45–57, 2019.

[173] Y. Wang, Z. Wang, and J. Liang. Global synchronization for delayed complex networks with randomly occurring nonlinearities and multiple stochastic disturbances. *Journal of Physics A: Mathematical and Theoretical*, 42:Art. No. 135101, 2009.

[174] Y. Wang, W. Zheng, and H. Zhang. Dynamic event-based control of nonlinear stochastic systems. *IEEE Transactions on Automatic Control*, 62(12):6544–6551, 2017.

[175] Z. Wang, D. W. C. Ho, Y. Liu, and X. Liu. Robust H_∞ control for a class of nonlinear discrete time-delay stochastic systems with missing measurements. *Automatica*, 45(3):684–691, 2009.

[176] Z. Wang, J. Lam, and X. Liu. Filtering for a class of nonlinear discrete-time stochastic systems with state delays. *Journal of Computational and Applied Mathematics*, 201(1):153–163, 2007.

[177] Z. Wang, J. Lam, G. Wei, K. Fraser, and X. Liu. Filtering for nonlinear genetic regulatory networks with stochastic disturbances. *IEEE Transactions on Automatic Control*, 53(10):2448–2457, 2008.

[178] S. Weerakkody, Y. Mo, B. Sinopoli, D. Han, and L. Shi. Multi-sensor scheduling for state estimation with event-based, stochastic triggers. *IEEE Transactions on Automaic Control*, 61(9):2695–2701, 2016.

[179] L. Wen, S. Yu, Y. Zhao, and Y. Yan. Leader-following consensus for multi-agent systems subject to cyber attacks: Dynamic event-triggered control. *ISA Transactions*, 2021. DOI: 10.1016/j.isatra.2021.09.002.

[180] S. Wen, Z. Zeng, and T. Huang. Robust H_∞ output tracking control for fuzzy networked systems with stochastic sampling and multiplicative noise. *Nonlinear Dynamics*, 70(2):1061–1077, 2012.

[181] J. Wu, X. Chen, and H. Gao. H_∞ filtering with stochastic sampling. *Signal Processing*, 90(4):1131–1145, 2010.

[182] J. Wu, Q.-S. Jia, K. H. Johansson, and L. Shi. Event-based sensor data scheduling: Trade-off between communication rate and estimation quality. *IEEE Transactions on Automatic Control*, 58(4):1041–1046, 2013.

[183] M. Wu, Y. He, J. She, and G. Liu. Delay-dependent criteria for robust stability of time-varying delay systems. *Automatica*, 40(8):1435–1439, 2004.

[184] Z. Wu, P. Shi, H. Su, and J. Chu. Reliable H_∞ control for discrete-time fuzzy systems with infinite-distributed delay. *IEEE Transactions on Fuzzy Systems*, 20(1):22–31, 2012.

[185] Z.-G. Wu, Y. Xu, R. Lu, Y. Wu, and T. Huang. Event-triggered control for consensus of multiagent systems with fixed/switching topologies. *IEEE Transactions on Systems, Man, and Cybernetics: Systems*, 48(10):1736–1746, 2018.

[186] W. Xia, S. Xu, J. Lu, Y. Li, Y. Chu, and Z. Zhang. Event-triggered filtering for discrete-time Markovian jump systems with additive time-varying delays. *Applied Mathematics and Computation*, 391:Art. No. 125630, 2021.

[187] W. Xia, W. Zheng, and S. Xu. Event-triggered filter design for Markovian jump delay systems with nonlinear perturbation using quantized measurement. *International Journal of Robust and Nonlinear Control*, 29(14):4644–4664, 2019.

[188] F. Xiao, Y. Shi, and T. Chen. Robust stability of networked linear control systems with asynchronous continuous- and discrete-time event-triggering schemes. *IEEE Transactions on Automatic Control*, 66(2):932–939, 2021.

[189] L. Xie, Y. C. Soh, and C. E. de Souza. Robust Kalman filtering for uncertain discrete-time systems. *IEEE Transactions on Automatic Control*, 39(6):1310–1314, 1994.

[190] S. Xu, J. Lam, and T. Chen. Robust H_∞ control for uncertain discrete stochastic time-delay systems. *Systems & Control Letters*, 51(3-4):203–215, 2004.

[191] W. Xu and D. W. C. Ho. Clustered event-triggered consensus analysis: An impulsive framework. *IEEE Transactions on Industrial Informatics*, 63(11):7133–7143, 2016.

[192] Y. Xu, R. Lu, P. Shi, J. Tao, and S. Xie. Robust estimation for neural networks with randomly occurring distributed delays and Markovian jump coupling. *IEEE Transactions on Neural Networks and Learning Systems*, 29(4):845–855, 2018.

[193] Y. Xu, H. Su, and Y.-J. Pan. Output feedback stabilization for Markov-based nonuniformly sampled-data networked control systems. *Systems & Control Letters*, 62(8):656–663, 2013.

[194] Y. Xu, Y. Wang, G. Zhuang, and J. Lu. Dynamic event-based asynchronous H_∞ control for TšCS fuzzy singular Markov jump systems with redundant channels. *IET Control Theory & Applications*, 13(14):2239–2251, 2019.

[195] Y. Yang, Y. Niu, and Z. Zhang. Dynamic event-triggered sliding mode control for interval Type-2 fuzzy systems with fading channels. *ISA Transactions*, 110:53–62, 2021.

[196] W. Yao, C. Wang, Y. Sun, C. Zhou, and H. Lin. Synchronization of inertial memristive neural networks with time-varying delays via static or dynamic event-triggered control. *Neurocomputing*, 404:367–380, 2020.

[197] D. Ye, E. Tian, and Q.-L. Han. A delay system method for designing event-triggered controllers of networked control systems. *IEEE Transactions on Automatic Control*, 58(2):475–481, 2013.

[198] H. Yu and F. Hao. Input-to-state stability of integral-based event-triggered control for linear plants. *Automatica*, 85:248–255, 2017.

[199] M. Yu, L Wang, and T. Chu. Sampled-data stabilisation of networked control systems with nonlinearity. *IEE Proceedings (Control Theory and Applications)*, 152(6):609–614, 2005.

[200] D. Yue, Q.-L. Han, and C. Peng. State feedback controller design of networked control systems. *IEEE Transactions on Circuits and Systems*, 51(11):640–644, 2004.

[201] D. Yue, E. Tian, and Q.-L. Han. A delay system method for designing event-triggered controllers of networked control systems. *IEEE Transactions on Automatic Control*, 58(2):475–481, 2013.

[202] D. Zeng, R. Zhang, X. Liu, S. Zhong, and K. Shi. Pinning stochastic sampled-data control for exponential synchronization of directed complex dynamical networks with sampled-data communications. *Applied Mathematics and Computation*, 337:102–118, 2018.

[203] L. Zha, J.-A. Fang, X. Li, and J. Liu. Event-triggered output feedback H_∞ control for networked Markovian jump systems with quantizations. *Nonlinear Analysis-Hybrid Systems*, 24:146–158, 2017.

[204] C. Zhang, J. Hu, J. Qiu, and Q. Chen. Event-triggered nonsynchronized H_∞ filtering for discrete-time T-S fuzzy systems based on piecewise Lyapunov functions. *IEEE Transactions on Systems, Man, and Cybernetics: Systems*, 47(8):2330–2341, 2017.

[205] D. Zhang, P. Shi, and W.-A. Zhang. Energy-efficient distributed filtering in sensor networks: A unified switched system approach. *IEEE Transactions on Cybernetics*, 47(7):1618–1629, 2017.

[206] H. Zhang, Y. Shi, and J. Wang. On energy-to-peak filtering for nonuniformly sampled nonlinear systems: A markovian jump system approach. *IEEE Transactions on Fuzzy Systems*, 2(1):212–222, 2014.

[207] J. Zhang and E. Fridman. Dynamic event-triggered control of networked stochastic systems with scheduling protocols. *IEEE Transactions on Automatic Control*, 66(12):6139–6147, 2021.

[208] R. Zhang, D. Zeng, S. Zhong, K. Shi, and J. Cui. New approach on designing stochastic sampled-data controller for exponential synchronization of chaotic Lur'e systems. *Nonlinear Analysis-Hybrid Systems*, 29:303–321, 2018.

[209] W. Zhang and L. Yu. Stabilization of sampled-data control systems with control inputs missing. *IEEE Transactions on Automatic Control*, 55(2):447–452, 2010.

[210] W.-A. Zhang, H. Dong, G. Guo, and L. Yu. Distributed sampled-data H_∞ filtering for sensor networks with nonuniform sampling periods. *IEEE Transactions on Industrial Informatics*, 10(2):871–881, 2014.

[211] W.-A. Zhang, S. Liu, and L. Yu. Fusion estimation for sensor networks with nonuniform estimation rates. *IEEE Transactions on Circuits and Systems I-Regular Papers*, 61(5):1485–1498, 2014.

[212] X.-M. Zhang and Q.-L. Han. A decentralized event-triggered dissipative control scheme for systems with multiple sensors to sample the system outputs. *IEEE Transactions on Cybernetics*, 46(12):2745–2757, 2009.

[213] X.-M. Zhang and Q.-L. Han. Event-triggered dynamic output feedback control for networked control systems. *IET Control Theory and Applications*, 8(4):226–234, 2014.

[214] X.-M. Zhang and Q.-L. Han. Event-based H_∞ filtering for sampled-data systems. *Automatica*, 51:55–69, 2015.

[215] X.-M. Zhang, Q.-L. Han, and B.-L. Zhang. An overview and deep investigation on sampled-data-based event-triggered control and filtering for networked systems. *IEEE Transactions on Industrial Informatics*, 13(1):4–16, 2017.

[216] Y. Zhang, Z. Wang, and F.-E. Alsaadi. Detection of intermittent faults for nonuniformly sampled multi-rate systems with dynamic quantisation and missing measurements. *International Journal of Control*, 93(4):898–909, 2020.

[217] D. Zhao, Z. Wang, D. Ding, G. Wei, and F. E. Alsaadi. H_∞ PID output-feedback control under event-triggered protocol. *International Journal of General Systems*, 47(5):454–467, 2018.

[218] G. Zhao, C. Hua, and X. Guan. Decentralized dynamic event-triggered H_∞ control for nonlinear systems with unreliable communication channel and limited bandwidth. *IEEE Transactions on Fuzzy Systems*, 29(4):757–771, 2021.

[219] X. Zhao, C. Liu, J. Liu, and E. Tian. Probabilistic-constrained reliable H_∞ tracking control for a class of stochastic nonlinear systems: An outlier-resistant event-triggered scheme. *Journal of the Franklin Institute*, 358(9):4741–4760, 2021.

[220] X. Zhao, C. Ma, X. Xing, and X. Zheng. A stochastic sampling consensus protocol of networked Euler-Lagrange systems with application to two-link manipulator. *IEEE Transactions on Industrial Informatics*, 11(4):907–914, 2015.

[221] L. Zou, Z. Wang, H. Gao, and X. Liu. Event-triggered state estimation for complex networks with mixed time delays via sampled data information: The continuous-time case. *IEEE Transactions on Cybernetics*, 45(12):2804–2815, 2015.

[222] W. Zou, P. Shi, Z. Xiang, and Y. Shig. Consensus tracking control of switched stochastic nonlinear multiagent systems via event-triggered strategy. *IEEE Transactions on Neural Networks and Learning Systems*, 31(3):1036–1045, 2020.

[223] Z. Zuo, P. Xie, and Y. Wang. Output-based dynamic event-triggering control for sensor saturated systems with external disturbance. *Applied Mathematics and Computation*, 374:Art. No. 125043, 2020.

Index

For Product Safety Concerns and Information please contact our EU
representative GPSR@taylorandfrancis.com
Taylor & Francis Verlag GmbH, Kaufingerstraße 24, 80331 München, Germany

www.ingramcontent.com/pod-product-compliance
Lightning Source LLC
Chambersburg PA
CBHW060337220326
41598CB00023B/2731
9781032310206